Eco-Materials and Green Energy for a Sustainable Future

Eco-Materials and Green Energy for a Sustainable Future emphasizes the synergy between eco-materials and green energy solutions, highlighting their combined power to reduce carbon emissions, conserve resources, and create a more resilient and sustainable future. It provides a detailed discussion on cutting-edge green energy technologies and their potential to transform the energy landscape.

Covering a range of applications and emerging technologies that are moving toward sustainable and green energy, this book includes topics on nano-batteries, nanoparticle treatments of toxic textile industry wastewater, and green building materials. It explores thin-film solar cells and luminescent materials in solar energy. This book considers green synthesis methods, such as plant extracts and microorganisms, with applications in regenerative medicine.

This book will interest researchers and senior undergraduate and graduate students studying renewable energy sources, green materials engineering and chemistry, and sustainability.

Eco-Materials and Green Energy for a Sustainable Future

Edited by
Amit Soni, Dharmendra Tripathi, Jagrati Sahariya, and Kamal Nayan Sharma

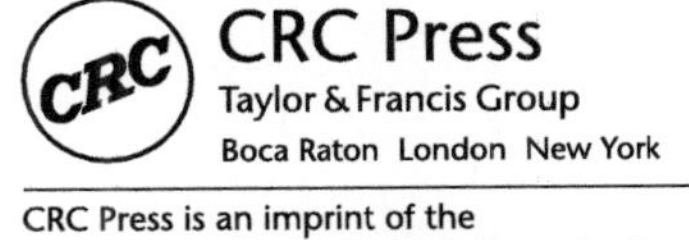

CRC Press
Taylor & Francis Group
Boca Raton London New York

CRC Press is an imprint of the
Taylor & Francis Group, an **informa** business

Cover image credit: Getty Images

First edition published 2025
by CRC Press
2385 NW Executive Center Drive, Suite 320, Boca Raton FL 33431

and by CRC Press
4 Park Square, Milton Park, Abingdon, Oxon, OX14 4RN

CRC Press is an imprint of Taylor & Francis Group, LLC

ISBN: 978-1-032-75380-5 (hbk)
ISBN: 978-1-032-75381-2 (pbk)
ISBN: 978-1-003-47374-9 (ebk)

DOI: 10.1201/9781003473749

Typeset in Times
by codeMantra

Contents

PART I Eco-Material Technologies

Rohit Jakhar and Sarika Jain

Anju Rani, Mrinal Dutta, Chander Shekhar, and Ayana Bhaduri

Khushbu Jain, Krishna Kumar Yadav, and Neeru Dabas

*Renuka Kanojia, Anand Salvi, Sonika Charak,
Manish Shandilya, and Joydeep Dutta*

Yogesh Kumar Sonia and Sapna Meena

PART II Green Energy Applications

*Prabha Gurawalia, Sudip Majumder,
Chandra Mohan Srivastava, Sonika Charak, and Manish Shandilya*

PART III Sustainable Future Approaches

Preface

In a world grappling with unparalleled environmental crises, the imperative for sustainable solutions has reached a pivotal juncture. We find ourselves at the crossroads of innovation and accountability, where the demand for profound and immediate change has never been more pressing. The book proposal, *Eco-Materials and Green Energy for a Sustainable Future*, aims to chart a course toward a world in which environmental harmony is the norm, all while capitalizing on the limitless opportunities presented by advancements in science and technology.

This book *Eco-Materials and Green Energy for a Sustainable Future* will address several important problems related to energy sustainability and environmental concerns for future sustainability. It will address different problems that include exponentially increased energy demands due to irrepressible factors such as growing population, industrialization, and fast-depleting fossil fuels. Hence, this directs us toward the urgent need for energy production as a new resource in this area which will play a significant role in future sustainability. Green energy employs non-toxic ingredients that are efficient and environment friendly and generates less waste products. Its contributions to research lie in synthesizing a wide range of information into a cohesive narrative, while its practical implications can drive the adoption of eco-friendly materials and technologies, fostering a more sustainable and resilient future.

The objectives of this book are subsequently presented chapter wise:

Chapter 1: Modern ocean engineering emphasizes flexible, compliant structures for floating cities and innovative energy harvesting. This chapter explores a vertical membrane structure's behavior, focusing on material properties and stress analysis.

Chapter 2: This chapter explores nanoparticle use (CuO, Al_2O_3, Ag, TiO_2) for improved heat transfer in hydromagnetic horizontal duct flow, particularly in nano-battery applications. It involves a three-layer duct and numerical analysis, finding that Al_2O_3 nanoparticles significantly influence transport properties.

Chapter 3: Luminescent materials offer potential solutions, particularly in solar energy applications. This chapter introduces luminescent materials and their diverse uses, emphasizing environmental benefits and future prospects.

Chapter 4: Over 85% of global energy comes from fossil fuels, contributing to climate change. Transitioning to renewable energy, particularly solar power, is essential. This review chapter compares single-crystal silicon (c-Si) and cadmium telluride (CdTe) thin-film solar cells (TFSCs) and explores CdTe solar cell performance using the SCAPS-1D program.

Chapter 5: Nanotechnology, with its diverse applications, holds promise for addressing environmental issues. This chapter highlights the green synthesis of metal oxide nanoparticles (NiO, MnO_2, MgO) using natural sources. These nanoparticles exhibit effective photocatalytic activity, offering potential eco-friendly solutions for treating toxic textile industry wastewater.

Chapter 6: One-dimensional TiO_2 (1D-TiO_2) nanomaterials have distinctive properties and versatile applications. This chapter focuses on TiO_2 nanotubes (TNT) synthesis, morphology changes based on reaction conditions, properties, engineering improvements, and potential applications in photocatalysis, solar cells, and energy storage.

Chapter 7: Global environmental pollution calls for immediate attention. This chapter presents the green synthesis methods for nanomaterials and advocates eco-friendly materials in various industries, with green chemistry emphasizing biodegradable and non-toxic substances for sustainable nanoscience and a greener future.

Chapter 8: This chapter showcases that green building materials are a valuable resource for sustainable construction. It covers the importance, types, benefits, sustainability, case studies, regulations, and implementation strategies, providing essential tools for eco-friendly construction practices.

Chapter 9: Ionic liquids (ILs), highly versatile due to tunable properties, find applications in science and technology. This chapter explores the use of ionic liquids, hybrid electrolytes derived from ionic liquids, and electrode materials based on ionic liquids in the development of energy-efficient smart devices.

Chapter 10: Hydroxyapatite (HAp) is biocompatible and used in biomedicine. Green synthesis methods, such as plant extracts and microorganisms, offer eco-friendly alternatives to traditional toxic processes, reducing environmental impact while providing control over particle properties. This has potential applications in healthcare and regenerative medicine.

Chapter 11: This book chapter tackles the global issue of sustainable waste management in the context of rising resource use and population growth. It advocates for a holistic approach, blending environmental and economic considerations. This chapter explores repurposing industrial solid waste materials within a circular economy model, promoting sustainable resource utilization and environmentally friendly waste management practices through ongoing research and innovation.

Chapter 12: Chitosan nanoparticles, sustainable and eco-friendly, have versatile uses in nanotechnology. They serve as nanocarriers for drug, protein, and gene delivery, enhance electrical conductivity, and find applications in nanoelectronics and nanosensors. This chapter outlines their synthesis, characteristics, and broad applications, highlighting their transformative potential in diverse fields and green nanotechnology.

Chapter 13: This book delves into lead-free inorganic double perovskite materials, highlighting their eco-friendliness and potential in modern energy technologies. Emphasizing their role in renewable energy, this chapter provides insights into their promising future for sustainable energy solutions, notably in solar cells and batteries.

Chapter 14: Climate change presents multifaceted challenges, including temperature-related issues, such as droughts and diseases. The CoCEB model, extended to an optimal control problem, seeks to minimize atmospheric carbon dioxide concentrations using chemical and mechanical methods, emphasizing their efficacy in controlling these levels.

Chapter 15: This chapter emphasizes the vital role of eco-materials in addressing environmental concerns and fostering sustainability, covering their types, design principles, and real-world applications. It briefly mentions the potential of nanomedicine and nanotechnology in healthcare.

About the Editors

Dr. Amit Soni has rich experience of 22 years with academic and administrative exposure of different organizations in various capacities, which includes 3 years of industrial experience in RRVPNL, Jaipur. He is currently working as a Professor in the Department of Electrical Engineering, Manipal University, Jaipur, and actively contributing to an administrative role as Director of SEEC (School of Electrical, Electronics & Communication). In addition to this, he also has an administrative role as In-Charge CEO at Atal Incubation Center (AIC), Manipal University, Jaipur. He had worked as the Director (International Collaborations) at Manipal University, Jaipur, and contributed toward MoU's/MoA's with top QS Ranked Universities, Faculty & Student exchanges, successfully organizing DoIC Flagship events. He had worked as the Director (Quality & Compliance) at Manipal University, Jaipur, and successfully contributed to NIRF Ranking 2021, AQAR-1, and NBA visit for five Engineering Departments. He has published 110 research papers that include reputed SCI international journals (31) with high impact Q1 & Q2 journals and Scopus indexed conferences (79) at national and international levels. He has published several book chapters and has contributed as 'Lead Editor' in LNEE Springer publications. He has successfully supervised five PhDs, and currently, six research scholars are working under his supervision. He has worked as PI for DST SERB and as Co-PI for CSIR funded research project in collaboration with MLSU, Udaipur, and NIT, Uttarakhand. He is a regular reviewer of research articles for many high-impact international journals. He is a life member of Optica, Solar Energy Society of India, Member ISTE, India, and IEEE, USA. His research interests include conventional power and renewable energy sources, power system stability, solar photovoltaic materials: material characterization, semiconductor device physics, bulk and thin film technology, photoluminescence, and power quality improvements.

Dr. Dharmendra Tripathi has been working as an Associate Professor in the Department of Mathematics, National Institute of Technology, Uttarakhand. Prior to joining NIT, Uttarakhand, he has worked for more than 12 years as a faculty member (Associate Professor, Assistant Professor) in various reputed institutions like Manipal University, Jaipur; NIT, Delhi; IIT, Ropar; and BITS Pilani, Hyderabad. He completed his PhD in Applied Mathematics (Mathematical Modelling of Physiological flows) in 2009 from the Indian Institute of Technology Banaras Hindu University and MSc in Mathematics from Banaras Hindu University.

He has supervised eight PhD students, and three PhD students are working under his supervision. He has also guided various BTech projects. He has published more than 200 papers in reputed international journals, three edited books in Springer, one edited book in CRC, and ten book chapters. He has presented more than 40 papers at international and national conferences. His research's h-index is 54 and i-10 index is 164, and his papers have more than 7870 citations as per Google Scholar.

He has delivered more than 100 lectures as an Invited Speaker, a Keynote Speaker, and a Resource Person in various conferences, workshops, FDP, STTP, STC, refresher courses, etc.

He has received the best faculty award in 2022 from NIT, Uttarakhand, for his outstanding contribution in teaching, research, and administration. He has been listed in the **top 2% researchers/scientists across the world** as per updated science-wide author databases of standardized citation indicators in the year 2020, year 2021, and year 2022. He has also received the President Award in 2017 by the Manipal University, Jaipur, for outstanding contribution. He is the recipient of Prof. PR Sharma Memorial Award from the International Academy for Physical Sciences (IAPS) in 2021. He has also been recognized by various reputed journals for reviewing the articles and editing special issues for the journals. He was awarded some prestigious INAE fellowships in 2015, 2016, 2017, and 2018, and postdoctoral fellowships (NBHM, Dr. D.S. Kothari, and Indo-EU) in 2010.

He has been discharging second-time additional administrative responsibilities as the Dean (Student Welfare) at NIT, Uttarakhand, since March 2022 and I/c Registrar since July 2022. He has already discharged many administrative responsibilities of the Institutes like I/c Registrar, Dean (R&C), CVO, Dean (SW), Chief Warden, and Chairman of various Institutes Committees.

His research work is focused on the mathematical modeling and simulation of biological flows in deformable domains, peristaltic flow of Newtonian and non-Newtonian fluids, dynamics of various infectious diseases, microfluidics, CFD, biomechanics, heat transfer, nanofluids, energy systems, numerical methods, etc.

Dr. Jagrati Sahariya is working as an Associate Professor in the Department of Physics, National Institute of Technology, Uttarakhand. She is having teaching experience of more than 11 years. She has been working in the field of γ-ray scattering and band structure calculations for the last 11 years. She received her PhD degree in 2012 in the field of electronic structure calculation and Compton scattering. As part of her PhD thesis, she has studied the electronic and magnetic properties of a variety of technologically important materials. She has sufficient expertise in using different band structure methods like full potential linearized augmented plane wave, linear combinations of atomic orbitals, and spin polarized relativistic Korringa-Kohn-Rostoker to compute electronic structure, optical and magnetic properties, and Compton profiles of a variety of materials.

She has published more than 100 papers in peer-reviewed international journals of high-impact factor and repute conference proceedings, ten book chapters, one edited book in CRC, and one edited book in Springer. She has supervised two PhD students, and currently, three PhD students are working under her guidance. She has also executed two research projects funded by SERB, New Delhi, and a research project sponsored by CSIR, New Delhi, is ongoing. She has organized several conferences, workshops, FDP, and STC, and has edited four Scopus-indexed conference proceedings. She is a reviewer for several reputed international journals of high-impact factors. Her research interests include renewable energy sources, solar photovoltaic materials: material characterization, semiconductor device physics, bulk and thin film technology, and charge and magnetic Compton scattering.

Dr. Kamal Nayan Sharma is working as an Assistant Professor of Chemistry at Amity School of Applied Sciences, Amity University Haryana, Gurugram. He has recently been honored with the prestigious SIRE (SERB International Research Experience) fellowship by the Science and Engineering Research Board (Govt. of India). Presently, he is on deputation as a SERB-SIRE awardee at the University of Waikato in Hamilton, New Zealand, for six months. During this six-month period, he is engaged in research collaboration with an Associate Prof. Dr. Graham C. Saunders. Before joining Amity University Haryana, Dr. Sharma served as an Assistant Professor at Vivekanand Global University in Jaipur, Rajasthan, India. He earned his PhD from the prestigious Indian Institute of Technology (IIT), Delhi, and subsequently completed a young scientist research project sponsored by SERB, New Delhi, at MNIT, Jaipur, as the Principal Investigator. His scholarly work is widely recognized, with 31 publications in esteemed international journals of high-impact factors, such as nanoscale, organometallics, chemical communications, Dalton transaction, organic and biomolecular chemistry, and tetrahedron letters. His research has garnered significant attention, reflected in an impressive h-index of 17 and an i-10 index of 21, along with more than 850 citations. He has also visited the University of Cape Town, Cape Town, for research work under the Department of Science and Technology (Govt. of India) sponsored Indo-South African Research Project. He has served as a reviewer in various international journals. He currently guides two PhD students and has supervised several master's students in their project work.

List of Contributors

Himanshu Arora
Department of Chemistry
University of Allahabad
Prayagraj, Uttar Pradesh, India

Manoj Kumar Banjare
MATS School of Sciences
MATS University
Pandri, Raipur (C.G.), India

Iman Bashtani
Faculty of Engineering, Mechanical
   Engineering Department
Ferdowsi University of Mashhad
Mashhad, Iran

O. Anwar Bég
Multi-Physical Engineering Sciences
   Group (MPESG), Department
   of Mechanical/Aeronautical
   Engineering
Salford University
Manchester, United Kingdom

Tasveer A. Bég
Engineering Mechanics Research
Israfil House
Manchester, United Kingdom

Kamalakanta Behera
Department of Chemistry
University of Allahabad
Prayagraj, Uttar Pradesh, India

Ayana Bhaduri
Department of Physics
Amity School of Applied Sciences,
   Amity University Haryana
Gurugram, Haryana, India

Gurpreet Singh Bhatia
Department of Mathematics
Lovely Professional University
Jalandhar, India

Martin L. Burby
Multi-Physical Engineering Sciences
   Group (MPESG), Department
   of Mechanical/Aeronautical
   Engineering
Salford University
Manchester, United Kingdom

Rajesh Kumar Chandrawat
School of Engineering and Information
   Technology
Manipal Academy of Higher Education
Dubai, United Arab Emirates

Sonika Charak
National Brain Research Centre
Manesar, Haryana, India

Neeru Dabas
CBFS, Amity School of Applied
   Sciences
Amity University Haryana
Gurugram, Haryana, India

Joydeep Dutta
Department of Chemistry,
   Biochemistry, and Forensic Science
Amity School of Applied Sciences,
   Amity University Haryana
Gurugram, Haryana, India

Mrinal Dutta
National Institute for Solar Energy
Gurugram, Haryana, India

Javad Abolfazli Esfahani
Faculty of Engineering, Mechanical
  Engineering Department
Ferdowsi University of Mashhad
Mashhad, Iran
and
Center of Excellence on Modelling and
  Control Systems (CEMCS)
Ferdowsi University of Mashhad
Mashhad, Iran

Ritika Gera
Department of Applied Chemistry
  (CBFS - ASAS)
Amity University Haryana
Gurugram, Haryana, India

Khushbu Jain
CBFS
Amity School of Applied Sciences,
  Amity University Haryana
Gurugram, Haryana, India

Sarika Jain
Amity School of Applied Sciences
Amity University Haryana
Gurugram, Haryana, India

Rohit Jakhar
Department of Civil Engineering
Amity University HaryanaHaryana,
  India

Hemant Joshi
Department of Chemistry
School of Chemical Sciences and
  Pharmacy, Central University of
  Rajasthan
Ajmer, Rajasthan, India

W.S. Jouri
MPESG, Department of Mechanical
  and Aeronautical Engineering
Salford University
Manchester, United Kingdom

Ali Kadir
MPESG, Department of Mechanical
  and Aeronautical Engineering
Salford University
Manchester, United Kingdom

Ishita Kapil
Department of Physics
Amity School of Applied Sciences,
  Amity University Haryana
Gurugram, Haryana, India

Bibi Shaguftah Khatoon
Department of Applied Chemistry
  (CBFS - ASAS)
Amity University Haryana
Gurugram, Haryana, India

S. Kuharat
Multi-Physical Engineering Sciences
  Group (MPESG), Department
  of Mechanical/Aeronautical
  Engineering
Salford University
Manchester, United Kingdom

Deepak Kumar
Department of Mathematics
Lovely Professional University
Jalandhar, India

Ritik Kumar
Department of Physics
Amity School of Applied Sciences,
  Amity University Haryana
Gurugram, Haryana, India

Henry J. Leonard
Multi-Physical Engineering Sciences
  Group (MPESG), Department
  of Mechanical/Aeronautical
  Engineering
Salford University
Manchester, United Kingdom

Sudip Majumder
International Institute of Innovation and
    Technology
Kolkata, West Bengal, India

Sapna Meena
Department of Chemistry
SMT. Madi Bai Mirdha Govt. Girls
    College
Nagaur, Rajasthan, India

Theo Morand
Aerospace Engineering Program, IPSA
Paris, France

J. Pattison
Hyde Group – Aerospace Engineering
Dukinfield, Manchester, United
    Kingdom

Prabha Gurawalia
Department of Chemistry,
    Biochemistry, and Forensic Science
Amity School of Applied Sciences,
    Amity University Haryana
Gurugram, Haryana, India

Anju Rani
Department of Physics
Amity School of Applied Sciences,
    Amity University Haryana
Gurugram, Haryana, India

Gyandshwar Kumar Rao
Department of Chemistry,
    Biochemistry, and Forensic Science
Amity School of Applied Sciences,
    Amity University Haryana
Gurugram, Haryana, India

Renuka Kanojia
Department of Chemistry,
    Biochemistry, and Forensic Science
Amity School of Applied Sciences,
    Amity University Haryana
Gurugram, Haryana, India

Partha Roy
Department of Chemistry
School of Chemical Sciences and
    Pharmacy, Central University of
    Rajasthan
Ajmer, Rajasthan, India

Anand Salvi
Department of Chemistry,
    Biochemistry, and Forensic Science
Amity School of Applied Sciences,
    Amity University Haryana
Gurugram, Haryana, India

Manish Shandilya
Department of Chemistry,
    Biochemistry, and Forensic Science
Amity School of Applied Sciences,
    Amity University Haryana
Gurugram, Haryana, India

Kamal Nayan Sharma
Department of Applied Chemistry
    (CBFS - ASAS)
Amity University Haryana
Gurugram, Haryana, India

Sachin Shaw
Department of Mathematics and
    Statistical Sciences
Botswana International University of
    Science and Technology
Palapye, Botswana

Chander Shekhar
Department of Physics
Amity School of Applied Sciences,
    Amity University Haryana
Gurugram, Haryana, India

Godfrey Shoko
Department of Mathematics and
    Statistical Sciences
Botswana International University of
    Science and Technology
Palapye, Botswana

Md Abrar Siddiquee
Centre for Interdisciplinary Research in
    Basic Science
Jamia Millia Islamia
New Delhi, India

Yogesh Kumar Sonia
Materials Electrochemical & Energy
    Storage Laboratory, Department of
    Chemistry
Malaviya National Institute of
    Technology
Jaipur, Rajasthan, India

Chandra Mohan Srivastava
Department of Chemistry,
    Biochemistry, and Forensic Science
Amity School of Applied Sciences,
    Amity University Haryana
Gurugram, Haryana, India

B. Vasu
Department of Mathematics
Motilal Nehru National Institute of
    Technology Allahabad
Prayagraj, Uttar Pradesh, India

Sachin Vats
Department of Physics
Amity School of Applied Sciences,
    Amity University Haryana
Gurugram, Haryana, India

Krishna Kumar Yadav
Department of Chemical Engineering
Sami Shamoon College of Engineering
Be'er Sheva, Israel

Nisha Yadav
Department of Chemistry,
    Biochemistry, and Forensic Science
Amity School of Applied Sciences,
    Amity University Haryana
Gurugram, Haryana, India

Pinky Yadav
Department of Physics
Amity School of Applied Sciences,
    Amity University Haryana
Gurugram, Haryana, India

Part I

Eco-Material Technologies

1 Exploring Green Materials for Building a Sustainable Future
Need, Challenges, and Strategies

Rohit Jakhar and Sarika Jain

1.1 INTRODUCTION

The construction industry is at a critical crossroads, facing the pressing need for environmental stewardship and the challenge of climate change. Green building materials represent a beacon of hope and transformation in this dynamic landscape. These materials are the cornerstones of sustainable architecture, the elemental choices that have the power to redefine cities, and the soul of conscientious construction. The journey ahead will explore the various facets of green building materials, including natural materials like wood, bamboo, and cork; recycled materials like salvaged metal, glass, and concrete; and innovative materials like aerogels, hempcrete, and bio-based plastics. Beyond the mere introduction to these materials lies a deeper exploration of their tangible benefits, such as enhanced energy efficiency, reduced carbon footprints, and improved indoor air quality. Sustainable sourcing is also highlighted, with responsible material extraction and ethical considerations paving the way for a brighter future. Real-world case studies and examples will show how these materials are making their mark on residential homes, commercial spaces, and even government initiatives. The impact of green materials on architectural design will be explored, diving into their influence on aesthetics, functionality, and overall building performance. Emerging trends and innovations, such as smart materials and 3D printing, will open our eyes to the cutting-edge possibilities that lie ahead. By embracing the potential of green building materials, we can create a sustainable future where buildings are not just structures but living entities that breathe life into our planet. Sustainable building is a growing movement that recognizes the pivotal role of construction in shaping our world's destiny. By understanding green building materials, we can make informed choices and take meaningful steps towards a more sustainable built environment. By incorporating environmental impact assessment tools, regulations, and certifications, we can build structures that not only serve as structures but also leave a legacy of stewardship, innovation, and hope. By embarking on this journey, we can create a better world and contribute to a sustainable future crafted with green building materials.

1.2 TYPES OF GREEN BUILDING MATERIALS

The diverse world of green building materials offers a spectrum of choices, each possessing unique qualities that contribute to sustainability and eco-friendliness. In this section, we will explore three main categories of green building materials: natural materials, recycled materials, and innovative materials [1].

1.2.1 NATURAL MATERIALS

1.2.1.1 Nature's Gift to Construction

Wood: Often regarded as the quintessential building material, wood stands as a testament to nature's capacity for creating both beauty and functionality. Its renewable nature and adaptability make it a top choice for sustainable construction. Timber sourced from responsibly managed forests forms the backbone of green architecture [2].

Benefits:

- Renewable Resource: Trees can be re-grown, making wood a renewable resource when managed responsibly.
- Energy Efficiency: Wood is an excellent insulator, reducing the need for additional heating and cooling.
- Aesthetic Appeal: Its natural beauty adds warmth and character to structures.

Bamboo: With its rapid growth rate and remarkable strength-to-weight ratio, bamboo is a sustainable alternative to traditional hardwoods. It is an emblem of eco-friendly construction and exemplifies the potential of renewable resources [2].

Benefits:

- Rapid Growth: Bamboo matures in a few years, unlike hardwood trees that can take decades.
- Strength: Despite its lightweight nature, bamboo exhibits impressive tensile strength.
- Versatility: Bamboo's versatility extends to flooring, furniture, and structural elements.

Cork: Derived from the bark of cork oak trees, cork is a natural wonder. Its resilience and insulation properties make it a compelling choice for green construction [2].

Benefits:

- Sustainability: Cork harvesting is sustainable, as trees are not cut down during the process.
- Thermal Insulation: Cork's cellular structure provides excellent thermal insulation.

- Acoustic Properties: It dampens sound and contributes to a quieter indoor environment.

1.2.2 RECYCLED MATERIALS

1.2.2.1 Reclaiming Resources for a Greener Tomorrow

Recycled Metal: Metals like steel and aluminium are infinitely recyclable. Incorporating recycled metal into construction significantly reduces the environmental impact of extracting and processing raw metals.

Benefits:

- Resource Conservation: Recycling metal conserves natural resources and energy.
- Strength and Durability: Recycled metal retains the strength and durability of virgin metal.
- Endless Recycling: Metal can be recycled indefinitely without quality loss.

Recycled Glass: Glass recycling not only diverts waste from landfills but also conserves the energy required for producing new glass. Recycled glass can be used in architectural elements like countertops and tiles.

Benefits:

- Energy Savings: Recycling glass consumes less energy compared to producing new glass.
- Colour Variations: Recycled glass offers unique colours and textures.
- Eco-Friendly Design: It adds an eco-conscious touch to architectural features.

Recycled Concrete: The demolition of old concrete structures generates vast amounts of waste. However, recycling concrete for use in new construction mitigates environmental impact.

Benefits:

- Waste Reduction: Recycling concrete diverts construction waste from landfills.
- Resource Efficiency: Crushed concrete can replace virgin aggregates in new concrete mixes.
- Cost Savings: It often comes at a lower cost than using new aggregates.

1.2.3 INNOVATIVE MATERIALS

1.2.3.1 Pushing the Boundaries of Sustainability

Aerogels: Aerogels are ultralight, highly porous materials known for their exceptional insulating properties. In construction, they are used to enhance energy efficiency [3].

Benefits:

- Superior Insulation: Aerogels provide high thermal resistance with minimal thickness.
- Energy Savings: Improved insulation reduces heating and cooling costs.
- Lightweight: Their low density minimizes structural load

Hempcrete: Hempcrete is a bio-composite material made from the inner fibres of the hemp plant, lime, and water. It offers a sustainable alternative to traditional concrete [3].

Benefits:

- Low Carbon Footprint: Hemp absorbs carbon dioxide during growth, offsetting emissions.
- Natural Insulation: Hempcrete provides natural thermal and acoustic insulation.
- Moisture Regulation: It helps maintain comfortable indoor humidity levels.

Bio-based Plastics: Derived from renewable resources like cornstarch or sugarcane, bio-based plastics have applications in sustainable construction.

Benefits:

- Renewable Source: They reduce reliance on fossil fuels for plastic production.
- Biodegradable: Some bio-based plastics are biodegradable, reducing long-term waste.

Each of these categories represents a conscious choice towards sustainability in construction. Green building materials have transcended their traditional counterparts, offering not only durability and functionality but also a commitment to a healthier, more sustainable planet. As architects, builders, and consumers increasingly prioritize eco-conscious construction, the utilization of these materials becomes not just a choice but a responsibility.

In the subsequent sections of this chapter, we will explore in-depth the advantages these materials offer, their role in sustainable sourcing, real-world case studies, and how they influence architectural design. Let's embark on this journey into the heart of sustainable construction, where the foundations of a greener future are laid with every choice of material.

1.3 BENEFITS AND ADVANTAGES OF GREEN BUILDING MATERIALS

Green building materials aren't just a nod to sustainability; they are the essential building blocks of a brighter, more eco-conscious future. They offer a myriad of benefits that extend far beyond their traditional counterparts. Understanding and embracing these advantages are crucial for making informed decisions in construction and contributing to a healthier planet [3].

1. **Energy Efficiency**

   Green building materials are renowned for their superior energy efficiency. Here's how they contribute to reduced energy consumption in buildings:

   Natural Insulation: Many green materials, such as cork and wool, have excellent insulating properties. They help maintain stable indoor temperatures, reducing the need for excessive heating or cooling.

   Thermal Mass: Materials like rammed earth and concrete possess high thermal mass, absorbing and slowly releasing heat. This stabilizes indoor temperatures, reducing the need for constant heating or cooling.

   Reflective Surfaces: Some materials, like cool roofing materials, reflect sunlight and heat, reducing the heat gain in buildings and lowering air conditioning requirements.

   Energy-efficient Lighting: Green materials also include energy-efficient lighting options, like LED fixtures, which reduce electricity consumption.

2. **Reduced Carbon Footprint**

   One of the most compelling advantages of green building materials is their ability to significantly reduce the carbon footprint of construction [4]:

   Lower Embodied Energy: Green materials are often produced with lower energy inputs compared to traditional materials. For example, recycled steel requires less energy to produce than virgin steel.

   Carbon Sequestration: Some materials, like wood, store carbon throughout their lifespan, effectively sequestering carbon dioxide and mitigating greenhouse gas emissions.

   Longevity: Green materials, known for their durability, reduce the need for frequent replacements and the associated energy and resource consumption.

3. **Improved Indoor Air Quality**

   Green building materials prioritize human health by promoting better indoor air quality:

   Low VOCs (Volatile Organic Compounds): Green materials are often low in VOCs, which are harmful chemicals that can off-gas from traditional materials and negatively affect indoor air quality.

   Natural Breathability: Materials like clay plaster and lime wash have natural breathability, allowing moisture to evaporate, reducing the risk of mild growth, and contributing to healthier indoor environments.

4. **Longevity and Durability**

   Green materials are chosen not only for their sustainability but also for their longevity and durability:

   Resilience: Many green materials are naturally resilient and can withstand harsh environmental conditions, reducing maintenance and replacement costs.

Reduced Lifecycle Costs: While some green materials may have slightly higher upfront costs, their longevity often translates into significant cost savings over the building's lifecycle.

5. **Cost Savings Over Time**

The initial investment in green materials can be justified by long-term cost savings.

Energy Cost Reduction: Lower energy consumption leads to reduced utility bills over the life of a building.

Maintenance Savings: The durability of green materials means fewer repairs and replacements, reducing maintenance costs.

Government Incentives: Some regions offer incentives, tax breaks, or rebates for sustainable construction practices, further enhancing cost-effectiveness.

Green building materials are practical choices that benefit the environment, save operating costs, improve occupant health, and contribute to a more resilient and sustainable future. They go beyond just satisfying sustainability targets. The benefits they provide, including as energy efficiency and a less carbon footprint, highlight their critical role in sustainable building. The use of these materials becomes more than simply a choice; it becomes a responsibility to create not just for the present but for future generations as architects, builders, and customers place an increasing emphasis on eco-conscious construction. We will examine how these materials might be obtained sustainably and incorporated into building designs in the next chapters, highlighting their useful applications and real-world success stories [5].

1.4   SUSTAINABLE SOURCING OF GREEN BUILDING MATERIALS

Sustainability in building involves all aspects of a material's lifespan, from extraction to disposal, and goes beyond the selection of that material. A crucial step in reducing the environmental effect of construction projects is the sustainable sourcing of green building materials. It makes sure that ethical and social standards are respected throughout the supply chain and that the materials utilized are ethically harvested and handled.

6. **Responsible Material Extraction**

The responsible extraction of materials forms the foundation of sustainable sourcing:

Timber Harvesting: When it comes to wood, sustainable forestry practices are key. This includes selective logging, reforestation efforts, and adherence to certification standards such as the Forest Stewardship Council (FSC) or Programme for the Endorsement of Forest Certification (PEFC).

Stone and Quarrying: For materials like stone, responsible quarrying practices involve minimizing habitat disruption, erosion control, and ensuring safe working conditions for labourers.

Mineral Mining: Ethical mineral extraction involves minimizing ecological disturbance, managing waste, and upholding labour standards.

Certification programmes like the Responsible Jewellery Council (RJC) address ethical sourcing of minerals.

7. **Ethical Considerations**

Sustainable sourcing extends to ethical considerations, including labour practices and social impacts:

Fair Labour Practices: Ethical sourcing requires ensuring fair wages, safe working conditions, and equitable treatment of workers throughout the supply chain.

Conflict Minerals: In cases involving minerals such as gold, tin, tungsten, and tantalum, responsible sourcing ensures that materials do not fund armed conflict or human rights abuses. Compliance with regulations like the Dodd-Frank Act is essential.

Community Engagement: Engaging with local communities to understand and address their concerns is a hallmark of ethical sourcing. This may involve land rights, community development, and environmental stewardship.

8. **Local vs. Global Sourcing**

The choice between sourcing materials locally or globally depends on various factors:

Transportation Emissions: Local sourcing often results in fewer transportation emissions, especially for heavy materials. However, global sourcing may provide access to unique sustainable options.

Resource Availability: Local availability of specific materials can influence sourcing decisions. Regions with abundant renewable resources like bamboo may favour local sourcing.

Cost and Availability: Cost considerations and the availability of green materials can vary between regions. Local materials may be more cost-effective in some cases.

9. **Certification and Verification**

Certification programmes and verification processes play a pivotal role in ensuring the sustainability and ethical sourcing of materials:

Forest Certification: Programmes like FSC and PEFC certify responsible forestry practices for wood products.

Fair Trade Certification: Fair Trade Certification ensures ethical labour practices and community benefits in various industries.

Conflict-Free Certification: Certification schemes and third-party audits verify that minerals are sourced without contributing to conflict.

10. **Building Material Transparency**

Transparency initiatives and tools, such as Environmental Product Declarations (EPDs) and Health Product Declarations (HPDs), provide information about the environmental and health impacts of building materials. This transparency allows architects, builders, and consumers to make informed choices.

11. **Design for Reuse and Recycling**

In addition to sourcing sustainable materials, sustainable construction also considers the end of life of materials. Designing buildings with materials that can be easily reused or recycled reduces waste and supports circular economy principles [6].

12. **Regulatory Compliance**

Sustainable sourcing often aligns with regional and international regulations aimed at environmental and social responsibilities. Complying with these regulations ensures that materials meet recognized standards of sustainability and ethical sourcing.

The construction sector is required by ethics and the environment to source green building materials sustainably. It necessitates a comprehensive strategy that takes into account ethical labour practices, responsible extraction, and the social effects of material sourcing. Environmental and economic factors must be taken into account while deciding between local and global sources. Tools like certification, openness, and regulatory compliance are crucial for guaranteeing that materials adhere to ethical and sustainable norms. The building sector makes tremendous progress towards lowering its environmental impact and promoting a more ethical and sustainable future by obtaining materials ethically. We will examine actual case studies and examples that show sustainable sourcing in action in the next sections of this chapter, showing how moral and responsible decisions have an impact on the construction industry.

1.5 CASE STUDIES AND EXAMPLES

Case Study 1: The Bullitt Centre, Seattle, Washington, USA

Key Takeaways:

Embrace Wood as a Sustainable Material: The Bullitt Centre's extensive use of wood, especially cross-laminated timber (CLT), demonstrates that wood can be a viable and sustainable alternative to conventional building materials. It showcases the importance of responsible timber sourcing and certification (e.g., FSC) for environmentally conscious construction.

Innovate for Energy Efficiency: Passive design strategies, coupled with the choice of green materials, can result in exceptional energy efficiency. The Bullitt Centre's commitment to energy reduction highlights the potential for green buildings to dramatically decrease energy consumption.

Case Study 2: The Edge, Amsterdam, Netherlands

Key Takeaways:

Smart Materials and Technology Integration: The Edge exemplifies the integration of smart materials, such as energy-efficient glass, with advanced technology to optimize energy use. This case study underscores the importance of incorporating cutting-edge solutions for energy efficiency in modern construction [7].

Recycled Materials for Sustainability: The building's use of recycled materials not only reduces waste but also showcases the feasibility of recycling in construction. It emphasizes the role of sustainable sourcing in minimizing the environmental footprint of building projects.

Case Study 3: The Change Initiative Building, Dubai, UAE

Key Takeaways:

Solar Energy Integration: The Change Initiative Building's use of solar panels highlights the potential for renewable energy sources in arid regions. It underscores the importance of considering local environmental conditions when selecting green materials and energy sources [8].

Resource Efficiency in Harsh Environments: Dubai's challenging environment necessitates innovative approaches to sustainability. The case study demonstrates the feasibility of sustainable sourcing and energy-efficient materials in regions with resource constraints.

Example 1: Bamboo in Traditional Building Practices

Key Takeaways:

Traditional Wisdom: Traditional practices, such as using bamboo in construction, often align with sustainable principles. This example highlights the wisdom of indigenous knowledge and its relevance in contemporary sustainable construction.

Rapid Renewable Resource: Bamboo's fast growth and versatility make it a remarkable renewable resource. This example emphasizes the potential for renewable materials to meet modern construction needs while minimizing environmental impact.

Example 2: Hempcrete in Residential Construction

Key Takeaways:

Natural Insulation: Hempcrete's insulation properties demonstrate the importance of considering materials that contribute to energy efficiency in residential construction. Its use in walls and foundations can reduce heating and cooling demands.

Sustainable Farming Practices: Hemp's minimal water and pesticide requirements showcase the potential for sustainable agriculture to support eco-friendly construction materials. This example underscores the importance of holistic sustainability, from sourcing to end-use.

In summary, these case studies and examples collectively emphasize several critical lessons:

Material Innovation: Green building materials can encompass both traditional and cutting-edge materials. Their selection and innovative use can lead to significant improvements in energy efficiency and sustainability.

Sustainable Sourcing: Responsible and ethical sourcing of materials is essential. Certification programmes, recycling, and local resourcing are effective strategies for reducing the environmental and social impact of construction.

Energy Efficiency: Green building materials contribute to energy-efficient buildings, reducing long-term energy consumption and operational costs [9].

Resource Efficiency: The use of recycled materials and rapidly renewable resources, such as bamboo and hemp, demonstrates a commitment to resource efficiency and a circular economy.

Adaptability to Local Conditions: Sustainable sourcing and material choices should consider local environmental conditions and constraints, demonstrating that green construction can thrive even in challenging environments.

These lessons can guide architects, builders, and policymakers in making informed choices and advocating for sustainable practices in the construction industry, ultimately contributing to a greener, more sustainable future.

1.6 DESIGN CONSIDERATIONS WITH GREEN BUILDING MATERIALS

13. Architectural Implications

- *Integration of Natural Elements*: Green building materials often have unique textures and colours that can enhance a building's aesthetic. Architects should consider how to incorporate these elements into the overall design to create visually appealing and harmonious structures.

- *Adaptive Design*: The choice of materials can influence the design's adaptability to changing environmental conditions. For example, designing with natural ventilation in mind can reduce the need for energy-intensive HVAC systems.

- *Flexibility*: The design should allow for future upgrades or changes that may involve additional green materials or sustainable technologies.

14. Aesthetic Considerations

- *Texture and Colour Harmony*: Architects should explore how the texture and colour of green materials complement the building's overall design and surrounding environment. Earthy tones, wood grains, and natural finishes can create a visually inviting space.

- *Daylighting and Views*: Design should prioritize maximizing natural light penetration and views to the outdoors, as these elements are not only visually pleasing but also contribute to occupant well-being and energy savings.

- *Sustainable Landscaping*: Integrating sustainable landscaping and green roofs into the design can further enhance the building's aesthetics while providing additional environmental benefits.

15. Compatibility with Various Building Types

- *Residential Design*: Green materials can be tailored to residential structures to improve energy efficiency, indoor air quality, and aesthetics. Examples include using sustainable flooring options like bamboo or incorporating energy-efficient windows and insulation.

- *Commercial and Industrial Design*: Green materials in commercial buildings can focus on optimizing energy performance, indoor air quality, and sustainability certifications like LEED. The design may involve advanced lighting systems, smart HVAC controls, and sustainable office layouts.

- *Institutional and Government Buildings*: Public buildings often have a responsibility to lead by example. Design considerations should emphasize transparency, sustainable practices, and user comfort. Green materials can be integrated into government offices, schools, and healthcare facilities.

16. **Energy-Efficient Design**
    - *Passive Design*: Passive design strategies, such as optimizing building orientation for natural light and heat gain, should be integrated into the overall building design. Green materials can work in synergy with passive design to enhance energy efficiency.
    - *Thermal Insulation*: The design should prioritize effective thermal insulation to reduce heating and cooling demands. Green materials like natural wool or recycled insulation can play a significant role in achieving this.
    - *Renewable Energy Integration*: The design should consider how to incorporate renewable energy sources, such as solar panels or wind turbines, seamlessly into the building's architecture.

17. **Sustainable Site Planning**
    - *Material Transport*: Minimizing the transportation distance of materials to the construction site reduces carbon emissions. Design should consider local sourcing of materials whenever possible.
    - *Water Efficiency*: Sustainable site design can include rainwater harvesting systems, permeable pavements, and drought-tolerant landscaping to conserve water resources.
    - *Biodiversity Preservation*: The site design should aim to protect and enhance local ecosystems and biodiversity, promoting a healthy natural environment.

18. **Building Envelope and Insulation**
    - *High-Performance Envelope*: The design should prioritize a high-performance building envelope that minimizes air leakage and thermal bridging. Green materials can be used for insulation, siding, and roofing to enhance energy efficiency.
    - *Air Tightness*: Attention to detail in design and construction can ensure an airtight building envelope, reducing drafts and heat loss.

19. **Life Cycle Assessment**
    - *Long-Term Sustainability*: The design should consider the long-term sustainability of materials. Conducting a life cycle assessment can help evaluate the environmental impact of materials throughout their entire lifespan, from production to disposal [5].

20. **Maintenance and Durability**
    - *Low-Maintenance Materials*: Whenever possible, select low-maintenance green materials that require minimal upkeep over time, reducing long-term maintenance costs.

- *Durability*: The design should prioritize materials known for their durability to ensure the building's longevity and minimize resource consumption [5].

21. **Building Certification and Compliance**
    - *Design for Certification*: If aiming for sustainability certifications like LEED, the design should align with the requirements of these programmes and consider how the chosen materials contribute to certification goals.

In conclusion, including green building materials in a construction project necessitates an all-encompassing design strategy. In order to ensure that these materials are smoothly incorporated into the overall concept for the project, architects and designers should take into account the aesthetic, functional, energy-efficient, and sustainable elements of these materials. Buildings may achieve increased sustainability and offer healthier, more pleasant places for people by addressing design factors that support green materials.

1.7 EMERGING TREND AND INNOVATIONS

Some emerging trends and innovations in the world of green building materials and construction, along with relevant examples that highlight these exciting developments, are given as follows:

22. **Smart Materials and Building Systems**

    Emerging Trend: The integration of smart materials and systems into construction is revolutionizing building performance and sustainability.

    Example: Dynamic Glass Windows – Dynamic glass windows, like those by View, can tint or change transparency in response to environmental conditions. They optimize natural light, reduce heat gain, and decrease the need for artificial lighting and cooling, contributing to energy savings and occupant comfort.

23. **3D Printing in Construction**

    Emerging Trend: 3D printing technology is being used to create entire building components, offering design flexibility, reduced waste.

    Example: Apis Cor's 3D-Printed House – Apis Cor, a construction tech company, 3D-printed a residential house in less than 24 hours. This innovation reduces construction waste and allows for design customization with sustainable materials.

24. **Transparent Solar Panels**

    Emerging Trend: Transparent solar panels are integrated into building facades and windows, harnessing sunlight without obstructing views.

    Example: Ubiquitous Energy's Transparent Solar Windows – Ubiquitous Energy has developed transparent solar cells that can be applied to windows to generate electricity while still allowing visible light to pass through. This technology can offset energy usage in buildings and reduce reliance on traditional power sources.

25. **Biodegradable and Bio-Based Materials**

    Emerging Trend: The use of biodegradable and bio-based materials is gaining traction, offering sustainable alternatives to traditional construction materials.

    Example: Mycelium-based Bricks – Companies like Ecovative Design use mycelium, the root structure of fungi, to create biodegradable building materials. Mycelium-based bricks are lightweight and strong and can decompose naturally, reducing long-term environmental impact.

26. **Carbon-Negative Materials**

    Emerging Trend: Some materials are engineered to capture and store more carbon dioxide than is emitted during their production.

    Example: Carbon Cure's Carbon-Negative Concrete – Carbon Cure's technology injects captured CO_2 into concrete during mixing. As the concrete cures, the CO_2 is mineralized, reducing the overall carbon footprint of the material.

27. **Sustainable Insulation Materials**

    Emerging Trend: Insulation materials are evolving to offer improved thermal performance while being eco-friendly.

    Example: Aerogel Insulation – Aerogels, like Aspen Aerogels' Space loft, are highly efficient insulators that are incredibly lightweight and composed of silica aerogel. They are used to enhance the energy efficiency of buildings while minimizing bulk.

28. **Recycled Ocean Plastic Materials**

    Emerging Trend: The growing concern over plastic pollution in oceans has led to the development of construction materials made from recycled ocean plastics.

    Example: The Ocean Cleanup's Interceptors – While not a construction material per se. The Ocean Cleanup's Interceptors are designed to collect plastic waste from rivers before it reaches the oceans. The collected plastics can potentially be used in recycled construction materials.

29. **Living Building Materials**

    Emerging Trend: Materials that can actively contribute to a building's sustainability, such as those with self-healing or self-cleaning properties.

    Example: Self-Healing Concrete – Researchers are developing concrete that can repair its own cracks using bacteria or other healing agents. This technology can extend the lifespan of structures and reduce maintenance requirements.

The construction sector is seeing exciting new developments as a result of these new trends and technologies, which are expanding the possibilities for design, energy efficiency, and sustainability. These technologies have the potential to transform the way we construct, making our buildings more environmentally friendly, robust, and responsive to the changing requirements of our world as they develop and become more widely available.

1.8   ENVIRONMENTAL IMPACT ASSESSMENT

30. **Reduction in Carbon Footprint**

Positive Impact: The utilization of green building materials, such as sustainably harvested wood, recycled materials, and carbon-negative substances, contributes to a significant reduction in the carbon footprint of construction projects. These materials sequester carbon, offsetting emissions generated during their production and transportation.

31. **Energy Efficiency and Reduction in Energy Consumption**

Positive Impact: The incorporation of green materials with superior insulation properties and energy-efficient attributes in building design leads to reduced energy consumption. This, in turn, lowers greenhouse gas emissions associated with electricity and heating, positively impacting the environment.

32. **Sustainable Sourcing Practices**

Positive Impact: The emphasis on sustainable sourcing of green materials, including responsible forestry practices and ethical mineral extraction, promotes environmentally responsible resource management. This reduces habitat disruption, conserves biodiversity, and minimizes the negative ecological impacts of extraction [3].

33. **Minimized Waste and Resource Conservation**

Positive Impact: The use of recycled materials and the emphasis on recyclability and reusability in green materials minimize construction waste. This conserves resources and reduces the burden on landfills, mitigating the environmental impact of construction.

34. **Improved Indoor Air Quality**

Positive Impact: Green materials are often low in volatile organic compounds (VOCs) and other harmful chemicals. Their use enhances indoor air quality, reducing health risks associated with poor indoor air. This promotes occupant well-being and comfort.

35. **Preservation of Ecosystems**

Positive Impact: Green materials that prioritize rapid renewable resources like bamboo help preserve natural ecosystems by reducing the pressure on old-growth forests. This approach protects diverse plant and animal species and contributes to overall ecosystem health.

36. **Reduced Water Consumption**

Positive Impact: Sustainable sourcing and construction practices often incorporate water-saving technologies and materials. This can include the use of rainwater harvesting systems and water-efficient landscaping, reducing overall water consumption and conserving this vital resource.

37. **Encouragement of Sustainable Agriculture**

Positive Impact: The use of bio-based materials, like hempcrete, supports sustainable agriculture. These materials require minimal water and pesticide use, promoting eco-friendly farming practices and reducing environmental pollution.

38. **Fostering Innovation and Awareness**

    Positive Impact: The exploration and adoption of emerging trends and innovations in green materials drive research and development efforts within the construction industry. This fosters innovation, encourages sustainable practices, and raises awareness about environmental responsibility.

39. **Long-Term Sustainability and Resilience**

    Positive Impact: The durability of green materials and their contribution to energy efficiency result in structures that are more resilient to environmental challenges. This, in turn, reduces the environmental impact associated with the frequent replacement or repair of building components.

In conclusion, using green building materials when building has a mainly good effect on the environment. These materials help with waste reduction, enhanced indoor air quality, responsible sourcing, carbon reduction, and energy efficiency. The construction sector contributes significantly to reducing its environmental impact and advancing the cause of global sustainability by encouraging sustainable practices and technologies.

1.9 REGULATION AND CERTIFICATIONS

There are undoubtedly several laws and certifications that help to ensure ecologically acceptable practices when it comes to green building materials and sustainable construction. Standards, rules, and compliance with sustainable construction methods are all aided by these laws and certifications. Here are a few noteworthy instances:

40. **LEED (Leadership in Energy and Environmental Design)**

    Certification: LEED Certification is one of the most widely recognized sustainability certifications for buildings worldwide.

    Regulatory Influence: LEED standards have influenced building codes and regulations in various regions. Many jurisdictions offer incentives, such as tax breaks or expedited permitting, for projects that achieve LEED certification.

41. **ENERGY STAR**

    Certification: ENERGY STAR certification is awarded to buildings and products that meet strict energy efficiency and environmental performance criteria.

    Regulatory Influence: ENERGY STAR ratings are often used by governments to set energy efficiency standards and requirements for buildings and appliances.

42. **Building Codes and Standards (International Building Code, ASHRAE, etc.)**

    Regulations: Building codes and standards, such as the International Building Code (IBC) and ASHRAE (American Society of Heating, Refrigerating and Air-Conditioning Engineers) standards, incorporate sustainability principles and guidelines.

Regulatory Influence: Governments adopt and update building codes and standards to promote sustainable construction practices and energy-efficient design.

43. **Forest Stewardship Council (FSC) Certification**

Certification: FSC Certification ensures responsible forest management and sustainable sourcing of wood products.

Regulatory Influence: Some regions require or encourage the use of FSC-certified wood in public construction projects to promote sustainable forestry practices.

44. **Cradle to Cradle (C2C) Certification**

Certification: Cradle to Cradle Certification evaluates products based on their material health, material reutilization, renewable energy use, water stewardship, and social fairness.

Regulatory Influence: While not a regulatory body, Cradle to Cradle principles have influenced product design and purchasing decisions in government and industry.

45. **Green Building Councils and Organizations (e.g., USGBC, IGBC)**

Regulatory Influence: Green Building Councils and organizations often collaborate with governments to develop and promote sustainable building practices. They may also offer certifications like LEED (USGBC) and Green Building Rating Systems (IGBC).

46. **EPEAT (Electronic Product Environmental Assessment Tool)**

Certification: EPEAT certification focuses on the environmental performance of electronic products.

Regulatory Influence: Some governments require or encourage the purchase of EPEAT-certified electronics for government use to reduce electronic waste and energy consumption.

47. **Regional and Local Green Building Programs**

Regulations: Many regions and localities have developed their own green building programmes and certifications that align with regional environmental priorities.

Regulatory Influence: These programmes may influence local building codes and regulations to encourage sustainable construction practices [10].

48. **Eco-labels and Product Certifications (e.g., Green Seal, Blue Angel)**

Certifications: Eco-labels and product certifications assess the environmental impact of building materials and products.

Regulatory Influence: Government procurement policies may require or prioritize the use of eco-labelled or certified products in public construction projects [10].

49. **International Agreements (e.g., Paris Agreement)**

Regulations: International agreements like the Paris Agreement set goals for reducing greenhouse gas emissions and addressing climate change.

Regulatory Influence: National and local governments may adopt policies and regulations to align with international commitments, influencing construction practices and material choices [10].

These regulations and certifications are essential tools for promoting sustainable construction and the use of green building materials. They encourage responsible sourcing, energy efficiency, reduced environmental impact, and healthier indoor environments, helping to create a more sustainable and resilient built environment [10].

1.10 PRACTICAL GUIDANCE FOR IMPLEMENTATION

Certainly, in the context of green building materials and sustainable construction, there are several regulations and certifications that play a significant role in ensuring environmentally responsible practices. These regulations and certifications help set standards, provide guidelines, and encourage compliance with sustainable building practices. Here are some notable examples:

50. **LEED (Leadership in Energy and Environmental Design)**

     Certification: LEED Certification is one of the most widely recognized sustainability certifications for buildings worldwide.

     Regulatory Influence: LEED standards have influenced building codes and regulations in various regions. Many jurisdictions offer incentives, such as tax breaks or expedited permitting, for projects that achieve LEED certification.

51. **ENERGY STAR**

     Certification: ENERGY STAR certification is awarded to buildings and products that meet strict energy efficiency and environmental performance criteria.

     Regulatory Influence: ENERGY STAR ratings are often used by governments to set energy efficiency standards and requirements for buildings and appliances.

52. **Building Codes and Standards (International Building Code, ASHRAE, etc.)**

     Regulations: Building codes and standards, such as the International Building Code (IBC) and ASHRAE (American Society of Heating, Refrigerating and Air-Conditioning Engineers) standards, incorporate sustainability principles and guidelines.

     Regulatory Influence: Governments adopt and update building codes and standards to promote sustainable construction practices and energy-efficient design.

53. **Forest Stewardship Council (FSC) Certification**

     Certification: FSC Certification ensures responsible forest management and sustainable sourcing of wood products.

     Regulatory Influence: Some regions require or encourage the use of FSC-certified wood in public construction projects to promote sustainable forestry practices.

54. **Cradle to Cradle (C2C) Certification**

     Certification: Cradle to Cradle Certification evaluates products based on their material health, material reutilization, renewable energy use, water stewardship, and social fairness.

Regulatory Influence: While not a regulatory body, Cradle to Cradle principles have influenced product design and purchasing decisions in government and industry.

55. **Green Building Councils and Organizations (e.g., USGBC, IGBC)**

    Regulatory Influence: Green Building Councils and organizations often collaborate with governments to develop and promote sustainable building practices. They may also offer certifications like LEED (USGBC) and Green Building Rating Systems (IGBC).

56. **EPEAT (Electronic Product Environmental Assessment Tool)**

    Certification: EPEAT certification focuses on the environmental performance of electronic products.

    Regulatory Influence: Some governments require or encourage the purchase of EPEAT-certified electronics for government use to reduce electronic waste and energy consumption.

57. **Regional and Local Green Building Programmes**

    Regulations: Many regions and localities have developed their own green building programmes and certifications that align with regional environmental priorities.

    Regulatory Influence: These programmes may influence local building codes and regulations to encourage sustainable construction practices.

58. **Eco-labels and Product Certifications (e.g., Green Seal, Blue Angel)**

    Certifications: Eco-labels and product certifications assess the environmental impact of building materials and products.

    Regulatory Influence: Government procurement policies may require or prioritize the use of eco-labelled or certified products in public construction projects.

59. **International Agreements (e.g., Paris Agreement)**

    Regulations: International agreements like the Paris Agreement set goals for reducing greenhouse gas emissions and addressing climate change.

    Regulatory Influence: National and local governments may adopt policies and regulations to align with international commitments, influencing construction practices and material choices.

These laws and certifications are crucial instruments for advancing green building practices and sustainable development. They promote ethical sourcing, energy efficiency, minimal environmental effect, and better indoor settings, all of which contribute to the development of a built environment that is more resilient and sustainable.

1.11　CONCLUSION

The exploration of the world of green building materials has revealed an environment rich in creativity, accountability, and potential. A new age of sustainability and environmental stewardship has begun as a result of our investigation into the transforming power of materials that go beyond traditional building methods.

Green building materials play a major role in determining the future of architecture, from the soaring buildings covered with dynamic glass windows that adjust to the rhythms of nature to the modest homes made from bamboo's embrace. These materials exemplify development because they combine modern technology with natural elegance to produce environments that are not only useful but also environmentally friendly.

Green building materials have an influence on many levels. By lowering carbon footprints, they pave the way for a time when buildings actively combat climate change. Materials are sourced carefully to guarantee forest health and moral mineral extraction. Reducing waste encourages a circular economy in which nothing is wasted.

Indoor spaces are turned into wellness havens that are free of dangerous chemicals and flooded with natural light. Rooted in the community, sustainable agriculture protects vulnerable environments. Energy efficiency is made the standard rather than the exception, and water is valued rather than wasted. But it's not only the materials; it's also the concept they represent, which is one of duty to the environment and to the next generation. "The belief in those of nature" , where human demands coexist with those of nature. It is the idea of a forward movement guided by moral principles and a determination to improve and better structures.

Let's bring the lessons we've learned ahead as we close this chapter. Let's build with a sustainable vision, where every material decision is a statement of our dedication to a greener, more sustainable future. Together, we can turn the page towards a more promising future that is based on green building materials, where accountability meets innovation and where development coexists with the environment.

As we set out to construct a sustainable future, one brick, one beam, and one green construction material at a time, the journey continues to be an exciting one.

REFERENCES

1. T. Theis, J. Tomkin (2013). *Sustainability: A Comprehensive Foundation*. University of Illinois at Urbana-Champaign.
2. S. M. Khoshnava, R. Rostami, R. Mohamad Zin, D. Štreimikienė, A. Mardani and M. Ismail (2020). The role of green building materials in reducing environmental and human health impacts. *International journal of environmental research and public health*, 17(7), p.2589.
3. W. T. Tsai (2017). Overview of green building material (GBM) policies and guidelines with relevance to indoor air quality management in Taiwan. *Environments*, 5(1), p.4.
4. I. J. Šenitkova (2017). Indoor air quality–buildings design. In *MATEC Web of Conferences* (Vol. 93, p. 03001). EDP Sciences.
5. P. Babu, G. Suthar (2020). Indoor air quality and thermal comfort in green building: a study for measurement, problem and solution strategies. In *Indoor Environmental Quality: Select Proceedings of the 1st ACIEQ* (pp. 139–146). Springer, Singapore.
6. Y. Wang, C.A. Brebbia (2018). *Sustainable Materials and Technologies*. WIT Press.
7. K. G. Dassios, G. C. Vayenas (2018). *Energy Storage: Fundamentals, Materials, and Applications*. CRC Press.
8. B. Sørensen (2018). *Renewable Energy: Physics, Engineering, Environmental Impacts, Economics & Planning*. Academic Press.
9. C. J. Kibert (2016). *Sustainable Construction: Green Building Design and Delivery*. John Wiley & Sons.
10. K. N. Sheth (2016, April). Sustainable building materials used in green buildings. In *9th International Conference on Engineering and Business Education (ICEBE) & 6th International Conference on Innovation and Entrepreneurship (ICIE)* (pp. 23–26).

2 Numerical Simulation Studies on CdTe-Based Solar Cells

Anju Rani, Mrinal Dutta, Chander Shekhar, and Ayana Bhaduri

2.1 INTRODUCTION

Earth's atmospheric carbon dioxide (CO_2) levels have been increasing at an unprecedented pace during the past century because of global industrialization. The year 2015 marked the first time in recorded history that the average monthly global CO_2 emission reached 400 parts per million. By the middle of the century, the world's total power consumption is predicted to reach 30 TW, and two-thirds of it must come from renewable energy sources like solar and wind if CO_2 levels are to be stabilized by 2050 [1].

Cadmium telluride (CdTe) is a highly effective thin film material for solar cell (TFSC) applications. Its 1.5 eV bandgap makes it almost perfect for solar energy transformation. The extraordinary optical absorption coefficient of the material absorbs nearly entire incident photons within 1–2 μm from the surface, with energy exceeding its bandgap. For thin-film CdTe solar cells, cadmium sulfide (CdS) is the n-type junction associate, which is normally heterojunctions [2].

CdTe technology was said to be roughly 40% less expensive than amorphous Si technology and roughly 30% less expensive than CIGS technology. Therefore, the highest ratio of "efficiency/cost" could be attained by using CdTe-based TFSCs [3].

2.2 CADMIUM TELLURIDE (CdTe)

The photovoltaic (PV) technology of cadmium telluride (CdTe) is based on the absorption and conversion of sunlight into electrical power by a thin layer of CdTe [4]. The second most common solar cell material worldwide after Si is CdTe, whose popularity is rising rapidly [5]. The sole thin film photovoltaic technology, CdTe solar panels, has surpassed crystalline silicon PV in terms of affordability for a substantial part of the PV market, namely, in multi-kilowatt system [5]. A stable crystalline substance made of tellurium and cadmium is called cadmium telluride, or CdTe. It is mostly utilized as an infrared optical window and as a semiconducting material in CdTe photovoltaic systems. Cadmium sulfide (CdS) is commonly used to initiate a p–n junction in CdTe PV cells [6].

DOI: 10.1201/9781003473749-3

CdTe has been regarded as one of the remarkably effective and affordable TFSC materials with a direct band gap of roughly 1.5 eV, which is almost optimum for the mono-junction solar cell [7]. For a long time, CdTe has been extensively employed in the production of solar cells due to its favorable band gap and elevated optical absorption coefficient [8]. Therefore, when absorbing photons with energy above the band gap, a nominal amount of CdTe is required [9].

2.2.1 Material Properties of CdTe

It is a group II-VI semiconductor as it is composed of cadmium (Cd), which is a II valence electron element, and tellurium (Te), which is a VI valence electron element. Each Cd atom in CdTe is bound to four Te atoms to produce the zinc blende lattice structure [10].

It is possible to produce n-doped CdTe by either swapping tellurium, which has a VI-valence electron, with an element with a VII-valence electron, like atoms of fluorine (F), chlorine (Cl), bromine (Br), or iodine (I), or by substituting the II-valence electron of Cd with III-valence electron atoms like aluminum (Al), gallium (Ga), or indium (In). Therefore, the III- and VII-elements are shallow donors. Similarly, a vacancy in tellurium serves as a donor [9].

Element substitution with an I-valence electron for cadmium can result in p-type doping in CdTe; examples of such elements include gold, copper, silver, and so on. An alternative method is to replace the tellurium atoms with atoms like nitrogen (N), phosphorus (P), or arsenic (As) that have V-valence electrons (shallow acceptors) [11]. Likewise, a vacancy in Cd can serve as an acceptor. Usually, P-doped CdTe is utilized in solar cells, but finding CdTe with high doping levels remains a challenge [12].

Figure 2.1a depicts a characteristic CdTe solar cell configuration. On the glass substrate, transparent front contact is first placed. Tin oxide or cadmium stannate, an alloy of cadmium and tin oxide, may be used in this. Cadmium sulfide n-layer is deposited on top of that, like the CIGS solar cells' n-buffer layer. Subsequently, a few micrometer thick absorber p-CdTe layer is deposited [13].

The limited selection of appropriate metals due to the material features of CdTe makes it difficult to make good back contact. The contact properties can be enhanced

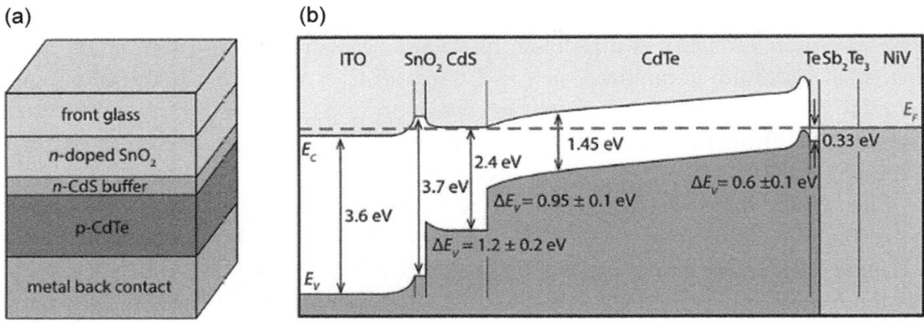

FIGURE 2.1 The characteristic CdTe solar cell (a) layer configuration and (b) band diagram [14]. (Permission taken from Elsevier.)

by employing significantly doped semiconductor material in the contact area; however, it is challenging to achieve highly doped CdTe material. Despite being used as back contacts, copper-containing contacts may eventually suffer instability issues due to Cu diffusion into the CdTe film and even up to CdS layer. Currently, molybdenum-infused stable antimony telluride layers are utilized [15].

Figure 2.1(b) displays a CdTe solar cell's band diagram. The bandgap of n-CdS and p-CdTe are 2.4 and 1.44 eV, respectively. After being stimulated by light, charge carriers split at the heterojunction, and minority electrons in the p-layer congregate at the transparent conducting front contact. By the rear contact, the holes are clustered [14].

2.2.2 Manufacturing CdTe Solar Cells

In general, closed-space sublimation is used for processing the CdS/CdTe layers. Under vacuum circumstances, the source and substrate are positioned close to one another, often at a distance of a few millimeters to several centimeters. CdTe powder or granulates can be used as the source. The temperature of the source is kept higher than that of the substrate despite their mutual heating. The precursors that are deposited on the substrate experience a driving force because of the temperature difference. As a result, bulk p-CdTe is produced. Within the deposition unit, an inert diluent gas, such as nitrogen or argon, may be employed [16].

The top American enterprise in CdTe technology is First Solar Inc. They employ closed space sublimation, like the German business Antec Solar. German-based Calyxo is another producer of CdTe solar cells. The largest manufacturer for CdTe based solar cells is First Solar by far. The yearly generation rate for First Solar in 2008 was 500 MW. It was already among the largest solar module producers in the world in 2006 and 2007 [17].

A CdTe-based record module with a surface area of 7,200 cm^2 has a 16.1% conversion efficiency as of 2013. The price of First Solar modules was ~0.68 to 0.70 USD/Wp (watt peak) and is projected to decrease further, maintaining a lower price per Wp with respect to c-Si wafers [13].

2.2.3 Limitations

The cadmium's toxicity is a problem that must be addressed at a significant level. Insoluble cadmium complexes such as CdTe and CdS are barely harmful as compared to elemental Cd. However, it is necessary to stop Cd from infiltrating the environment. It is imperative to consider if CdTe modules have the potential to enhance Cd pollution [17].

First Solar would take up around 2% of the overall industrial Cd usage. Nevertheless, CdTe solar module recycling programs have been established. For example, First Solar offers a recycling program with a 5 cent per Wp deposit to pay for the recycling expenses when the module reaches the end of its life [13].

For the CdTe technology, tellurium (Te) supplies might provide an even greater challenge. Tellurium, with a 1 g/kg abundance analogous to platinum, is one of the most rare stable solid elements found in Earth's crust [18]. Te is scarce, which could prevent CdTe PV technology from being expanded to terawatt-scale levels. However, there have only been a few numbers of usage of tellurium as a raw material. Therefore, no specific mining has been developed so far for Te. New deposits of ores high in tellurium were also discovered at Xinju, China [13]. The availability of tellurium may hinder the development of CdTe PV technology; however, this is presently unclear [19].

2.2.4 IDENTIFICATION OF ISSUES AND SCOPE OF IMPROVEMENTS

With a record-breaking laboratory efficiency of 22.1% and modules with an efficiency of 18.6%, cadmium telluride is the finest thin film PV technology accessible commercially [20]. The Shockley-Queisser limit (32%), however, has a lot of potential for growth. The finest CdTe devices have near to maximum short-circuit current, but there are still greater chances for improvement in open-circuit voltage. Any improvement is going to depend largely on the optimization of back contact. For example, significant degrees of visible and/or near-infrared transparency are necessary for their suitability for complicated structures like tandem and bifacial CdTe solar cells along with their electrical characteristics in addition to their electrical properties. Several back contact materials and approaches have been employed by the CdTe research community to realize them.

Thus far, there has not been a single solution established for the perfect rear contact for p-CdTe layer. Even though back contact quality is not the sole factor influencing the efficiency of TFSC with CdTe absorbers, using poor contact materials can eventually lead to degradation or lower photovoltaic conversion efficiency.

To achieve optimal efficiency, either metallic or almost metallic conductivity is essential. To prevent too much carrier recombination, the interface also needs to have a low defect density. Electron reflectivity is exhibited by optimal back contact. High carrier lifetime within CdTe and low defect density at the rear interface and grain boundaries are prerequisites for carriers to reach the rear contact.

The CdTe rear contact concern and the hole transport layer (HTL) in perovskite solar cells (PSC) share similarities [15]. The contact material's thermal stability, encapsulation, optimizing transport layer conductance, dopant diffusion, and material cost are issues that both CdTe and PSC encounter. High near-infrared transparency is required for wide-gap perovskites because they are frequently used for tandem cell designs. A variety of inorganic materials, such as carbon allotropes, MoO_x, CuO_x, NiO, MoS_2, NiS, V_2O_5, CuSCN, CuPc, and CuI, have been employed as rear contacts and HTL for CdTe solar cells and PSC, respectively [21] (Figure 2.2).

Improved p-CdTe can be formed by infusing group V elements surpassing the 10^{16} cm^{-3} level, eliminating the need for copper-containing compounds in the back contact structure, and a wider range of materials are available for this.

In the end, material prices are important, as are any costly materials eliminated from a process. Tellurium is going to be the most expensive impure element when it comes to the thin-film CdTe solar cell material cost. Pricey metals like gold (Au) or platinum (Pt) are usually employed for making contacts and are beneficial for basic research, yet too expensive to produce. Tellurium is a somewhat rare element, and hence, any material used in the CdTe cell had to be much less expensive than Te.

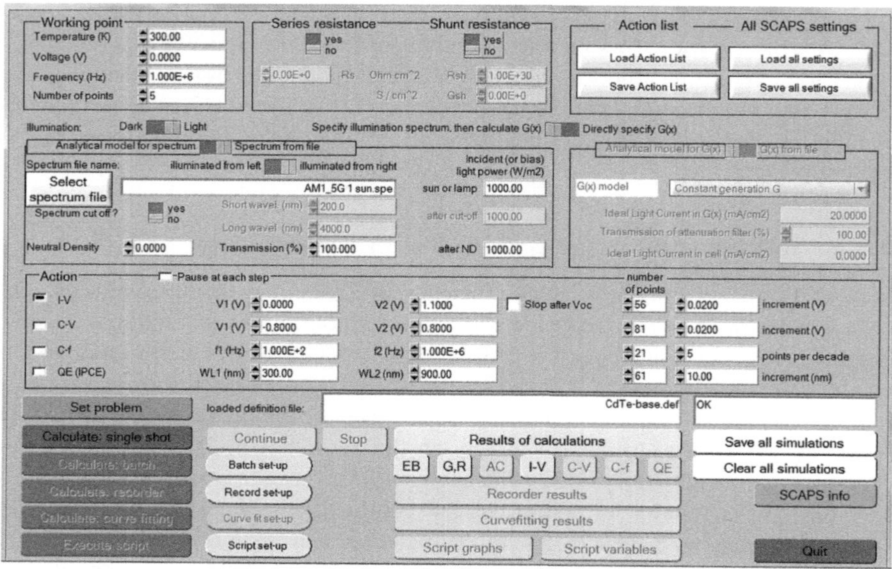

FIGURE 2.2 SCAPS action panel snapshot.

Favorable materials are those that are stable in thin interlayers and would not disperse to dope CdTe. The most tamable layers are organic ones, but they must be designed to remain stable at higher processing temperatures. A larger range of perovskite-like halides have been explored up to this point, and further recommendations for future testing include MoS_2/Ni, copper-free thiocyanates and phthalocyanines, and Bi_2Te_3 on arsenic-doped CdTe [22].

2.3 BENEFITS OF USING SOLAR CELLS SIMULATION SOFTWARE

Energy professionals and engineers may easily execute a broad range of tasks and estimations by employing solar simulation software. These computations would have taken an extended amount of time without them. Easy automation is another benefit that enables the solar community to quickly receive customer feedback. For large-scale solar manufacturing plants facing various obstacles, solar simulation software has the potential to greatly simplify and optimize the engineering and sales processes.

2.3.1 OVERVIEW ON SCAPS SOFTWARE

SCAPS-1D is a "one-dimensional solar cell capacitance simulator" developed by Stefaan Degrave, Koen Decock, Johan Verschraegen, Marc Burgelman, and Alex Niemegeers [23,24]. Photovoltaic researchers across the globe, including academic and industrial users, have free access to the program. It requires roughly 50 MB of disk space to run on a different Windows operating system such as Windows 7 or lower [24]. SCAPS was designed originally for $CuInSe_2$ and CdTe cell architectures.

However, several extensions have augmented its capabilities, making it relevant to both amorphous (silicon and micromorphous silicon) and crystalline (silicon and gallium arsenide family) solar cells. Here is a summary of its key attributes.

2.3.2 SALIENT FEATURES OF SCAPS PROGRAM

Some of the salient features of SCAPS Program are listed below:

- As many as seven semiconducting material layers can be simulated.
- Nearly all physical parameters concerned with the layered material can be optimized for the given study.
- The SRH-type, Auger, and direct recombination of charge carrier mechanism are employed.
- Defect levels present at the interface or in bulk; recombination and charge state are taken into consideration.
- Levels of defect. different types of charges, e.g., monovalent, divalent, user defined multivalent, and not having any charge (idealization), single donor, single acceptor, double donor, double acceptor, amphoteric.
- Defect levels, various energy distributions such as single level, uniform distribution, Gauss, tail, or any combinations of any of these energy levels.
- Optical property, defect levels: it is possible to excite a light source directly (impurity level, photovoltaic effect, impurity photovoltaic effect).
- Defect levels and metastable transitions between different types of defects.
- Contacts: optical property (transmission filter reflection); work function or flat band.
- Tunneling within the valence band and/or inside the conduction band tunneling to and from interfacial states.
- Generation: a user-supplied $g(x)$ file or from an internal computation can be used.
- Illumination: a range of standard and non-standard spectra (such as AM0, AM1.5D, AM1.5G, AM1.5G edition2, monochromatic, and white) can be added.
- Illumination: attenuation and spectrum cut-off, from the p-side or n-side.
- Voltage, frequency, and temperature as the working points for computations.
- The program computes solar cell response as a function of light or bias voltage, also for different energy bands, concentrations of defects, currents at a working point, J-V characteristics, and alternating current characteristics (C and G as functions of V and/or f).
- Ability to do computations in batches and display settings and results based on batch parameters.
- Loading and storing all configurations; configuring SCAPS at startup; a scripting language with an open-ended user feature.
- An extremely user-friendly interface.
- A feature in SCAPS allows the launch of the program from "script file" using script language to calculate and plot all internal variables.
- Integrated facility for curve fitting.
- An option in the panel for interpretation of admittance measurement [24].

One can optimize various parameters like band gap [25] and thickness of the layers of solar cell [26] and also study the effect of doping [27], temperature [28], etc. on the efficiency of solar cell [27].

2.3.3 MATHEMATICAL MODEL EMPLOYED IN SCAPS SIMULATION SOFTWARE

It is possible to evaluate performance parameters such as the current density (J) – voltage (V) curve, energy band gap, and power conversion efficiency by solving the Poisson equation employing continuity equation for different charge carriers given below [29]:

$$\frac{d}{dz}\left[-e(z)\frac{d\varphi}{dZ}\right] = q\left[p(z) - n(z) + N_D^+(z) - N_A^-(z) + p_t(z) - n_t(z)\right] \qquad (2.1)$$

$$\frac{dp_n}{dt} = G_p - \frac{p_n - p_{no}}{\tau_p} + p_n\mu_p\frac{d\varepsilon}{dz} + \mu_p\varepsilon\frac{dp_n}{dz} + D_p\frac{d^2p_n}{dz^2} \qquad (2.2)$$

$$\frac{dp_n}{dt} = G_n - \frac{n_p - n_{po}}{\tau_n} + n_p\mu_n\frac{d\varepsilon}{dz} + \mu_n\varepsilon\frac{dn_p}{dz} + D_n\frac{d^2n_p}{dz^2} \qquad (2.3)$$

where the rate of generation (G), the lifetime of an electron (τ_n) the lifetime of a hole (τ_p), coefficient of diffusion (D), electronic charge (q), e-mobility (μ_n), h-mobility (μ_p), free e-concentration $(n(z))$, free h-concentration $(p(z))$, trapped e-concentration $(n_t(z))$, trapped h-concentration $(p_t(z))$, ionized acceptor concentration $(N_A^-(z))$, ionized donor concentration $(N_D^+(z))$, electric field (ξ), and z denotes the direction.

The short circuit current density (J_{SC}) is mostly determined by the spectrum of incident light and its optical properties like reflection, recombination, and absorption. Equation (2.4) can be used to get the J_{SC} for the solar cell [19]:

$$J_{SC} = q\sum T(\lambda)\frac{\phi_i(\lambda_i)}{hv_i}\eta(\lambda_i)\Delta\lambda_i \qquad (2.4)$$

Here, the electron charge is denoted by q, the optical transmission is represented by $T(\lambda)$, the spectral power density is represented by ϕ_i, and the wavelength gap between two consecutive values is shown by $\Delta\lambda_i$.

The open-circuit voltage, or V_{OC}, is the voltage we get when the solar cell is not receiving any net current. It may be expressed as follows in Equation (2.5):

$$V_{OC} = \frac{nkT}{q}\ln\left(\frac{J_{SC}}{J_o} + 1\right) \qquad (2.5)$$

Equation (2.6) provides the formula for the fill factor percent, or $FF\%$, which is used to determine the solar cell's maximum power:

$$FF\% = \frac{V_{OC} - \ln(V_{OC} + 0.72)}{V_{OC} + 1}$$

(2.6)

The determination of *J-V* characteristics is performed using AM1.5G (the global standard solar cell spectra) as incident illumination. The power conversion efficiency ($\eta\%$) of the solar cell can be expressed by Equation (2.7):

$$\eta\% = \frac{V_{OC} \times J_{SC} \times FF\%}{p_{in}}$$

(2.7)

With a focus on polycrystalline cell architectures of the CdTe and CuInSe$_2$ communities, the simulation tool is accessible without charge to the scientific community. Several factors, such as thin films, large band gaps, and various interfaces, are managed by this simulation tool. Over time, the software evolved to include additional processes such as Auger recombination, tunnelling and multiple enhancements. This simulation tool was created specifically with the assistance of the Newton Raphson approach and the Gummel scheme for CdTe solar cells and CIGS. The action panel of the simulation tool is shown in Figure 2.3.

This package has so far been used for micro amorphous Si solar cells and related applications, as well as crystalline solar cells, a-Si: H and GaAs cells [6].

2.3.4 IMPROVEMENTS IN CdTe SOLAR CELL PERFORMANCES PREDICTED BY SCAPS

CdS-based thin film buffer layers have been utilized as n-type heterojunction partner in both developing and existing thin film photovoltaic systems. Employing numerical simulation, the best simulation parameters were ascertained by varying the

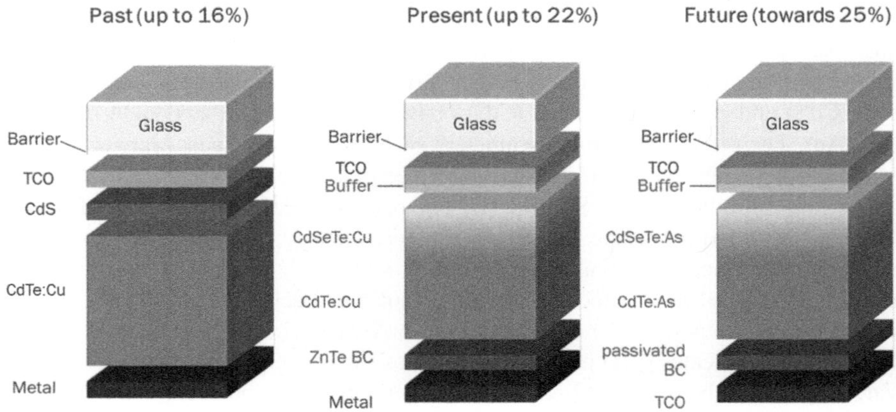

FIGURE 2.3 Comparisons of the material stacks employed in CdTe-based solar cells: past, present, and future [20]. (Permission taken from Elsevier.)

concentrations of charge carrier in the buffer layer (N_D) and absorber layer (N_A), as well as a suitable non-toxic material (ZnS, In_2S_3, ZnSe, and ZnO) in place of CdS. In thin film-based solar cells, potential replacements for the CdS buffer layer were discovered to be ZnO and ZnSe. The buffer (ZnSe) and absorber (CdTe) layers were selected to enhance efficiency by determining the ideal density of charge carriers (acceptor and donor), and the results provided useful insights for building highly efficient metal oxide-based photovoltaic solar cells in the research laboratory. Taking into account the results of the simulation using SCAPS-1D, it can be concluded that ZnX (X=Se and O) has the potential to replace CdX (X= S and Te) as the buffer layer in solar cells [30].

Research indicates that employing In_2S_3 as the window layer and $FeSi_2$ as the secondary absorber layer can enhance photovoltaic performance parameters. Due to the increased absorption at longer wavelengths (λ) in the solar spectrum, the $FeSi_2$ layer addition to the cell structure has led to a considerable gain in quantum efficiency [31].

CdTe solar cells' performance could be improved by utilizing a 70–100 nm-thick Mg-doped ZnO(ZMO):In window layer with dopant concentrations of x = 0.0625 in $Zn_{1-x}Mg_xO$ for both Mg and In. CdTe solar cells' short-circuit current density is mostly impacted by ZMO:In thickness. The findings of this simulation give valuable insights into window layer design and theoretical direction for producing CdTe solar cells with high efficiency [32].

The photovoltaic performance of a solar cell device array based on CdTe employing WO_3 as a window layer as Al/ZnO:Al(AZO)/WO_3/CdTe/NiO/Ni was simulated and analyzed using the SCAPS 1-D software. The results indicated that WO_3 can be a potential material as the alternative window material for the CdTe-based solar cells. This is primarily since it can favor the increase in J_{SC} owing to its wide band gap and is also a non-toxic and inexpensive material. This shows that the WO_3/CdTe heterojunction is a fascinating replacement for the future development of CdTe-based photovoltaic solar cells [33].

To optimize the photovoltaic performance of CdS and CdTe-based solar cells, the hole transport layer (HTL) transparent conducting oxide (TCO) layers have been added with the use of NiO and ZnO, respectively, in the simulation of cadmium sulfide (CdS) and cadmium telluride (CdTe) based solar cell such as (ZnO/CdS/CdTe/NiO/Au). The effect of various parameters such as layer thickness, carrier concentration, defect density, and CdS/CdTe interface defect density on solar cell performance was simulated using SCAPS-1D. The results revealed that NiO as HTL and ZnO as TCO had a positive impact on the performance of CdS/CdTe-based solar cells. The parameters such as an open circuit voltage (V_{OC}), a short circuit current density (J_{SC}), a fill factor, and the optimized solar cell achieved a maximum power conversion efficiency (PCE) of 1.09 V, 29.09 mA/cm², 87.84%, and 28.04% respectively, which indicates great promise for low-cost solar energy harvesting. Therefore, ZnO TCO could be a good substitute for ITO and fluorine-doped tin oxide (FTO) in the development of low-cost, high-efficiency CdS/CdTe-based solar cells. NiO is an encouraging hole transport material (HTL) that has the possibility for additional efficiency increase of the CdS/CdTe-based solar cells [34]. Table 2.1 provides glimpses

TABLE 2.1

Summary of the Theoretical and Experimental Works

Study Type	Cell Structure	V_{OC} (V)	J_{SC} (mA/cm²)	FF%	η%	Ref.
Experimental	ITO/SnO$_2$/CdS/CdTe/graphite paste/AgSC	0.700	18.50	52.5667	6.8184	[35]
Experimental	Glass/FTO/CdS/CdTe	0.63	38.5	0.33	8	[36]
Experimental	FTO/SnO$_2$/CdS/CdTe	0.81	21.8	67	11.8	[37]
Experimental	FTO/SnO$_2$/CdSe$_x$Te$_{1-x}$/CdTe	0.824	26	66	14.1	[37]
Experimental	Glass/n-SnO$_2$/n-CdS/p-CdTe	0.835	23.67	69.1	13.66	[38]
Theoretical	CdTe/CdS/SnO$_2$	0.9198	23.153579	66.81	14.23	[30]
Theoretical	Glass/SnO$_2$/Zn2SnO$_4$/CdS/CdTe/ZnTe	0.90	24.92	0.70	15.8	[39]
Theoretical	CdTe/ZnS/SnO$_2$	0.9121	23.260166	74.84	15.88	[30]
Theoretical	CdTe/ZnO/SnO$_2$	0.9142	23.303926	76.37	16.27	[30]
Theoretical	i-ZnO/CdS/CdTe/ZnTe	0.90	26.42	0.782	16.9	[39]
Theoretical	Glass/SnO$_2$/Zn$_2$SnO$_4$/CdS/CdTe/Sb$_2$Te$_3$	0.91	24.94	0.75	17.2	[39]
Theoretical	CdTe/ZnSe/SnO$_2$	0.9112	23.484037	82.38	17.42	[30]
Theoretical	CdTe/CdS/SnO$_2$	0.9113	23.44973	81.41	17.43	[30]
Theoretical	Glass/SnO$_2$/Zn$_2$SnO$_4$/CdS/CdTe	0.90	24.60	0.80	17.8	[39]
Theoretical	Glass/SnO$_2$/Zn$_2$SnO$_4$/CdS/CdTe/As2Te$_3$	0.92	24.97	0.81	18.6	[39]
Theoretical	ZnO/CdS/CdTe	1.06	24.56	86.46	22.42	[34]
Theoretical	ZnO/CdS/CdTe/ZnTe	0.946	34.40	75.72	24.66	[40]
Theoretical	CdSe$_x$Te$_{1-x}$/CdTe	0.923	31.421	83.985	24.354	[41]
Theoretical	FTO/TiO$_2$/ZnO/CdS/CdTe/V$_2$O$_5$	0.811	38.51	80	25	[42]
Theoretical	FTO/CdS/CdTe/NiO	1.09	27.38	87.85	26.35	[34]
Theoretical	ITO/n-CdS/p-CdTe/p+-CdSe	1.15	30.66	88.57	31.11	[43]
Theoretical	ITO/n-CdS/p-CdTe/p+-CdSe	1.05	49.23	85.71	44.14	[43]
Theoretical	ZnO/CdS/CdTe/NiO	1.09	29.09	87.84	28.04	[34]
Theoretical	CdS/FeSi$_2$/BaSi$_2$	0.958	51	83	38.93	[44]
Theoretical	FTO/In$_2$S$_3$/CdTe	0.6312	25.505526	82.37	13.26	[31]
Theoretical	FTO/In$_2$S$_3$/CdTe/FeSi$_2$	0.656	49.776	83.68	27.35	[31]

of experimental and SCAPS simulation-related research works in recent times. Many more initiatives could be taken to simulate, and improvements could be done in CdTe solar cells experimental work.

2.4 RELEVANCE OF THE RESEARCH AND EXPECTED IMPACT ON ACADEMICS/INDUSTRY

For a long time, thin film CdTe/CdS has been considered an appropriate alternative for the production of reliable and inexpensive solar cells [17]. This technique has a very minimal environmental effect, as demonstrated by several scholarly papers. Even unfavorable reviews indicate that it is quite unlikely that damaged modules would cause significant cadmium leakage in the soil. Furthermore, in the event of a domestic fire, no cadmium emissions are possible. Moreover, this fantastic technology is completely clean due to the CdTe modules' full recyclable nature [45].

Further improvements in cell efficiency may be achieved through disruptive modifications and the adoption of new device designs [46].

In spite of still being plenty of opportunity for improvement, solar technology based on CdTe has reached the state-of-the-art field stability and one of the most affordable prices of power among different energy sources [11]. For thin-film based photovoltaic solar cells, cadmium telluride is a potential photovoltaic material [47].

2.5 CONCLUSION

Solar energy appears as a sustainable and practical way to address future energy demands responsibly in the modern era of technological breakthroughs. This chapter explores the fundamentals of CdTe solar cells and SCAPS simulation software-related studies to this eco-friendly technology, emphasizing unique approaches, novel structures, and device simulation.

These simulators are useful for both predicting the effects of device alterations and analyzing manufactured cells. As the field of solar energy technology advances, solar simulation software becomes increasingly important in helping to build environmentally acceptable solar energy systems and determine future solar PV systems.

AUTHOR CONTRIBUTIONS

Anju Rani: Literature survey and writing the original draft; **Mrinal Dutta**: Revision; **Chander Shekhar**: Conceptualization, revision, and supervision; **Ayana Bhaduri**: Conceptualization, revision, and editing of the manuscript and supervision.

REFERENCES

1. N. K. Singh, A. Agarwal, and T. Kanumuri, "Effect of MoS_2 as a buffer layer on CdTe photovoltaic cell through numerical simulation," *J. Eng. Res.*, vol. 9, pp. 89–98, 2021, doi: 10.36909/jer.EMSME.13879.

2. C. S. Ferekides et al., "High efficiency CSS CdTe solar cells," *Thin Solid Films*, vol. 362, pp. 520–526, 2000.

3. L. I. Nykyruy, R. S. Yavorskyi, Z. R. Zapukhlyak, G. Wisz, and P. Potera, "Evaluation of CdS/CdTe thin film solar cells: SCAPS thickness simulation and analysis of optical properties," *Opt. Mater. (Amst).*, vol. 92, no. April, pp. 319–329, 2019, doi: 10.1016/j.optmat.2019.04.029.

4. S. Sharma, K. K. Jain, and A. Sharma, "Solar cells: In research and applications-a review," *Mater. Sci. Appl.*, vol. 06, no. 12, pp. 1145–1155, 2015, doi: 10.4236/msa.2015.612113.

5. K. E. Sarah, "A review of solar photovoltaic technologies," *Int. J. Eng. Res.*, vol. V9, no. 07, pp. 741–749, 2020, doi: 10.17577/ijertv9is070244.

6. B. M. Sakunde, N. B. Chaure, S. Patole, S. R. Jadkar, and H. M. Pathan, "Numerical modeling to improve the efficiency of cadmium sulfide/copper indium sulfide thin film-based solar cells," *ES Energy Environ.*, vol. 18, pp. 111–121, 2022, doi: 10.30919/esee8c784.

7. V. Shukla and G. Panda, "Effect of BSF layer on the performance of CdTe solar cell," *Mater. Today Proc.*, vol. 44, pp. 2300–2303, 2021, doi: 10.1016/j.matpr.2020.12.394.

8. B. M. Basol and B. McCandless, "Brief review of cadmium telluride-based photovoltaic technologies," *J. Photonics Energy*, vol. 4, no. 1, p. 040996, 2014, doi: 10.1117/1.jpe.4.040996.

9. A. D. Husainat, "Simulation and Design Implementation of Low-Cost and High-Efficiency Perovskite Solar Cells," 2020, [Online]. Available: https://digitalcommons.pvamu.edu/pvamu-dissertations/3/.

10. J. L. Gray, *The Physics of the Solar Cell*. In Handbook of Photovoltaic Science and Engineering, Second Edition, (ed.) Antonio Luque and Steven Hegedus (2011) John Wiley & Sons, Ltd. 2011.

11. W. K. Metzger et al., "Exceeding 20% efficiency with in situ group V doping in polycrystalline CdTe solar cells," *Nat. Energy*, vol. 4, no. 10, pp. 837–845, 2019, doi: 10.1038/s41560-019-0446-7.

12. S. S. Hussain et al., "Numerical modeling and optimization of lead-free hybrid double Perovskite solar cell by using SCAPS-1D," *J. Renew. Energy*, vol. 2021, pp. 1–12, 2021, doi: 10.1155/2021/6668687.

13. T. S. C. Kesterites, C. Pv, and T. Iii -, *Solar_Energy_Section_13_4_2*. In book Solar Energy: The physics and engineering of photovoltaic conversion Technologies and Systems By Arno H.M. Smets, Klaus Jäger, Olindo Isabella, René A.C.M.M. van Swaaij Miro Zeman UIT Cambridge LTD, 2016, vol. 22, pp. 205–207.

14. J. Fritsche, D. Kraft, A. Thißen, T. Mayer, A. Klein, and W. Jaegermann, "Band energy diagram of CdTe thin film solar cells," *Thin Solid Films*, vol. 403–404, pp. 252–257, 2002, doi: 10.1016/S0040-6090(01)01528-0.

15. M. Moustafa and T. AlZoubi, "Effect of the n-MoTe$_2$ interfacial layer in cadmium telluride solar cells using SCAPS," *Optik (Stuttg).*, vol. 170, no. April, pp. 101–105, 2018, doi: 10.1016/j.ijleo.2018.05.112.

16. P. Gorai, D. Krasikov, S. Grover, G. Xiong, W. K. Metzger, and V. Stevanović, "A search for new back contacts for CdTe solar cells," *Sci. Adv.*, vol. 9, no. 8, 2023, doi: 10.1126/sciadv.ade3761.

17. Z. Fang, X. C. Wang, H. C. Wu, and C. Z. Zhao, "Achievements and challenges of CdS," *CdTe Solar Cells*, vol. 2011, no. 1, 2011, doi: 10.1155/2011/297350.

18. I. Montoya De Los Santos et al., "Towards a CdTe solar cell efficiency promotion: The role of ZnO:Al and CuSCN nanolayers," *Nanomaterials*, vol. 13, no. 8, pp. 1–16, 2023, doi: 10.3390/nano13081335.

19. S. H. Zyoud and A. H. Zyoud, "Effect of absorber (acceptor) and buffer (donor) layers thickness on Mo/Cdte/Cds/ITO thin film solar cell performance: SCAPS-1D simulation aspect," *Int. Rev. Model. Simulations*, vol. 14, no. 1, pp. 10–17, 2021, https://doi.org/10.15866/iremos.v14i1.19953

20. M. A. Scarpulla et al., "CdTe-based thin film photovoltaics: Recent advances, current challenges and future prospects," *Sol. Energy Mater. Sol. Cells*, vol. 255, no. March, p. 112289, 2023, doi: 10.1016/j.solmat.2023.112289.

21. M. K. Hossain, M. H. K. Rubel, G. F. I. Toki, I. Alam, M. F. Rahman, and H. Bencherif, "Effect of various electron and hole transport layers on the performance of $CsPbI_3$-based perovskite solar cells: A numerical investigation in DFT, SCAPS-1D, and wxAMPS frameworks," *ACS Omega*, vol. 7, no. 47, 2022. doi: 10.1021/acsomega.2c05912.

22. R. S. Hall, D. Lamb, S. James, and C. Irvine, "Back contacts materials used in thin film CdTe solar cells - A review," *Energy Sci. Eng.*, vol. 9, no. 5, pp. 606–632, 2021, doi: 10.1002/ese3.843.

23. C. College and A. Irinjalakuda, "Optimisation of Inorganic Hole Transport Layers For Perovskite Solar Cells Using SCAPS-1D" Submitted by Department of Physics, no. June, pp. 1–42, 2021.

24. M. Burgelman, K. Decock, A. Niemegeers, J. Verschraegen, and S. Degrave, "Marc Burgelman Koen Decock, Alex Niemegeers, Johan Verschraegen, Stefaan Degrave Version: 18-12-2020," 2020.

25. N. Touafek, R. Mahamdi, and C. Dridi, "Boosting the performance of planar inverted perovskite solar cells employing graphene oxide as HTL," *Dig. J. Nanomater. Biostructures*, vol. 16, no. 2, pp. 705–712, 2021.

26. M. Mostefaoui, H. Mazari, S. Khelifi, A. Bouraiou, and R. Dabou, "Simulation of high efficiency CIGS solar cells with SCAPS-1D software," *Energy Procedia*, vol. 74, pp. 736–744, 2015, doi: 10.1016/j.egypro.2015.07.809.

27. H. Elfarri, M. Bouachri, A. Frimane, M. Fahoume, O. Daoudi, and M. Battas, "Optimization of simulations of thickness layers, temperature and defect density of CIGS based solar cells, with SCAPS-1D software, for photovoltaic application," *Chalcogenide Lett.*, vol. 18, no. 4, pp. 201–213, 2021.

28. C. Zuo, H. J. Bolink, H. Han, J. Huang, D. Cahen, and L. Ding, "Advances in perovskite solar cells," *Adv. Sci.*, vol. 3, no. 7, pp. 1–16, 2016, doi: 10.1002/advs.201500324.

29. P. K. Jha et al., "Study of eco-friendly organic-inorganic heterostructure $CH_3NH_3SnI_3$ perovskite solar cell via SCAPS Simulation," *J. Electron. Mater.*, vol. 52, no. 7, 2023, doi: 10.1007/s11664-023-10267-3.

30. S. H. Zyoud, A. H. Zyoud, N. M. Ahmed, and A. F. I. Abdelkader, "Numerical modelling analysis for carrier concentration level optimization of CdTe heterojunction thin film - based solar cell with different non - toxic metal chalcogenide buffer layers replacements: Using SCAPS - 1D software," *Crystals*, vol. 11, no. 12, p. 1454, 2021.

31. M. F. Rahman et al., "Design and numerical investigation of cadmium telluride (CdTe) and iron silicide ($FeSi_2$) based double absorber solar cells to enhance power conversion efficiency," *AIP Adv.*, vol. 12, no. 10, 2022, doi: 10.1063/5.0108459.

32. X. He, S. Yuehan, L. Wu, J. Zhang, and L. Feng, "Simulation of high-efficiency CdTe solar cells with $Zn_{1-x}Mg_xO$ window layer by SCAPS software," *ACS Appl. Mater. Interfaces*, vol. 10, pp. 22408–22418, 2018.

33. J. C. Zepeda Medina et al., "Performance simulation of solar cell based on AZO/CdTe heterostructure by SCAPS 1D software," *Heliyon*, vol. 9, no. 3, 2023, doi: 10.1016/j.heliyon.2023.e14547.

34. S. Ahmmed, A. Aktar, F. Rahman, and J. Hossain, "Optik A numerical simulation of high-efficiency CdS / CdTe based solar cell using NiO HTL and ZnO TCO," *Optik (Stuttg).*, vol. 223, no. May, p. 165625, 2020, doi: 10.1016/j.ijleo.2020.165625.

35. T. M. Razykov, K. M. Kuchkarov, B. A. Ergashev, and S. A. Esanov, "Fabrication of thin-film solar cells based on CdTe films and investigation of their photoelectrical properties," *Appl. Solar Energy*, vol. 56, no. 2, pp. 94–98, 2020, doi: 10.3103/S0003701X20020097.

36. I. M. Dharmadasa et al., "Fabrication of CdS/CdTe-based thin film solar cells using an electrochemical technique," *Coatings*, pp. 380–415, 2014, doi: 10.3390/coatings4030380.

37. E. Artegiani et al., "Effects of CdTe selenization on the electrical properties of the absorber for the fabrication of $CdSe_xTe_{1-x}$ / CdTe based solar cells," *Sol. Energy*, vol. 227, no. August, pp. 8–12, 2021, doi: 10.1016/j.solener.2021.08.070.

38. V. D. Novruzov et al., "CdTe thin film solar cells prepared by a low-temperature deposition method," *physica status solidi (a)*, vol. 733, no. 3, pp. 730–733, 2010, doi: 10.1002/pssa.200925217.

39. N. Amin, M. A. Matin, M. M. Aliyu, M. A. Alghoul, M. R. Karim, and K. Sopian, "Prospects of back surface field effect in ultra-thin high-efficiency CdS / CdTe solar cells from numerical modeling," *Int. J. Photoenergy*, vol. 2010, 2010, doi: 10.1155/2010/578580.

40. R. S. Sultana and A. N. Bahar, "Numerical modeling of a CdS / CdTe photovoltaic cell based on ZnTe BSF layer with optimum thickness of absorber layer Numerical modeling of a CdS / CdTe photovoltaic cell based on ZnTe BSF layer with optimum thickness of absorber layer," *Cogent Eng.*, vol. 2010, no. May, 2017, doi: 10.1080/23311916.2017.1318459.

41. X. Yin and L. Wu, "Numerical investigations on the working mechanisms of solar cells with a CdTe-based composite-absorber layer," *Chalcogenide Lett.*, vol. 17, no. 9, pp. 439–455, 2020.

42. A. Kuddus, F. Rahman, J. Hossain, and A. B. Ismail, "Enhancement of the performance of CdS / CdTe heterojunction solar cell using TiO2 / ZnO bi-layer ARC and V_2O_5 BSF layers: A simulation approach," *Eur. Phys. J. Appl. Phys.*, vol. 92, no. 2, p. 20901, 2020.

43. A. Kuddus, A. Bakar, and J. Hossain, "Design of a highly efficient CdTe-based dual-heterojunction solar cell with 44% predicted efficiency," *Sol. Energy*, vol. 221, no. November 2020, pp. 488–501, 2021, doi: 10.1016/j.solener.2021.04.062.

44. N. Light, M. A. Moon, H. Ali, F. Rahman, J. Hossain, and A. B. Ismail, "Design and simulation of $FeSi_2$-based novel heterojunction solar cells for harnessing visible and near-infrared light," *physica status solidi (a)*, vol. 217, no. 6, p. 1900921, 2020, doi: 10.1002/pssa.201900921.

45. A. Romeo and E. Artegiani, "CdTe-based thin film solar cells: Past, present and future," *Energies*, vol. 14, no. 6, p. 1684, 2021.

46. M. Gloeckler, I. Sankin, and Z. Zhao, "CdTe solar cells at the threshold to 20% efficiency," *IEEE J. Photovolt.*, vol. 3, no. 4, pp. 1389–1393, 2013.

47. S. Banerjee, "High-efficiency CdTe / CdS thin film solar cell," *Int. J. Eng. Res.*, vol. 4, no. 09, pp. 700–703, 2015.

Green Synthesis of Nanomaterials and Their Applications for Sustainable Environment

Khushbu Jain, Krishna Kumar Yadav, and Neeru Dabas

3.1 INTRODUCTION

With the advent of time, the use of nanotechnology is spreading across all fields including energy, environment, medical, and electronics.

Nanotechnology involves methods to get various materials in nanoscale dimensions. Nanoparticles refer to the group of atoms or molecules size in 1–100 nm. The physical and chemical properties of nanomaterials are greatly affected by their size, and so, tuning the size of materials is often desirable to get the diversified properties and applications to solve various problems in our environment today.

Conventional methods of synthesis of nanomaterials involve several hazardous chemicals that adversely impact our environment. Conventional methods may employ corrosive acids such as H_2SO_4, HNO_3, and HCl; bases such as $NaOH$ and NH_3; reducing agents such as $LiAlH_4$ and $NaBH_4$ for the synthesis of nanoparticles. These corrosive acids, bases, and chemical reducing agents harm our environment and should be replaced with safer chemicals/methods, e.g., silver nanoparticles can be synthesized from silver nitrate using nitrate reductase enzyme or plants extracts instead of conventional agents such as $NaBH_4$, $LiAlH_4$, $NaOH$, or ammonia. Moreover, a conventional synthesis that uses solvents that are flammable and toxic must be replaced with green solvents like water, ionic liquids, and supercritical CO_2, e.g., graphene synthesis is reported from benzene, which is flammable. Moreover, benzene vapor exposure is linked with a high risk of blood cancer. So, such compounds must be avoided, and safer, eco-friendly chemicals must be adopted for the synthesis of useful nanomaterials.

Listed below are some techniques/methods for the synthesis of nanoparticles and the uses of reagents along with the disadvantages/hazards associated with them that may involve:

DOI: 10.1201/9781003473749-4

a. **Precipitation Method**: Precipitation method is frequently employed in the synthesis of nanomaterials from the precursor soluble metal salts by adding a precipitating agent. Common precipitating agents include sodium hydroxide, ammonium hydroxide, and various acids, which are corrosive and toxic.

b. **Microemulsion Technique**: Herein, surfactants are added to stabilize the nanoparticles in a continuous phase. Surfactants often persist in the environment and water bodies and deteriorate water quality, adversely effecting aquatic life.

c. **Chemical Vapor Deposition (CVD)**: The method is used for the fabrication of high-quality pure solid thin film 2-D material. However, the disadvantage of the method is that precursors used in CVD, such as metal carbonyls and organometallic compounds, can be toxic or hazardous, and byproducts such as CO and HCl are also toxic and hazardous in nature.

d. **Sol-Gel Process**: The process is used to prepare ceramic and glass materials by hydrolysis of metal alkoxides or other metal compounds such as metal chlorides, metal sulfates, and metal nitrides. Alkoxides are flammable and may release toxic fumes upon hydrolysis. Moreover, sometimes it comprises several steps in synthetic procedure and may be time consuming.

e. **Hydrothermal/Solvothermal Synthesis**: These methods involve the use of high-temperature and high-pressure conditions to promote the formation of nanomaterials from precursors dissolved in water or organic solvents. These conditions can pose safety risks, and the organic solvents used may be flammable or toxic.

f. **Electrochemical Deposition**: Nanoparticle deposition or coating of various metals or metal oxides from solution onto the substrate using electrical energy is a rapid synthetic approach for preparing nanoparticles with controlled morphology. Electroplating solutions often contain hazardous substances such as cyanides, which are highly toxic.

g. **Mechanical Milling**: This method involves the mechanical grinding of bulk materials to produce nanoscale particles. Although it does not involve as many hazardous chemicals as other methods, it can still generate fine dust particles, which may pose inhalation risks.

Most of the methods described involve hazardous chemicals or raw materials or corrosive liquids or solvents at one or other step of the nanoparticle synthesis. Although some of the processes or reagents involved in the synthesis of nanoparticles can lead to deleterious impacts on the environment, the synthesis of newer materials is always desirable and crucial for strengthening technological development.

Development is a continuous process and need of the current society. Although development is necessary, it should be in a sustainable manner so it does not harm our environment. However, it has been witnessed that in the process of development, our environment is deteriorating day by day. So, various countries across the globe, developed as well as developing, are striving to save our environment by adopting several policies for the restoration of healthy and clean air, water, and land.

Scientists across the globe follow strategies, for example, elimination of the hazardous and toxic materials with eco-friendly green materials. Moreover, the process design involved in the synthesis of smart nanomaterials also involves corrosive materials and lengthy steps including unnecessary derivatization and use of protecting groups. It is desirable to reduce the unnecessary steps and also adopt the usage of reagents that are naturally available and biodegradable. Different ranges of metal and metal oxide NPs are synthesized from their precursor metal ions by reduction or precipitation. Green methods employed for the synthesis of nanoparticles are highly advantageous as they are cost-effective, time-efficient, eco-friendly, and highly effective to achieve amorphous as well as crystalline nanoparticles with a variety of shapes (spherical, cubical, hexagonal, triangular, tetrahedral) and sizes. Green synthetic methods involve the use of safer chemicals or renewable feedstock to get the desired products in high yield. More often, synthesis of metal nanoparticles is carried in organic solvents from precursor salts by using strong reducing agents. Nowadays, reduction and precipitation by classical corrosive reagents are replaced by plant-based natural products.

Not only plants but also microorganisms, for example, bacteria, fungi, yeast, and algae, are employed for the synthesis of valuable nanoparticles of varying shapes, sizes, properties, and applications. In addition, the use of waste material to synthesize valuable nanomaterials is also interesting and is eco-friendly as it reduces waste accumulation and burden, and its proper utilization without any need for separate disposal. Waste to value-added nanomaterial is a diversified area of research under waste management category and is extensively explored to design valuable nanomaterial with advanced applications in energy and environment.

Green nanomaterials for sustainable environment are the need of hour as rapid industrialization pollution of surface and groundwater as well as air has grown enormously. Removal of toxic pollutants from the land, air, and water is the key focus to manage all kinds of pollution. Nanotechnology has great potential and provides the solution to unwanted problems faced by humans. Various types of green synthetic routes for the synthesis of nanoparticles/nanocomposites in an eco-friendly and sustainable manner are illustrated in Figure 3.1.

3.1.1 Green Synthesis of Nanoparticles/Composites by Plant Extracts

The production of nanoparticles from plant sources is popular worldwide because plants provide an ecofriendly source that can replace harmful chemical reagents often used in chemical synthesis.

The use of various plant products/parts for the synthesis of nanoparticles has tremendous applications in the field of environment. For a sustainable environment, it is highly essential that metals of high economic importance must be synthesized in a facile, pollution-free, less-energy intensive process and with minimal synthetic steps. Green synthesis involving phytochemicals derived from plants has fulfilled the criteria for the synthesis of variable valuable metal catalysts such as gold (Au), silver (Ag), copper (Cu), palladium (Pd), platinum (Pt), and rhodium (Rh) [1–4].

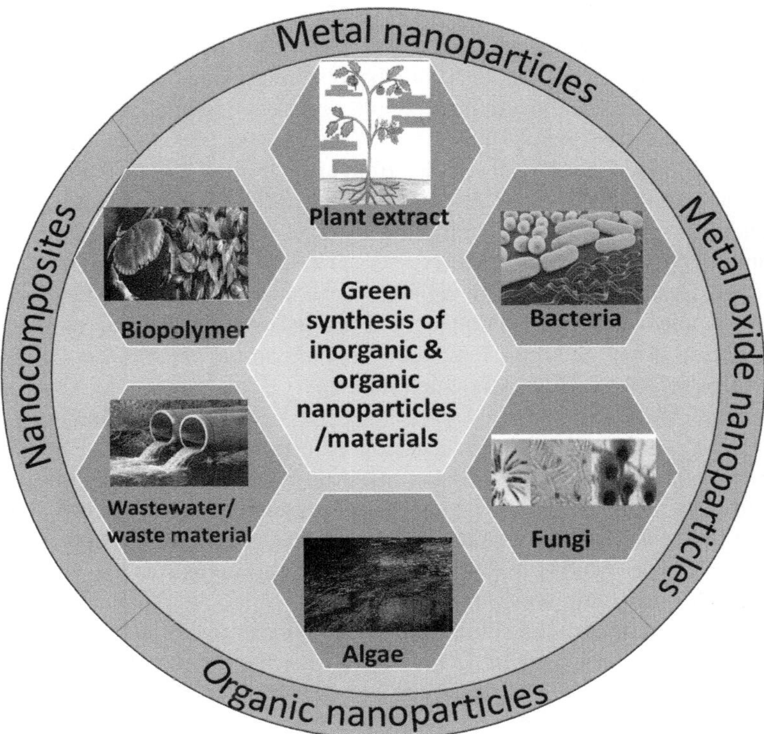

FIGURE 3.1 Various types of nanoparticles synthesized by green methods (using biopolymer, plant extract, microorganisms, wastewater, and waste material).

Not only precious metals but also metal oxides with good photocatalytic activity such as zinc oxide, copper oxide, iron oxide, nickel oxide, titanium dioxide, and cerium oxide have been synthesized by using phytochemicals [5].

Various metal and metal oxide nanoparticles are synthesized by green methods using phytochemicals derived from different parts of plants such as leaves, roots, stems, and flowers. The use of bioactive materials derived from plants for the synthesis of metal and metal oxide nanoparticles is growing extensively as it is a safe and environmentally friendly solution to the synthesis of valuable nanoparticles. Leaf extract of plants is very popular and frequently used to synthesize inorganic nanoparticles. Almost all classes of inorganic oxides belonging to transition, inner transition, and main group metals have been synthesized by the use of phytochemicals [6]. For example, ZnO nanoparticles have been synthesized by the leaf extracts of several plant species such as *Ocimum tenuiflorum*, *Syzygium cumini*, *Eucalyptus* spp., *Calotropis procera*, *Solanum rantonnetii*, *Mangifera indica*, *Lagerstroemia speciosa*, *Azadirachta indica*, *Solanum nigrum*, *Pinus brutia*, *Corymbia citriodora*, *Mimosa pudica*, *Laurus nobilis*, *Plectranthus amboinicus*, wheat seedling, Green tea, and Aloe vera. Phytochemicals present in different parts of the plants act as capping, as well as reducing, agent, and using them is a safe and environmentally benign approach to synthesize the nanoparticles. Leaves extract containing various natural products such as alkaloids, flavonoids, terpenoids, amino acids, saponins, tannins, citric acids, heterocyclic, and polyphenolic compounds react with metal salts. Polyphenols and flavonoids are the reducing agents, while alkaloids and terpenoids stabilize the nanoparticles.

Neem (*Azadirachta indica*) is known from ancient times and has been used in the field of medicine from the ages. Each and every part of the neem plant has bioactive components that have medicinal value and are used as antifungal, anti-inflammatory, anti-tumor, and anti-bacterial agents.

Allouche et al. reported a batch method for the removal of methyl orange from the aqueous sample using chitosan polymer [7]. The adsorption of dye over the polymer was best fitted as per the Freundlich model, and maximum efficiency was reported at pH 3 with adsorption reaching the maximum up to 29 mg/g in 1 hour.

Spherical MnO nanoparticles of average size 80 ± 0.5 nm synthesized from *Abutilon indicum* are useful antibacterial and cytotoxic agents effective against HeLa cancerous cells [8]. For the synthesis of NPs, the leaves extract of *Abutilon indicum* was used along with manganese precursor sulfate salt and base sodium hydroxide to maintain the basic pH. Calcination process further gives MnO nanoparticles with desirable properties needed in clinical uses and environmental applications like dye degradation and chromium (VI) removal.

Sesbania grandiflora, a legume tree, has extensive use in medicinal chemistry. The leaves of *Sesbania grandiflora* are highly enriched in phenolic acids, polyphenol, and flavonoids, and the leaf extract of this leguminous family tree was reported for the synthesis of zinc oxide and iron oxide nanoparticles. Both metal oxide NPS were reported as potentially useful in treating seafood industry effluents having a high chemical oxygen demand of 912 ppm [9].

Compared to microorganisms, the plant-based synthesis of inorganic nanoparticles is more popular because it is time effective and doesn't require longer incubation time like that done previously to synthesize the nanoparticles [5].

Nanocomposites made up of metal and polymers are also prepared by using the principles of green chemistry. Polymers used for the synthesis of composites can be natural as well as synthetic. Biopolymers such as chitosan, lignin, cellulose, silk fibroin, and alginate have been used extensively to synthesize green nanocomposites [10]. Polymers such as polyvinyl alcohol, polyacrylamide, hydrogel, polystyrene, and polyurethane are also used. There is an increasing use of green synthetic routes for composites made up of reduced graphene oxide (RGO) and carbon nanotubes (CNT) as they have advanced features with diverse applications.

3.1.2 Green Synthesis of Nanoparticles by Microorganisms

The use of microorganisms (bacteria, fungi, algae) is an alternative sustainable method to synthesize a range of nanoparticles. Green synthesis of nanoparticles by the microorganisms is widely accepted as it is a less energy demanding, non-toxic, economically viable method with better control over the synthesis process. There are several reports on the synthesis of inorganic nanoparticles of metals (gold, silver, palladium, platinum, selenium, tellurium), alloys, metal oxides (silica, titania, zirconia, magnetite, and uraninite), and quantum dots by bacteria, fungi, molds, viruses, and yeast.

The rich biodiversity of microbes and techniques of microbial culture, altogether, expand the scope to synthesize advanced metal nanoparticles in a non-polluting, environment friendly way.

Redox active bacteria are often used to synthesize metal NPs in a reduced state from their source metal ions. Reduction process in such bacteria is catalyzed by enzyme/redox active proteins. Accumulation of metals in some of the bacterial strains inspired microbial-assisted recovery of metals from the waste/spent ore. It is known that silver acts as an antimicrobial agent but some bacterial strain remains unaffected by the metal. Surprisingly, the strain accumulates silver and immobilizes it to inhibit the toxic effect. Klaus et al. in 1999 reported biosynthesis of silver nanoparticles up to size 200 nm using *Pseudomonas stutzeri* within the periplasmic space of the cell [11]. The bacterial strain is resistant to silver and its possible use was suggested for recovery of Ag metal from the ore.

Bacterial synthesis of gold and silver nanoparticles has been known since a long time ago, and both gram-positive and gram-negative bacteria have been used for the synthesis. Some of the bacteria used in the past for the synthesis of Ag and Au nanoparticles are *Bacillus subtilis, Brevibacterium casei, Pseudomonas strutzeri, Geobacillus stearothermophilus, Marinobacter Pelagius, Lactobacillus, Microbacterium resistens*, and *Sporosarcina koreensis* DC4.

Recently, it has been reported that *Bacillus pumilus*, a silver tolerant species, synthesized highly stable Ag Nps of size less than 25 nm, and it remains stable for as long as 180 days. The Ag NPs were effective against microorganisms *S. aureus, E. coli, P. aeruginosa, and A. baumannii* in the concentration range from 1.4 to 5.6 µg/mL.

Yeast like *Candida guilliermondii* was reported in 2011 for the synthesis of gold and silver nanoparticles [12].

The fungus produces metal nanoparticles by producing enzymes and other compounds that break down the metal salt to produce nanoparticles [13]. Biological species gives an opportunity to produce metal nanoparticles of well-defined shape, size, properties, and applications [14]. In a novel biosynthetic approach, fungus *Verticillium* was used to prepare Ag nanoparticles of ~25 nm [15]. Formation of Ag particles on the cell wall occurs due to the reduction of Ag salt by the enzyme present on the cell membrane of *Verticillium*.

3.1.3 Green Synthesis of Nanoparticles by Biopolymers

Biopolymers are naturally occurring polymers, which are increasingly used in the green synthesis of various nanoparticles. Biopolymers such as starch, cellulose, and chitosan are abundantly available and are low-cost materials used to remove water pollutants. They offer several advantages over the chemical methods as they are biodegradable, economical, and non-toxic. They have desirable physical and chemical properties essentially required for the treatment of water contaminants. Biosorption is a popular method wherein biopolymers are used to remove the organic and inorganic contaminants present in water. The use of biopolymers has been extensively explored in wastewater treatment and provides a sustainable solution to treat water fed with a range of contaminants. Moreover, their use in the synthesis of metal nanoparticles for application as active catalysts in organic synthesis for carbon-carbon bond formation is also appreciable.

3.1.3.1 Cellulose for Synthesis of Metal Nanoparticles

Cellulose-supported synthesis of metal nanoparticles is a promising approach for the synthesis of metal nanoparticles, which interestingly catalyzes organic transformations and also comprises green solvents like H_2O, supercritical CO_2, and ionic liquids. [16].

Thielemans et al. reported the synthesis of Pd nanoparticles supported on cellulose nanocrystals in a single step from palladium complex, Pd(hexafluoroacetylacetonate)$_2$ using cellulose nanocrystal and green solvent, supercritical CO_2 [17]. Nanostructured material derived from biological reductants offers huge scope in the field of green synthesis involving nanoscience and technology. Nanocrystalline cellulose is used as a reducing agent to prepare Pt nanoparticles from the aqueous solution for application in the field of electrocatalysis [18].

3.1.3.2 Starch for Synthesis of Metal Nanoparticles

Starch has been used as another green reductant and stabilizer for precious metal nanoparticles for versatile applications. Antibacterial, Ag nanoparticles were synthesized from silver nitrate and starch using a green one-pot microwave-assisted synthesis method. The role of starch is of reductant and stabilizer. In another example, degraded pueraria starch was used for the synthesis of Au/Ag bimetallic nanoparticles [19].

3.1.3.3 Chitosan for Synthesis of Metal Nanoparticles

Chitosan and its derivatives are known for various biomedical applications such as antibacterial, antifungal, and antiparasitic. Chitosan serves multiple roles in the preparation of metal nanoparticles such as of reducing, stabilizing, and capping agents. Cu nanoparticles are known for their high thermal and electrical conductivity and can be synthesized in eco-friendly manner using chitosan. Cu nanoparticles of high stability and longer life were synthesized by one-pot synthesis method using chitosan, sodium phosphinate, and ascorbic acid [20]. Sodium phosphinate acts as a co-reductant, while ascorbic acid is an antioxidant and does not let Cu nanoparticles to oxidize further, enhancing the stability of the particles. Interestingly, chitosan capped Cu nanocomposite synthesized from an aqueous medium finds application in the agriculture sector where it exhibits activity against common plant pathogens, viz., *Fusarium* sp., *Aspergillus* sp., *Alternaria alternata*, *Pythium* sp., and *Bacillus cereushas* [21].

3.1.3.4 Peptide for Synthesis of Ag Nanoparticles

Researchers have also used biomolecules such as proteins and peptides for the synthesis of metal and metal oxide nanoparticles. The use of biomolecules is an attractive approach for the synthesis of inorganic nanomaterials as it provides a sustainable strategy to obtain the nanomaterials of high value with enormous applications and unique properties. Naik et al. in 2002 strategically designed variable-shaped silver nanoparticles in the size range 60–150 nm using a peptide [22]. Amino acid sequence of the peptide serves as a template to bind and reduce Ag(I) to Ag(0). Interestingly, nano-sized silver crystals exhibit fluorescence upon shining a mercury lamp onto them.

3.1.4 WASTE TO VALUABLE NANOMATERIALS: GREEN APPROACH FOR SYNTHESIS OF NANOMATERIALS AND APPLICATIONS

It has been reported that around 84% of global energy has been fulfilled by fossil fuels [23]. At the same time, the waste to eco materials for energy has a huge potential for sustainable and renewable energy generation [24–30]. This shift of waste to energy has been shown in various studies where economic viability and environmental sustainability of waste to energy has been discussed deeply. The waste to energy requires sustainable waste management, and at the same time, their economic feasibility should also be provided. Several studies emphasize the energy recovery from waste incineration by processing the waste into eco materials [26,31]. The high operational and maintenance cost are the main drawbacks for waste to energy recovery [32]. In essence, waste to energy technologies provide a means to extract energy from waste materials in the form of usable heat, electricity (direct or indirect), or fuel; therefore, waste to energy is gaining recognition as a highly suitable option for addressing waste-related issues. In recent days, several studies indicate the preference for recycling over direct energy recovery [33,34]. In most of the countries where energy recovery increases, there recycling also increases. In contrast, most developing countries' wastes are managed by landfills, and they often exhibit lower recycling rates [35]. In 2015, Arafat et al. assessed the recoverable energy content of various components of municipal solid waste (MSW) using different waste-to-energy (WTE) technologies and suggested that anaerobic digestion is well-suited for food and yard waste, while gasification is effective for plastic waste [36]. At the same time, incineration is a viable option for energy recovery from a wide range of waste streams. Challenges in WTE technology identification persist, including social opposition due to potential toxic emissions [37] and unfavorable characteristics such as high costs and funding difficulties [38]. Moreover, protests from local communities, particularly in densely populated developing countries, pose a significant challenge [39,40]. For successful WTE facility implementation, local community acceptance is crucial [41]. Developed countries have recognized the potential of WTE options and have successfully integrated them into their waste management strategies. A few types of municipal waste and their treatment have been shown in Figure 3.2 adopted from reference [42].

It is important to mention that the industrial sector plays a pivotal role in a country's economy and overall prosperity. However, the process of industrialization has led to a pressing concern—environmental pollution [43]. This issue has prompted significant research efforts to address industrial waste remediation. Of particular concern is industrial pollution stemming from thermal power plants and ceramic industries, which demands urgent attention. Ceramic industries are known for generating substantial quantities of corrosive waste, while coal-based power plants emit

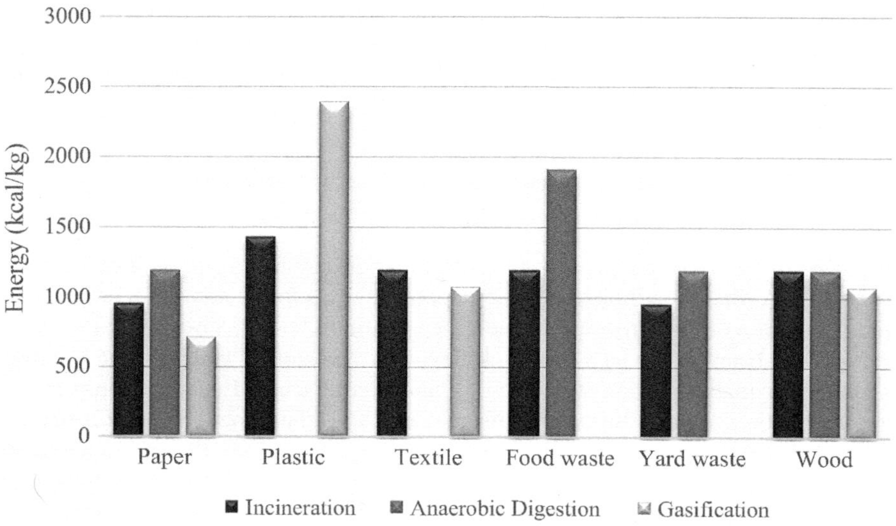

FIGURE 3.2 Energy recovery from various components of municipal waste [42].

carbon dioxide (CO_2) in gaseous form [43]. Both liquid and gaseous effluents from these sources pose a serious threat to our environment and the well-being of living organisms. Corrosive waste includes substances like industrial cleansers, effluents from paper and pulp industries, and byproducts of chemical reactions, such as spent catalysts or sludge from wastewater treatment systems. To combat environmental pollution caused by industries, key strategies involve reducing the consumption of polluting products, wastewater treatment, and implementing proper discharge and disposal procedures. Based on various parameters, the waste can be mainly categorized as biowaste, electronic waste, nuclear waste, agrowaste, and so on. In this chapter, we will mainly focus on materials recovery for electronic waste and their utilization in energy and water purification.

E-waste composition

Electronic waste is commonly referred to as e-waste, and most of the e-waste is generated from electrical and electronic equipment as well as chargeable and rechargeable batteries. E-waste contains the following components. **Valuable Metals**: In e-waste, most of the eco materials are in the form of metals such as gold (Au), silver (Ag), platinum (Pt), palladium (Pd), copper (Cu), aluminum (Al), nickel (Ni), tin (Sn), zinc (Zn), iron (Fe), and other [24]. Out of these metals few metals are regarded as metals of concern and hazardous in nature, e.g., metals such as mercury (Hg), beryllium (Be), indium (In), lead (Pb), cadmium (Cd), arsenic (As), and antimony (Sb). These materials collectively form the intricate composition of electronic waste, reflecting the diverse nature of discarded electronic devices. It's important to note that the specific materials present in e-waste can vary based on the source and

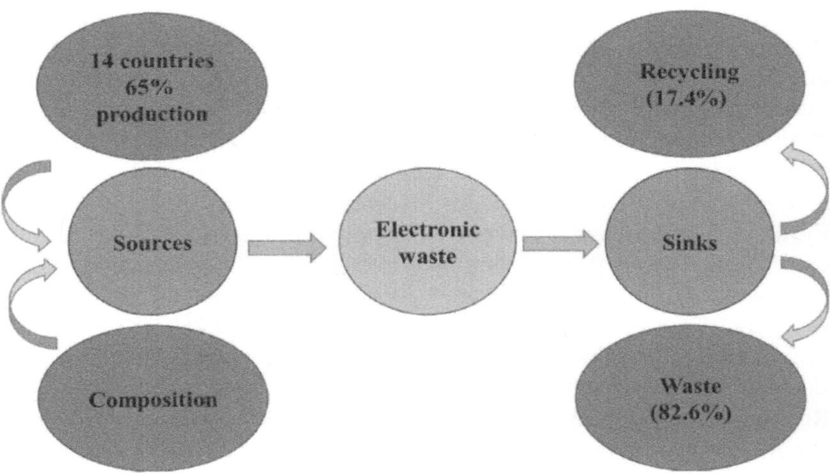

FIGURE 3.3 E-waste components and their source [24].

type of electronic equipment. The composition and recycling of e-waste has been shown in Figure 3.3 [24]. Out of various e-waste, we focused on battery waste for recovery of materials in this chapter.

Conventional methods for treatment of e-waste (battery) and their application in energy

Nowadays portable batteries have gained popularity as power sources for household items like toys, radios, cameras, torches, and more. The most widely used batteries are zinc-carbon dry cells and zinc-manganese dioxide batteries. However, a significant issue arises when these batteries are casually discarded after they are depleted, leading to environmental risks [44]. In Mexico alone, a staggering 635,000 tons of batteries have been disposed of in the environment over the past 43 years [45]. The disposed batteries predominantly contain metallic components such as cadmium (Cd), mercury (Hg), and lead (Pb), posing threats to living organisms and causing pollution in natural reservoirs such as soil and groundwater. To address the issue of battery waste, the most effective approach involves repairing, reusing, or recycling the valuable materials it contains. However, this can be challenging due to the toxic components involved in electronic waste. The ideal principle for e-waste management revolves around recycling, proper transportation, and safe disposal to minimize health and environmental impacts. Various e-waste management methods are employed worldwide, with differences between developed and developing nations. For example, China faces challenges in processing printed circuit boards (PCBs) and handling toxic substances in waste, while the European Union has established efficient collection centers for recycling many metals. Nevertheless, basic recycling methods may not efficiently recover all metals, including rare ones like indium, ruthenium, and palladium [44]. The following are the conventional methods for recycling e-waste with their merits and demerits.

Landfilling: Landfilling is a commonly employed method for all types of waste, especially e-waste, disposal due to its simplicity and ease of operation. This approach entails the open dumping of electronic waste in landfills, resulting in the creation of leachate in the landfill site. E-waste disposed of in landfills can release toxic metals and polyhalogenated organic contaminants, posing risks to both human health and the environment [44]. These harmful metals, which resist decomposition through physical processes, persist as pollutants for extended periods.

Pyrolysis: Pyrolysis and incineration require the burning of waste to get the useful metals. This method is fast and eliminates the requirement of laborious separation and recycling of the metals. The byproducts obtained from the incineration process not only contribute to the sustainability of the operation but also serve as potential sources of both chemical compounds and energy; however, at the same time, the end product has been dependent mostly on the temperature of pyrolysis [46].

Repair: The e-waste can be reduced by repairing the product that was damaged; however, we can delay the e-waste generation by this method, but in the end, it will go into waste, and we have to utilize it from various methods.

Recycling: This method is a widely used and highly effective method that reduces carbon emissions and limits the presence of valuable or hazardous substances. In 2019, only a small fraction (17.4%) of total e-waste was collected through recycling [24], which implies that this method has a huge potential in e-waste management. It has been suggested that approximately 38% ferrous metals, 28% non-ferrous metals, 19% plastics, 4% glass, 1% wood, and 10% other materials are present in the e-waste sector. Printed circuit boards within electronic devices are notably rich in metals, containing significantly more gold and copper than natural ores.

Hydrometrical Processing: Hydrometallurgy is a highly effective method for treating e-waste, surpassing smelting in terms of precision, predictability, and its positive environmental impact [47]. This process involves converting e-waste into a granular form, followed by separation, purification, and metal recovery using alkaline or acidic solutions to leach metals from the waste. Commonly used leaching agents in hydrometallurgy include nitric acid for base metals, sulfuric acid and aqua regia for copper, cyanide for gold and silver, and sodium chlorate for palladium [47]. However, one drawback of hydrometallurgical processing is the use of cyanide as a leaching agent for gold, which poses toxicity risks and environmental contamination, especially in water bodies [48]. Safer alternatives to cyanide include thiosulfate, thiourea, and halides, which offer efficient gold recovery, cost-effectiveness, and reduced environmental impact [49]. Aqua regia, a mixture of concentrated hydrochloric and nitric acids, can also dissolve gold, but the use of chlorine can generate toxic gases and strong corrosive solutions [49].

3.1.4.1   Application of Waste to Wealth

3.1.4.1.1 Extraction of ZnO from Waste Spent Battery

The demand for portable devices increases; therefore, battery demands should be also increased significantly, out of which lots of spent batteries are received [50]. Electronic devices in the last decade have led to a significant rise in electronic waste generation. It has been reported that ZnO can be recovered from spent (discharge) Zn-Mn primary alkaline batteries [51]. According to the United States Environmental Protection Agency (USEPA), each U.U.A. citizen uses ~8 batteries per year, translating to around 2.4 billion batteries disposed of annually due to the country's large population [52]. Alkaline Zn-Mn batteries are widely used in small modern gadgets like recorders, remote controls, and toys. Various studies have attempted the recycling of alkaline batteries, with the most common method being hydrometallurgy. This process involves using corrosive leachants like sulfuric acid, hydrochloric acid, and potassium hydroxide to recover pure Zn and Mn from waste batteries [53]. In a recent study, efforts were made to develop an eco-friendly process to extract various ZnO nanostructures from waste alkaline primary zinc-manganese batteries. This study also investigated the chemical and morphological changes that occur during the discharge process of primary batteries, revealing the conversion of bulk zinc into zinc oxide rods during battery operation. The production of a ZnO-based field emitter involved applying a layer of ZnO onto a silicon substrate through a scalable spin-coating method. The activation threshold for ZnO rods was 4.9 V/μm, accompanied by an impressive field enhancement factor of 4,655 in high electric field conditions. This innovative technique holds great appeal for researchers in the fields of materials chemistry and environmental waste management. A method describing the extraction of ZnO and its utilization for field emission has been shown in Figure 3.4.

3.1.4.1.2 Spent Battery to Energy

Sapna et al. also utilized spent Zn-Mn battery for the extraction of Zn-Mn-based oxide and performed the electrochemical ethanol oxidation [54]. The disposal of these used electronic batteries in landfills results in hazardous waste, contributing to environmental pollution. Considering that spent batteries contain valuable elements, there is growing interest in recovering these resources to meet the material demands of sustainable energy production. Sapna et al. synthesized Ni_6MnO_8 nanoparticles extracted from the spent battery through a straightforward chemical process [54]. She utilized manganese chloride obtained from the cathode portion of spent batteries as a precursor. The resulting Ni_6MnO_8 material was then employed in electrochemical ethanol oxidation. Notably, Ni_6MnO_8 exhibited robust electroactivity for ethanol oxidation, with a high current density ($13.69\,mA/cm^2$), a low oxidation peak potential ($1.53\,V$), a smaller Tafel slope value ($61\,mV/dec$), and a high current density ratio (J_f/J_b) of 1.01. Furthermore, the surface area of the Ni_6MnO_8 sheet, approximately $136\,m^2/g$, exceeded the maximum reported surface area to date ($91.32\,m^2/g$). This research highlights the potential of non-noble metal-based mixed metal oxides derived from waste as promising anodic materials for direct alkaline fuel cells.

3.1.4.1.3 Tin Can to Iron Chloride

The rapid growth of the global population, coupled with urbanization and improved living standards in both developed and developing nations, has resulted in a significant increase in solid waste production. Solid waste encompasses discarded or unused materials from various sectors, including industry, commerce, and residential areas. Among several waste types, industrial waste, including iron and other metal waste, plays a significant role in waste generation in countries like India, where

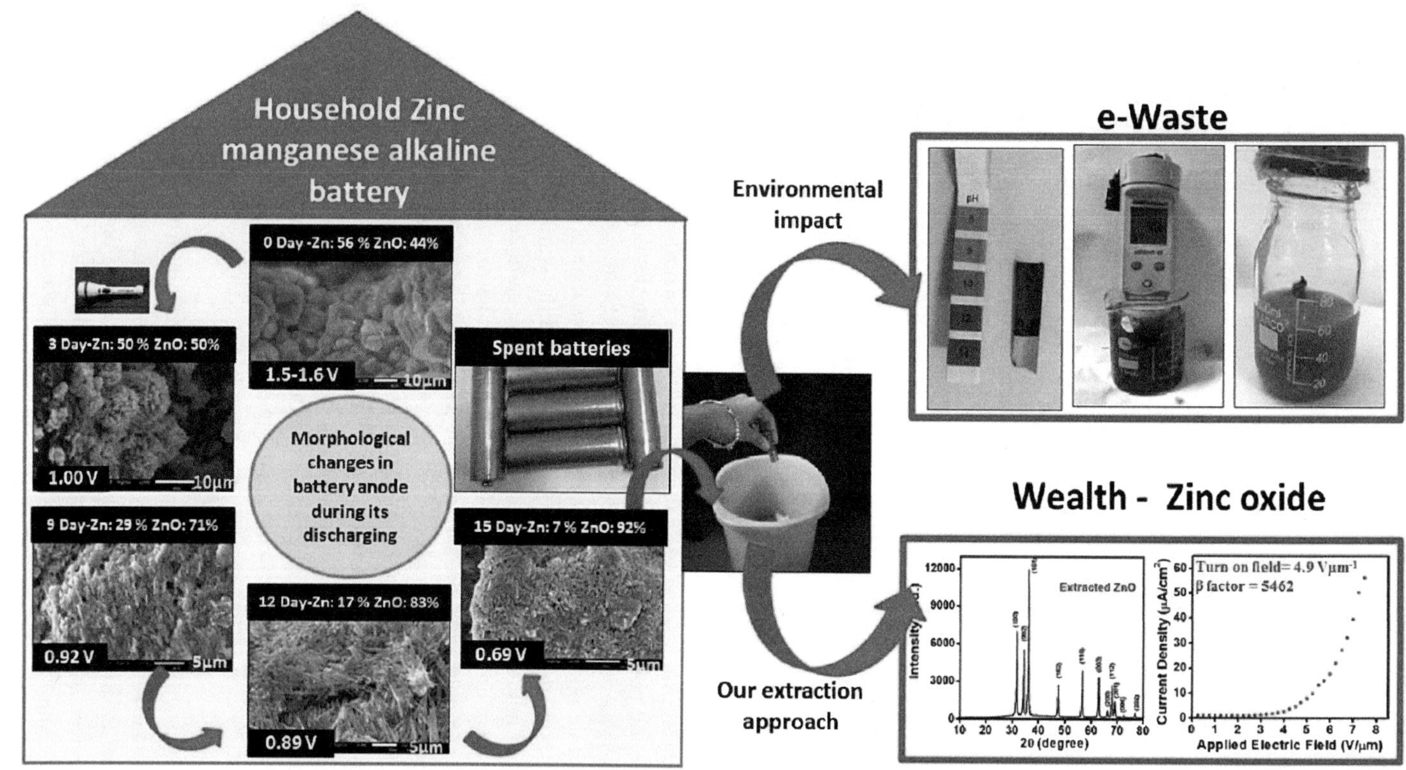

FIGURE 3.4   Extraction of ZnO from spent battery and its utilization for field emissions [53].

an estimated 290 million tons per annum are generated [55]. These iron-based and non-iron-based solid metal wastes find applications in building materials, such as bricks, tiles, lightweight aggregates, reinforced composites, and ceramics produced from iron ore tailings [56]. Within the category of ferrous metals, items like cast iron, tin cans, stainless steel, industrial scrap, car bodies, and household appliances contribute to the waste stream. In particular, tin-plated iron containers, often used in food packaging, are prevalent and their disposal in landfills contributes significantly to solid waste and environmental pollution [57].

For the synthesis of iron chloride, a waste tin container was immersed in concentrated hydrochloric acid (HCl) at room temperature for 10 minutes followed by cutting the waste tin can into pieces. The as-obtained tin-free pieces were placed into HCl containing beaker followed by heating for making soluble iron. The soluble waste can was slowly evaporated at a constant temperature of 60°C for 27 hours. This process led to the formation of solid green crystals. The excess green crystals were then filtered using Whatman filter paper, dried under vacuum conditions, and stored in an inert atmosphere for further analysis. The as-obtained sample is $FeCl_2 \cdot 4H_2O$.

Water purification from waste-derived materials

Water is a fundamental necessity for all life on Earth. Water is of paramount importance in sustaining life and plays a critical role in maintaining ecosystems [58]. However, in recent years, providing clean water to a significant portion of the population remains a challenge. Factors such as rapid population growth, extensive industrialization, increased agricultural activities to meet the growing food demand, and environmental changes have led to a rise in water pollution [59]. Consequently, water quality is deteriorating on a global scale. Wastewater treatment has become a major challenge for scientists, water regulatory authorities, and government agencies. Water pollutants are broadly categorized as inorganic, organic, and biological pollutants [60]. Shockingly, around 80% of industrial and domestic effluents are directly discharged into freshwater bodies, leading to health, environmental, and climate-related disasters. Developed countries have wastewater treatment capacity at approximately 70%, while developing countries lag significantly at around 8%. This disparity exacerbates water-related issues in developing nations. Urbanization's rapid growth has accelerated water pollution, causing eutrophication and human health problems, and contributing to greenhouse gas emissions like nitrous oxide and methane. Proper wastewater treatment is essential to maintain sanitation and prevent diseases. Managing community wastewater and sewage is crucial to prevent contamination of local environments and drinking water supplies, reducing the risk of disease transmission. Access to clean water, sanitation, and hygiene education can decrease disease-related health issues and deaths, ultimately leading to better global health and socioeconomic development. However, many countries face challenges in managing wastewater due to limited resources, funds, infrastructure, technology, and space, leaving their populations at risk of contaminated water, poor sanitation, and hygiene-related diseases. Different sources of water contamination are illustrated in Figure 3.5.

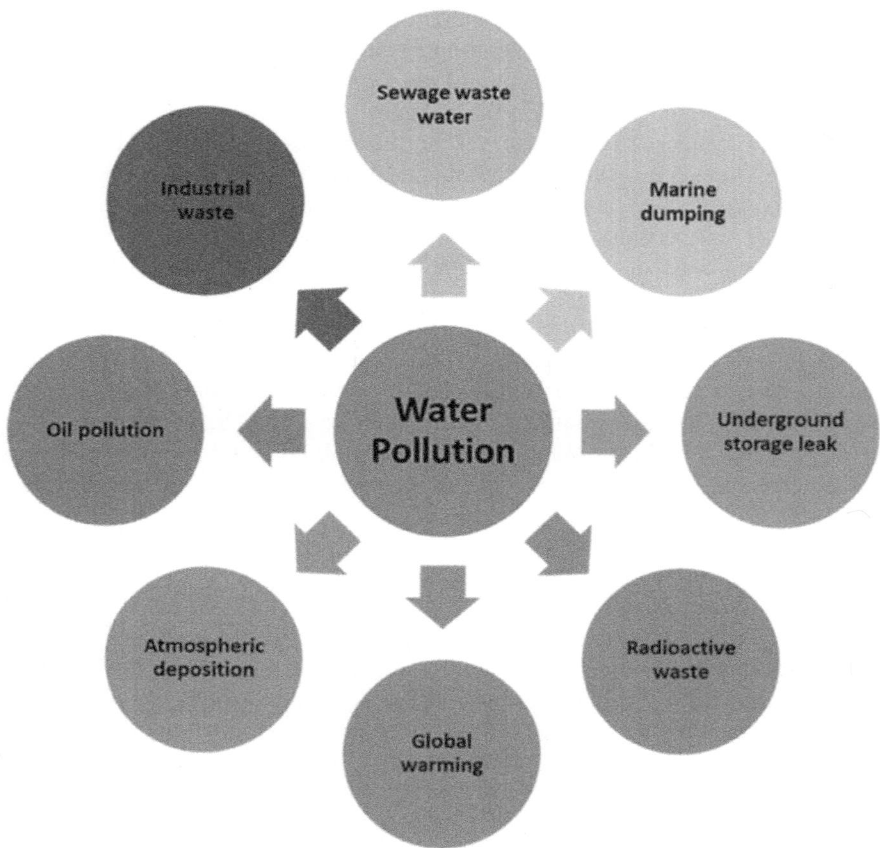

FIGURE 3.5 Source of water contamination [25].

Arora et al. reported the extraction of iron(II) oxide from scrap iron received from building materials and utilized for the pharmaceutical pollutant ciprofloxacin (CIP) [61]. The iron was reclaimed from scrap metal wire using a hydrometallurgical procedure, converting it into highly pure ferrous chloride crystals. These ferrous chloride crystals served as the source of iron for stabilizing iron oxalate nanorods through a micellar approach. The iron oxide nanorods and nanoparticles were employed for the efficient photodegradation of CIP. The iron oxide produced through calcination at 250°C exhibited a reasonably high surface area of 133 m^2/g. Detailed investigations into the mechanistic aspects of Fe$_2$O$_3$ nanorods for CIP degradation revealed that e$^-$ (electrons) and ˙OH (hydroxyl radicals) were the primary reactive species involved in the degradation of CIP. The photocatalyst displayed robust durability and could be recycled for more than three consecutive reaction cycles. Furthermore, this study showcases a viable approach to convert scrap iron into a valuable product, addressing both disposal concerns and aiding in wastewater remediation.

REFERENCES

1. R.W. Raut, A.S.M. Haroon, Y.S. Malghe, B.T. Nikam, S.B. Kashid, Rapid biosynthesis of platinum and palladium metal nanoparticles using root extract of asparagus racemosus linn, *Adv Mater Lett.* 4 (2013) 650–654. https://doi.org/10.5185/amlett.2012.11470.

2. A.M. El Shafey, Green synthesis of metal and metal oxide nanoparticles from plant leaf extracts and their applications: A review, *Green Process Synth.* 9 (2020) 304–339. https://doi.org/10.1515/gps-2020-0031.

3. E. Ismail, M. Kenfouch, and M. Dhlamini, S. Dube, M. Maaza, Green biosynthesis of rhodium nanoparticles via aspalathus linearis natural extract, *J Nanomater Mol Nanotechnol.* 06 (2017). https://www.scitechnol.com/peer-review/green-biosynthesis-of-rhodium-nanoparticles-via-aspalathus-linearis-natural-extract-ycsW.php?article_id=5957.

4. S. Chouhan, S. Guleria, Green synthesis of AgNPs using Cannabis sativa leaf extract: Characterization, antibacterial, anti-yeast and α-amylase inhibitory activity, *Mater Sci Energy Technol.* 3 (2020) 536–544. https://doi.org/10.1016/j.mset.2020.05.004.

5. J. Singh, T. Dutta, K.H. Kim, M. Rawat, P. Samddar, P. Kumar, "Green" synthesis of metals and their oxide nanoparticles: Applications for environmental remediation, *J Nanobiotechnol.* 16 (2018) 1–24. https://link.springer.com/article/10.1186/s12951-018-0408-4.

6. Priya, Naveen, K. Kaur, A.K. Sidhu, Green synthesis: An eco-friendly route for the synthesis of iron oxide nanoparticles, *Front Nanotechnol.* 3 (2021). https://doi.org/10.3389/fnano.2021.655062.

7. F.-N. Allouche, N. Yassaa, H. Lounici, Sorption of methyl orange from aqueous solution on chitosan biomass, *Procedia Earth Planet Sci.* 15 (2015) 596–601. https://doi.org/10.1016/j.proeps.2015.08.109.

8. S.A. Khan, S. Shahid, B. Shahid, U. Fatima, S.A. Abbasi, Green synthesis of MNO nanoparticles using abutilon indicum leaf extract for biological, photocatalytic, and adsorption activities, *Biomolecules.* 10 (2020). https://doi.org/10.3390/biom10050785.

9. S.P. Rajendran, K. Sengodan, Synthesis and characterization of zinc oxide and iron oxide nanoparticles using Sesbania grandiflora leaf extract as reducing agent, *J Nanosci.* 2017 (2017) 1–7. https://doi.org/10.1155/2017/8348507.

10. E. Matei, A.M. Predescu, M. Râpă, A.A. Ţurcanu, I. Mateș, N. Constantin, C. Predescu, Natural polymers and their nanocomposites used for environmental applications, *Nanomaterials.* 12 (2022). https://doi.org/10.3390/nano12101707.

11. T. Klaus, R. Joerger, E. Olsson, C.G. Granqvist, Silver-based crystalline nanoparticles, microbially fabricated, *Proc Natl Acad Sci USA.* 96 (1999) 13611–13614. https://doi.org/10.1073/pnas.96.24.13611.

12. Mishra, Amrita, S.K. Tripathy, S.I. Yun, Bio-synthesis of gold and silver nanoparticles from Candida guilliermondii and their antimicrobial effect against pathogenic bacteria, *J Nanosci Nanotechnol.* 11(1) (2011) 243–248. https://doi.org/10.1166/jnn.2011.3265.

13. K.S. Siddiqi, A. Husen, Fabrication of metal nanoparticles from fungi and metal salts: Scope and application. *Nanoscale Res Lett.* 11 (2016) 1–15. https://link.springer.com/article/10.1186/s11671-016-1311-2.

14. G.S. Dhillon, S.K. Brar, S. Kaur, M. Verma, Green approach for nanoparticle biosynthesis by fungi: Current trends and applications. *Crit. Rev. Biotechnol.* 32 (2012) 49–73. https://doi.org/10.3109/07388551.2010.550568.

15. P. Mukherjee, A. Ahmad, D. Mandal, S. Senapati, S.R. Sainkar, M. I., Khan, R. Parishcha, P.V. AjayKumar, M. Alam, R. Kumar, M. Sastry, Fungus-mediated synthesis of silver nanoparticles and their immobilization in the mycelial matrix: A novel biological approach to nanoparticle synthesis. *Nano Lett.* 1 (2001) 515–519. https://pubs.acs.org/doi/10.1021/nl0155274.

16. S. Kamel, T.A. Khattab, Recent advances in cellulose supported metal nanoparticles as green and sustainable catalysis for organic synthesis. *Cellulose.* 28 (2021) 4545–4574. https://link.springer.com/article/10.1007/s10570-021-03839-1.

17. M. Rezayat, R.K. Blundell, J.E. Camp, D.A. Walsh, W. Thielemans, Green one-step synthesis of catalytically active palladium nanoparticles supported on cellulose nanocrystals. *ACS Sustain Chem Eng.* 2 (2014) 1241–1250. https://pubs.acs.org/doi/abs/10.1021/sc500079q.

18. L. Johnson, W. Thielemans, D.A. Walsh, Synthesis of carbon-supported Pt nanoparticle electrocatalysts using nanocrystalline cellulose as reducing agent, *Green Chem.* 13 (2011) 1686–1693. https://doi.org/10.1039/C0GC00881H.

19. B. Xia, F. He, L. Li, Preparation of bimetallic nanoparticles using a facile green synthesis method and their application, *Langmuir.* 29 (2013), 4901–4907. https://pubs.acs.org/doi/10.1021/la400355u.

20. K. Tokarek, J.L. Hueso, P. Kuśtrowski, G. Stochel, A. Kyzioł. Green synthesis of chitosan-stabilized copper nanoparticles, *Eur J Inorg Chem.* 28 (2013) 4940–4947. https://doi.org/10.1002/ejic.201300594.

21. M.R. Mehta, H.P. Mahajan, A.U. Hivrale, Green synthesis of chitosan capped-copper nano biocomposites: Synthesis, characterization, and biological activity against plant pathogens. *BioNanoScience* 11 (2021) 417–427. https://link.springer.com/article/10.1007/s12668-021-00823-8.

22. R.R. Naik, S.J. Stringer, G. Agarwal, S.E. Jones, M.O. Stone, Biomimetic synthesis and patterning of silver nanoparticles, *Nat Mater.* 1 (2002) 169–172. https://doi.org/10.1038/nmat758.

23. S. Shafiee, E. Topal, When will fossil fuel reserves be diminished? *Energy Policy.* 37 (2009) 181–189. https://doi.org/10.1016/j.enpol.2008.08.016.

24. M. Jha, Sunaina, S. Devi, N. Khan, Insight into Current Scenario of Electronic Waste to Nanomaterials Conversion, In: *Waste to Profit*, CRC Press, Boca Raton, FL, 2023: pp. 133–146. https://www.taylorfrancis.com/chapters/edit/10.1201/9781003334415-10.

25. M. Jha, Sunaina, A. Arora, K. Sood, Wastewater Treatment Using Nanoadsorbents Derived from Waste Materials, In: *Waste to Profit*, CRC Press, Boca Raton, FL, 2023: pp. 147–164. https://www.taylorfrancis.com/chapters/edit/10.1201/9781003334415-11.

26. B. Baran, M.S. Mamis, B.B. Alagoz, Utilization of energy from waste potential in Turkey as distributed secondary renewable energy source, *Renew Energy.* 90 (2016) 493–500. https://doi.org/10.1016/j.renene.2015.12.070.

27. B.Ž. Bajić, S.N. Dodić, D.G. Vučurović, J.M. Dodić, J.A. Grahovac, Waste-to-energy status in Serbia, *Renew Sustain Energy Rev.* 50 (2015) 1437–1444. https://doi.org/10.1016/j.rser.2015.05.079.

28. K.K. Yadav, M. Sharma, M. Jha, Extraction of nanostructured sodium nitrate from industrial effluent and their thermal properties, *Water Environ Res.* 92 (2020) 1123–1130. https://doi.org/10.1002/wer.1307.

29. K.K. Yadav, R. Wadhwa, N. Khan, M. Jha, Efficient metal-free supercapacitor based on graphene oxide derived from waste rice, *Curr Res Green Sustain Chem.* 4 (2021) 100075. https://doi.org/10.1016/j.crgsc.2021.100075.

30. K.K. Yadav, H. Singh, S. Rana, Sunaina, H. Sammi, S.T. Nishanthi, R. Wadhwa, N. Khan, M. Jha, Utilization of waste coir fibre architecture to synthesize porous graphene oxide and their derivatives: An efficient energy storage material, *J Clean Prod.* (2020) 124240. https://doi.org/10.1016/j.jclepro.2020.124240.

31. S.K. Guchhait, K.K. Yadav, Sunaina, S. Khatana, R.K. Saini, Pranay, U.K. Arora, R. Satyakam, R. Bajaj, M. Jha, Conversion of gaseous effluents of power plant to sodium carbonate: A value-added material for powder detergent, *Cleaner Waste Systems.* 3 (2022) 100042. https://doi.org/10.1016/j.clwas.2022.100042.

32. A.S. Erses Yay, Application of life cycle assessment (LCA) for municipal solid waste management: A case study of Sakarya, *J Clean Prod.* 94 (2015) 284–293. https://doi.org/10.1016/j.jclepro.2015.01.089.

33. O.K.M. Ouda, S.A. Raza, A.S. Nizami, M. Rehan, R. Al-Waked, N.E. Korres, Waste to energy potential: A case study of Saudi Arabia, *Renew Sustain Energy Rev.* 61 (2016) 328–340. https://doi.org/10.1016/j.rser.2016.04.005.

34. S.T. Tan, H. Hashim, J.S. Lim, W.S. Ho, C.T. Lee, J. Yan, Energy and emissions benefits of renewable energy derived from municipal solid waste: Analysis of a low carbon scenario in Malaysia, *Appl Energy.* 136 (2014) 797–804. https://doi.org/10.1016/j.apenergy.2014.06.003.

35. Ch. Achillas, Ch. Vlachokostas, N. Moussiopoulos, G. Banias, G. Kafetzopoulos, A. Karagiannidis, Social acceptance for the development of a waste-to-energy plant in an urban area, *Resour Conserv Recycl.* 55 (2011) 857–863. https://doi.org/10.1016/j.resconrec.2011.04.012.

36. H.A. Arafat, K. Jijakli, A. Ahsan, Environmental performance and energy recovery potential of five processes for municipal solid waste treatment, *J Clean Prod.* 105 (2015) 233–240. https://doi.org/10.1016/j.jclepro.2013.11.071.

37. X. Zhao, G. Jiang, A. Li, L. Wang, Economic analysis of waste-to-energy industry in China, *Waste Manage.* 48 (2016) 604–618. https://doi.org/10.1016/j.wasman.2015.10.014.

38. D.Q. Zhang, S.K. Tan, R.M. Gersberg, Municipal solid waste management in China: Status, problems and challenges, *J Environ Manage.* 91 (2010) 1623–1633. https://doi.org/10.1016/j.jenvman.2010.03.012.

39. K.A. Kalyani, K.K. Pandey, Waste to energy status in India: A short review, *Renew Sustain Energy Rev.* 31 (2014) 113–120. https://doi.org/10.1016/j.rser.2013.11.020.

40. X. Ren, Y. Che, K. Yang, Y. Tao, Risk perception and public acceptance toward a highly protested Waste-to-Energy facility, *Waste Manage.* 48 (2016) 528–539. https://doi.org/10.1016/j.wasman.2015.10.036.

41. R. Kikuchi, R. Gerardo, More than a decade of conflict between hazardous waste management and public resistance: A case study of NIMBY syndrome in Souselas (Portugal), *J Hazard Mater.* 172 (2009) 1681–1685. https://doi.org/10.1016/j.jhazmat.2009.07.062.

42. A. Kumar, S.R. Samadder, A review on technological options of waste to energy for effective management of municipal solid waste, *Waste Manage.* 69 (2017) 407–422. https://doi.org/10.1016/j.wasman.2017.08.046.

43. N. Dabas, K.K. Yadav, A.K. Ganguli, M. Jha, New process for conversion of hazardous industrial effluent of ceramic industry into nanostructured sodium carbonate and their application in textile industry, *J Environ Manage.* 240 (2019) 352–358. https://doi.org/10.1016/j.jenvman.2019.03.066.

44. H. Ghimire, P.A. Ariya, E-wastes: Bridging the knowledge gaps in global production budgets, composition, recycling and sustainability implications, *Sustain Chem.* 1 (2020) 154–182. https://doi.org/10.3390/suschem1020012.

45. M. Bartolozzi, The recovery of metals from spent alkaline-manganese batteries: A review of patent literature, *Resour Conserv Recycl.* 4 (1990) 233–240. https://doi.org/10.1016/0921-3449(90)90004-N.

46. A. Kumar, M. Holuszko, D.C.R. Espinosa, E-waste: An overview on generation, collection, legislation and recycling practices, *Resour Conserv Recycl.* 122 (2017) 32–42. https://doi.org/10.1016/j.resconrec.2017.01.018.

47. S. Ilyas, C. Ruan, H.N. Bhatti, M.A. Ghauri, M.A. Anwar, Column bioleaching of metals from electronic scrap, *Hydrometallurgy.* 101 (2010) 135–140. https://doi.org/10.1016/j.hydromet.2009.12.007.

48. I. Birloaga, I. De Michelis, F. Ferella, M. Buzatu, F. Vegliò, Study on the influence of various factors in the hydrometallurgical processing of waste printed circuit boards for copper and gold recovery, *Waste Manage.* 33 (2013) 935–941. https://doi.org/10.1016/j. wasman.2013.01.003.

49. J. Cui, L. Zhang, Metallurgical recovery of metals from electronic waste: A review, *J Hazard Mater.* 158 (2008) 228–256. https://doi.org/10.1016/j.jhazmat.2008.02.001.

50. M.M. Devi, Sunaina, H. Singh, K. Kaur, A. Gupta, A. das, S.T. Nishanthi, C. Bera, A.K. Ganguli, M. Jha, New approach for the transformation of metallic waste into nanostructured Fe3O4 and SnO2-Fe3O4 heterostructure and their application in treatment of organic pollutant, *Waste Manage.* 87 (2019) 719–730. https://doi.org/10.1016/j. wasman.2019.03.007.

51. A. Salgado, Recovery of zinc and manganese from spent alkaline batteries by liquid-liquid extraction with Cyanex 272, *J Power Sources.* 115 (2003) 367–373. https:// doi.org/10.1016/S0378-7753(03)00025-9.

52. J.-P. Wiaux, J.-P. Waefler, Recycling zinc batteries: An economical challenge in consumer waste management, *J Power Sources.* 57 (1995) 61–65. https://doi. org/10.1016/0378-7753(95)02242-2.

53. Sunaina, M. Sreekanth, M. Manolata Devi, V. Sethi, S. Ghosh, S.K. Mehta, A.K. Ganguli, M. Jha, New approach for fabrication of vertically oriented ZnO based field emitter derived from waste primary batteries, *Mater Sci Eng B Solid State Mater Adv Technol.* 274 (2021). https://doi.org/10.1016/j.mseb.2021.115480.

54. S. Devi, Sunaina, R. Wadhwa, K.K. Yadav, M. Jha, Understanding the origin of ethanol oxidation from ultrafine nickel manganese oxide nanosheets derived from spent alkaline batteries, *J Clean Prod.* 376 (2022) 134147. https://doi.org/10.1016/j. jclepro.2022.134147.

55. S.K. Guchhait, H. Sammi, K.K. Yadav, S. Rana, M. Jha, New hydrometallurgical approach to obtain uniform antiferromagnetic ferrous chloride cubes from waste tin cans, *J Mater Sci: Mater Electron.* 32 (2021) 2965–2972. https://link.springer.com/ article/10.1007/s10854-020-05048-1.

56. B. Gorai, R.K. Jana, Premchand, Characteristics and utilisation of copper slag-a review, *Resour Conserv Recycl.* 39 (2003) 299–313. https://doi.org/10.1016/S0921-3449(02)00171-4.

57. R.M. Yoada, D. Chirawurah, P.B. Adongo, Domestic waste disposal practice and perceptions of private sector waste management in urban Accra, *BMC Public Health.* 14 (2014) 697. https://bmcpublichealth.biomedcentral.com/articles/10.1186/1471-2458-14-697.

58. Z.W. Kundzewicz, Water resources for sustainable development. *Hydrolog Sci J.* 42 (1997), 467–480. https://doi.org/10.1080/02626669709492047.

59. M.E.A. Raouf, N.E. Maysour, R.K. Farag, A.M. Abdul-Raheim, Wastewater treatment methodologies, review article. *Int J Environ Agri Sci.* 3 (2019), 018.

60. N. Abdel-Raouf, A.A. Al-Homaidan, I.B.M. Ibraheem, Microalgae and wastewater treatment. *Saudi J Biol Sci.* 19 (2012), 257–275. https://doi.org/10.1016/j.sjbs.2012. 04.005.

61. A. Arora, Sunaina, R. Wadhwa, M. Jha, Conversion of scrap iron into ultrafine α-Fe$_2$O$_3$ nanorods for the efficient visible light photodegradation of ciprofloxacin, *New J Chem.* 46 (2022) 5861–5868. https://doi.org/10.1039/D2NJ00245K.

4 Chitosan Nanoparticles
Sustainable and Eco-Friendly Materials for Advancing Nanotechnological Applications

Renuka Kanojia, Anand Salvi, Sonika Charak, Manish Shandilya, and Joydeep Dutta

4.1 INTRODUCTION

In the fast-changing world of nanotechnology, everyone is looking for materials that are sustainable and good for environment. The idea of sustainable development has received significant political and social attention in recent years. Prioritizing the development of green technologies and applications of products formed from it in order to promote sustainability and less environmental degradation is necessary (Gomes et al., 2017). Nanotechnology nowadays with the help of different structures has been used to prepare innovative food substitutes and different advanced products that provide significant benefits to human health (Magnuson et al., 2011). Natural biopolymers such as chitosan and its derivatives have potential applications like improving human health and production of various drugs, vaccines, and many food constituents because of their inherent characteristics like biocompatibility, biodegradability, and low immunogenicity; they are considered as safe and are also readily accepted by many health institutions (Gomes et al., 2017; Elmoghayer et al., 2024). Chitosan is a natural biopolymer derived from chitin, which is a structural component of the exoskeleton of crustaceans and fungi (de Souza et al., 2018; Wijayadi & Rusli, 2019). Chitosan consists of both deacetylated and acetylated units. The acetylated units are composed of N-acetyl-D-glucosamine, while the deacetylated unit is composed of β-(1,4)-D-glucosamine (Wijayadi & Rusli, 2019; Sachdeva et al., 2023). Chitosan possesses inherent anti-allergenic characteristics and exhibits natural antibacterial activity, rendering it a valuable agent for promoting wound healing. Chitosan is anti-allergenic in nature and also has natural antibacterial properties, which makes it a useful wound healing agent (Nasir, 2010; Wijayadi & Rusli, 2019). It is insoluble in water and can only be dissolved in dilute acetic acid, formic acid, and other acids (Jiang et al., 2014). The combination of nanotechnology and polymers has generated significant interest in various fields such as drug development and therapeutic innovation in recent times (Agarwal et al., 2018).

DOI: 10.1201/9781003473749-5

Nanotechnology is such a broad field that includes engineering, material science, physics, chemistry, and life science. It is an emerging area of science that includes the study of structures with a size between 1 and 100 nm and their applications (Sharifi-Rad et al., 2021). Nanomaterials include a diverse range of materials with nanoscale structures, such as nano-brushes, nano-pins, nano-rods, nano-particles, nano-sheets, nano-reactors, nano-fibers, and nano-clusters (Shukla et al., 2013). At this very time, and throughout the history of mankind, nanoparticles are all around us (Aljebory et al., 2017). These particles have small size and large surface area and consist of unique physical and chemical properties (Agarwal et al., 2018). Polymeric nanoparticles are becoming more and more popular, especially in the field of nano-biotechnology (Shukla et al., 2013). Chitosan nanoparticles and their uses are at the center of an increasing number of scientific papers and patents, which shows an extremely advanced level of study on this biopolymer (Gomes et al., 2017). Chitosan has been considered as a choice of material for the production of nanoparticles because of its biocompatibility and biodegradability (Ochekpe et al., 2009). The synthesis of nanoparticles from synthetic polymers involves the application of heat, organic solvents, and high shear force, which can potentially compromise the stability of drugs. On the other hand, nanoparticles derived from natural polymers such as chitosan use easy and mild techniques that do not require the use of any organic solvents or high shear force (Agarwal et al., 2018). The initial findings have proved that the use of these nanoparticles is in the favor of human beings as well as animals and is also considered as safe for environmental applications (Gomes et al., 2017). In addition to this, the nanoparticles prepared from chitosan have fascinating properties like color stabilization, emulsification, and dietary fiber like properties and facilitate the trapping of fat and water; due to these multifunctional health benefits, the use of these chitosan derivatives has increased much in food industries (Gomes et al., 2017). Chitosan nanoparticles are a very promising and versatile method for overcoming the bioavailability and stability concerns of a majority of active ingredients. They carry the inherent properties of polymer with adaptable size and the possibility of surface modification according to individual needs (Mohammed et al., 2017). Many drugs have the issue of poor stability, low selectivity, water insolubility, and other side effects. Good carriers such as chitosan nanoparticles play a crucial role in solving these problems. Chitosan nanoparticles are a promising drug carrier, and their small size allows them to pass through biological barriers (Wang et al., 2011). Nanotechnology also has a fascinating role in agriculture production to improve soil health or crop health against harmful environmental conditions such as drought, salt, UV rays, heavy metal, flooding, and biotic stress; nanoparticles can be placed in the soil or delivered directly to crops (Hoang et al., 2022).

This chapter aims to shed light on the synthesis, properties, and different applications of chitosan nanoparticles that are environment friendly. Additionally, we explore the drastic need for sustainable materials in nanotechnology and introduce chitosan nanoparticles as an ecofriendly solution to this problem. In order to improve technology while also protecting the environment, we want this chapter to be a useful resource for scientists, engineers, and researchers.

4.2 CHITOSAN AS A SUSTAINABLE MATERIAL: SOURCE AND EXTRACTION OF CHITOSAN

Due to depleting natural resources and environmental damage caused by petroleum products, biomaterials are gaining more attention (Hameed et al., 2022). Chitosan is the second largest marine polysaccharide and is sourced and extracted from its natural origin, primarily found in the shells of crustaceans like shrimp, crab, and lobsters (Azuma et al., 2015). This cationic polysaccharide makes a greater contribution to biomedical research and exhibits impermeable pharmacological applications because it is easily accessible, biocompatible, biodegradable, non-toxic, and of low immunogenic in nature (Hameed et al., 2022). This has led to increased interest in various pharmaceutical and biological applications (Aljebory et al., 2017). According to some researchers, the global food sector produces between 60,000 and 80,000 tons of shellfish waste each year, and the transformation of byproducts into useful chitin or chitosan is very beneficial for environmental protection and resource recycling (Yu et al., 2022). It is also helpful in reducing the absorption of fat and cholesterol from the food we consume, and also it has been used in various applications like purifying contaminated drinking water, which has an adverse effect on human health (Hameed et al., 2022). Heavy metals, dyes, and other impurities can be removed from water during the purifying process using chitosan (Hameed et al., 2022). Crabs and shrimps are the commonly found sources cited in the literature which are used as the raw materials to prepare chitosan; however, other species like crayfish, oyster, and lobster have also been used (Yao et al., 2012; Kou et al., 2021). Other than crustaceans, seafood also contain 40%–50% of chitin of their total weight (Ehrlich et al., 2018; Iber et al., 2022). It has been found that there are mainly three steps in the chemical extraction of chitin, which are deproteinization, demineralization, and discoloration. Chitosan is derived from chitin by a process called deacetylation (Shaala et al., 2019; Iber et al., 2022). The enzymatic process of deacetylation is carried out using chitin deacetylase, which is derived from fungi, but is rare due to its lack of commercial availability (Kozma et al., 2022). Proteins are removed during the deproteinization process through the utilization of a strong base like sodium hydroxide. However, researchers are currently investigating alternative bases with less harmful cations, such as potassium hydroxide. On the other hand, demineralization involves the reduction of calcium carbonates and calcium phosphates by utilizing a strong acid (Iber et al., 2022). It is typically performed at room temperature over a period of 2–3 h, but some studies suggest it may be too long for some purposes (Kozma et al., 2022); this is known as acid-alkali method of chitin extraction (Kou et al., 2021). Decolorization can be performed using an alcohol like ethanol, and deacetylation involves the use of a strong alkali like NaOH (Iber et al., 2022). The extraction of green chitin has been enhanced through the utilization of ionic liquids, deep eutectic solvents, and other innovative methods (Morais et al., 2020; Kozma et al., 2022). Biological extraction of chitin includes different methods like enzymatic methods and fermentation methods (Kozma et al., 2022). Enzymatic methods involve proteinases to eliminate shell proteins and extract chitin, usually obtained

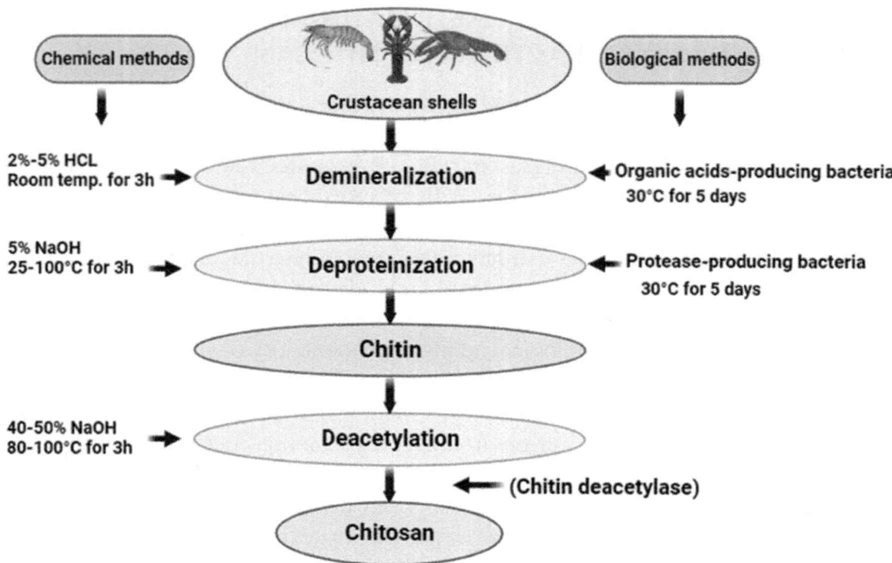

FIGURE 4.1 Sources and method of extraction of chitosan.

from bacteria or marine organisms (Hongkulsup et al., 2016). These techniques use specific enzymes and gentle reaction conditions, resulting in minimum deacetylation and harm to the chitin chain. Reaction timings exhibit similarities to chemical deproteination. However, the other rarely exceeds 90% as a result of active site depletion (Hamdi et al., 2017) (Figure 4.1).

Fermentation techniques exchange enzymes with bacterial cultures, reducing costs and incorporating demineralization through acid-producing bacteria (Tan et al., 2020). Lactic acid fermentation is commonly used for chitin extraction, while protease-producing cultures can improve purity (Sedaghat et al., 2017). Fermentation requires media preparation and a carbon source, such as glucose, for better results (Liu et al., 2020; Kozma et al., 2022).

4.3 DIFFERENT ROUTES TO PREPARE CHITOSAN NANOPARTICLES

Chitosan nanoparticles were first prepared in 1994 by Ohya et al. for their use in the intravenous delivery of 5-fluorouracil, an anticancer drug (Ohya et al, 1994). These nanoparticles were produced by the emulsification and cross-linking method (Grenha, 2012). Since then, these nanoparticles have been used in drug delivery systems and other applications (Calvo et al., 1997; Grenha, 2012). For more than 50 years, the pharmaceutical industries have been using processes including encapsulation and coating to introduce polymers into bioactivity (Guadarrama-Escobar et al., 2023). Chitosan nanoparticles can be synthesized utilizing simple methods that do not require organic solvents or significant shear forces due to their inherent hydrophilicity (Rajitha et al., 2016; Jha & Mayanovic, 2023). Chitosan is particularly useful in producing nanoparticles due to its distinct properties that can improve their attraction to negatively charged biological membranes and allow specific targeting within living organisms (Rajitha et al., 2016; Rodrigues et al., 2012). These particles possess the ability to regulate the release of active substances. Chitosan exhibits solubility in aqueous acidic solutions, hence enabling the production of particles without the need for potentially harmful chemical solvents (Aljebory et al., 2017). There are different methods developed by the researchers which include emulsion droplet coalescence, cross-linking and emulsification (Ohya et al., 1994), emulsion solvent diffusion, reverse micellization (Mitra et al., 2001), ionotropic gelation (Calvo et al., 1997), polyelectrolyte complexation (Sarmento et al., 2006), and modified ionic gelation with radical polymerization desolvation (Tian & Groves, 2010). In addition to this, plant sources, including leaves, flowers, seeds, stems, fruits, peels, and roots, have been successfully used for nanoparticle synthesis due to their eco-friendliness and non-toxic nature, offering stability, biocompatibility, and enhanced biological interactions (Beulah et al., 2019). These organic nanoparticles have better advantageous properties as compared to other types such as high permeability, non-toxicity, better biodegradability and ecofriendly (Youssef et al., 2017). The properties of these nanoparticles like charge, size, and rigidity determine their interaction with the immune system (Liu et al., 2017).

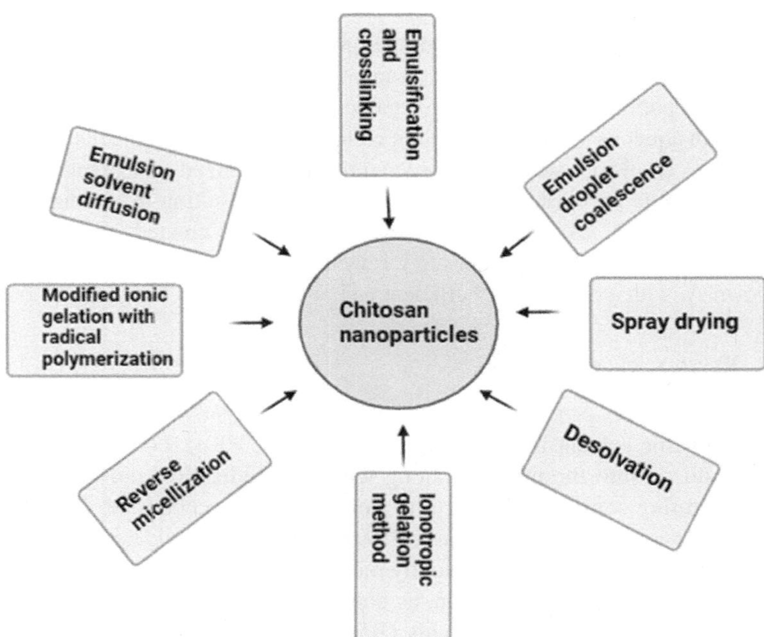

FIGURE 4.2 Different methods to synthesize chitosan nanoparticles.

4.3.1 EMULSIFICATION AND CROSS-LINKING

This method involves the initial preparation of chitosan nanoparticles, followed by the addition of a cross-linking agent that causes the droplets to freeze. The amino groups of chitosan react with glutaraldehyde groups to form covalent cross-links. The delivery of 5-fluorouracil was achieved using this technique, which was subsequently adapted for use with liquid paraffin and petroleum ether (Ohya et al., 1994). However, the method had drawbacks like uninteresting procedures and harsh cross-linking agents, which can cause over toxicity and compromise drug integrity (Agnihotri et al., 2004). As a result, its application was limited to a few works.

4.3.2 EMULSION DROPLET COALESCENCE

This technique of synthesis of chitosan nanoparticles requires loading with gadolinium used for the neutron-capture therapy of cancer: the process involves dissolving chitosan in gadolinium, adding it to liquid paraffin containing sorbitansesquiolate, stirring with a high-speed homogenizer, and mixing with NaOH to form droplet coalescence (Grenha, 2012). This solidifies the chitosan particles, which are then washed and centrifuged. This method was also used to encapsulate 5-fluorouracil. It has been shown that after decreasing the deacetylation degree of chitosan, the size of the particle increases (Grenha, 2012). Variations in chitosan concentration do not affect encapsulation efficiency or release profile.

4.3.3 EMULSION SOLVENT DIFFUSION

This preparation method depends on the phenomenon of incomplete miscibility between water and an organic solvent (Nagpal et al., 2010a). The procedure involves adding a hydrophobic drug-containing organic phase, such as methylene chloride and acetone, to an aqueous solution containing chitosan and a stabilizer. This results in the creation of an emulsion, which is subsequently homogenized and extracted. During this phase, acetone permeates into the aqueous phase, resulting in a reduction in the solubility of chitosan. As a consequence, nanoparticles are created when the polymer precipitates. This method is appropriate for enclosing hydrophobic medicines, such as cyclosporin-A, with a high level of efficiency in encapsulation (Grenha, 2012).

4.3.4 REVERSE MICELLIZATION

This method involves producing a microemulsion by dissolving a lipophilic surfactant in an organic solvent like n-hexane (Mitra et al., 2001). Following that, chitosan is introduced into the mixture, along with the medication and glutaraldehyde, while maintaining constant stirring. The nanoparticles are obtained via the process of solvent evaporation. The reverse micellization method provides the advantage of producing ultrafine nanoparticles, usually measuring around 100 nm or smaller, with a more concentrated size distribution, as compared to the larger particles produced by alternative emulsion-based methods (Banerjee et al., 2002; Tang et al., 2007). This method has been widely used in anticancer therapy and enzyme immobilization (Tang et al., 2007; Yang et al., 2010).

4.3.5 Ionotropic Gelation Method

The ionotropic gelation method of preparing chitosan nanoparticles was first described by Calvo et al in 1997, and it is currently undergoing extensive research and development (Calvo et al., 1997; Janes et al., 2001). The amine group of chitosan interacts electrostatically with negatively charged polyanion groups of tripolyphosphate (Janes et al., 2001). This is an easy and mild method of preparation in an aqueous environment. Chitosan is firstly dissolved in acetic acid solution in the presence or absence of any stabilizing agent. Polyanion and anionic polymers were added to the chitosan solution dropwise under vigorous mechanical stirring at room temperature. Particle size and charge on the surface of the particle can be changed by varying the ratio of chitosan and stabilizer (Nagpal et al., 2010a). It is the simple and mild method of preparation of chitosan nanoparticles and does not involve the use of any organic solvents and high temperatures, and that's why it is suitable for the use of thermosensitive moieties like proteins, peptides, vaccines, and hormones (Rajitha et al., 2016).

4.3.6 Modified Ionic Gelation with Radical Polymerization

This method employs the ionic gelation process, but with a modification in which the gelation of chitosan and the polymerization of acrylic acid take place at the same time (Grenha, 2012). The first step involves mixing an aqueous solution of either acrylic or methacrylic acid monomer with an aqueous chitosan solution that has opposite charges (Grenha, 2012). Polyethylene glycol or polyether may be introduced into the reaction medium in certain instances, either independently into the monomer solution or following its combination with chitosan. The combination of chitosan and methacrylic/acrylic acid results in the formation of nanoparticles, which are subsequently subjected to radical polymerization using potassium persulfate. The procedure requires a duration of 6h, followed by an overnight settling period, during which any unreacted monomers are eliminated. The nanoparticles have been used for the delivery of insulin, bovine serum albumin, and silk peptide (Hu et al., 2002). The study reveals that chitosan molecular weight increases with polyacrylic acid molecular weight in nanoparticles, indicating a template polymerization. The characteristics of nanoparticles are substantially influenced by parameters such as the ratio of chitosan to acrylic monomers and the concentration of the polymer (Hu et al., 2002).

4.3.7 Desolvation

This process, also referred to as simple coacervation or phase separation, involves the clustering of large molecules due to the partial removal of solvent from fully dissolved molecules (Psimadas et al., 2012; Grenha, 2012). Desolvating agents were first used for micron-sized carrier preparation in 1996, but are now frequently used for producing chitosan nanoparticles (Berthold et al., 1996). Substances like sodium sulfate and non-solvents like acetone are proposed as precipitating agents, with acetone being more commonly used (Mao et al., 2001). This method for preparing chitosan nanoparticles involves adding a solvent agent like sodium sulfate, leading to the gradual elimination of solvation water surrounding the chitosan solution, resulting in polymer solubilization and precipitation (Alonso & Sánchez, 2010). The favorable interactions between water and salt result in the partial removal of solvent molecules from chitosan, which in turn promotes stronger interaction between chitosan molecules, resulting in the formation of nanocarriers. Polysorbate80 is frequently employed for the purpose of stabilizing solutions of nanoparticles, which are then hardened through cross-linking with glutaraldehyde. The characteristics of nanoparticles are significantly influenced by various factors, including the molecular weight of chitosan, its concentration, the desolvating agent used, and the stirring rate. Therefore, it is necessary to optimize these factors (Borges et al., 2007).

4.3.8 Spray Drying

Spray drying is a technique used to produce chitosan nanoparticles. This process involves dissolving chitosan in water using glacial acetic acid and then splitting it down into small droplets by atomization (Mikušová & Mikuš, 2021). The liquid phase is vaporized through the amalgamation of droplets with a desiccating gas, leading to the creation of chitosan nanoparticles. The procedure is uncomplicated and single phase and is applicable to medications that are resistant to heat, sensitive to heat, soluble in water, or insoluble in water. Spray drying is a protein-compatible method for producing chitosan nanoparticles with a high protein content. This approach is a straightforward and effective way to create particles, and it serves as a system for delivering drugs to the lungs (Rodrigues et al., 2012; Pourshahab et al., 2011).

4.4 ADVANTAGES

4.4.1 Enhanced Surface Area

Because of their high surface area to volume ratio, ultra-fine chitosan nanoparticles interact with drugs more effectively, leading to improved drug loading capacity.

4.4.2 Increased Bioavailability

The smaller size of nanoparticles promotes their absorption by cells, resulting in improved bioavailability for drugs.

4.4.3 Targeted Drug Delivery

These nanoparticles can be surface modified to target specific cells or tissues for targeted drug delivery. It enhances the efficacy of drugs and minimizes the side effects.

4.4.4 Stability

Ultrafine nanoparticles have improved stability to encapsulate drugs that extend their shelf life and protect them from degradation.

4.4.5 Biocompatibility

Chitosan is a naturally occurring and biodegradable polymer and can be used in biomedical applications. The ultrafine nanoparticles prepared from chitosan have very less toxicity and are easily accepted by the body when incorporated with drugs.

4.5 DRAWBACKS

4.5.1 Agglomeration

Chitosan nanoparticles may aggregate during the formation process, which could result in uneven particle size distribution and reduced efficacy in drug delivery.

4.5.2 Complex Synthesis Process

The method of producing ultra-fine chitosan nanoparticles can be more complex and expensive as it often requires specified equipment.

4.5.3 Potential Toxicity

Although chitosan is usually regarded as safe, there may be concerns over the possible toxicity of nanoparticles, particularly when exposed to higher concentrations or for a prolonged amount of time.

4.5.4 Limited Suitability for Drugs

There is a possibility that certain drugs do not interact well with chitosan nanoparticle encapsulation, as their stability may be reduced, and this could limit the possible applications.

4.6 FACTORS AFFECTING SIZE, SHAPE, AND STABILITY

4.6.1 Chitosan Concentration

The chitosan concentration plays a key role in the development of nanoparticles. As there are more polymer chains available for particle production, higher concentrations of chitosan may lead to larger nanoparticle sizes. On the other hand, lower concentrations usually result in smaller nanoparticles (Grenha, 2012).

4.6.2 Molecular Weight and Degree of Deacetylationof Polymer

The degree of deacetylation and molecular weight significantly affect the formation of nanoparticles (Mohammed et al., 2017). Due to higher chain entanglement, chitosan with a higher molecular weight might produce larger nanoparticles (Paolicelli et al., 2009), whereas chitosan with a lower molecular weight may produce smaller nanoparticles. This can influence the particle size and stability (Garavand et al., 2022).

4.6.3 pH AND IONIC STRENGTH

The size and stability of nanoparticles are highly affected by the pH and ionic strength (Shukla et al., 2013). Chitosan is more soluble at lower pH values, and at this pH, smaller nanoparticles are produced (Katas & Alpar, 2006). Higher pH values can result in larger particles. Ionic strength can influence particle aggregation and stability (Garavand et al., 2022).

4.6.4 STIRRING SPEED AND TIME

The size and shape of chitosan nanoparticles can be influenced by the speed and duration or stirring during nanoparticle production (Garavand et al., 2022; Grenha, 2012).

4.6.5 TEMPERATURE

Size and stability of nanoparticles can be influenced by temperature during synthesis. In some circumstances, higher temperatures might cause faster agglomeration and larger particles (Garavand et al., 2022; Bugnicourt & Ladavière, 2016).

4.6.6 CROSS-LINKING AGENTS

Chitosan nanoparticles are frequently stabilized using cross-linking agents like tripolyphosphate or glutaraldehyde. The concentration and type of cross-linking agent can impact nanoparticle stability and shape (Garavand et al., 2022; Bugnicourt & Ladavière, 2016).

4.6.7 MIXING METHOD

The method used for mixing chitosan and other components (e.g., drug and cross-linking agents) can affect nanoparticle size and shape. Techniques like emulsion, coacervation, or ionic gelation are commonly used. The size distribution and particle shape can be affected by the method of preparation (Garavand et al., 2022) (Salar et al., 2017).

4.7 CHARACTERIZATION OF CHITOSAN NANOPARTICLES

Chitosan nanoparticle characterization is a crucial step in determining their structural, chemical, and physical characteristics. These characterization techniques help scientists and researchers to understand the nature of nanoparticles and to confirm their suitability for specific applications and assure their quality. Following are some characterization techniques of chitosan nanoparticles:

4.7.1 DYNAMIC LIGHT SCATTERING (DLS)

Dynamic light scattering (DLS) is a frequently used method for determining the hydrodynamic size of nanoparticles in a solution. It provides information about the mean size and size distribution of chitosan nanoparticles (Yang et al., 2010).

The participation of a low concentration of nanoparticles in DLS prevents the occurrence of a multiple scattering effect. The size values obtained from dynamic light scattering (DLS) tests are frequently influenced by several parameters, such as particle morphology and concentration, colloidal stability, and potential surface coating of nanoparticles (Garavand et al., 2022).

4.7.2 TRANSMISSION ELECTRON MICROSCOPY (TEM) AND SCANNING ELECTRON MICROSCOPY (SEM)

The morphology and surface of nanoparticles are evaluated through scanning electron microscopy (SEM), which scans the particles through an intensified electron beam. The device produces various signals that are detected by an extremely sensitive detector. This detector offers detailed information on the surface of the sample, including a high-resolution image and data on the smoothness and roughness of the structures (Aljebory et al., 2017). Another technique used for the analysis of chitosan-based particles is transmission electron microscopy (TEM) (Devi, 2013). This process involves utilizing a high-energy electron beam that is accelerated to traverse through a nanomaterial sample. In order to prepare the samples for transmission electron microscopy (TEM), chitosan-based nanomaterials are placed onto a copper grid that is coated with a layer of carbon film followed by drying of the samples (Ma et al., 2002).

4.7.3 ZETA POTENTIAL

The zeta potential magnitude measures the intensity of the electrostatic repulsions between charged particles. A high zeta potential, irrespective of its polarity, signifies that the particles in a system have a low tendency to aggregate, indicating the stability of the colloidal system. Zeta potential provides information about the surface charge on the chitosan nanoparticles. This is essential for understanding their stability and potential interactions with other particles or surfaces (Garavand et al., 2022).

4.7.4 FOURIER TRANSFORM INFRARED SPECTROSCOPY (FTIR)

The interaction between chitosan chains and different cross-linking groups at a molecular level is evidenced by FTIR spectroscopy (Rampino et al., 2013). It is used to study the chemical functional group present in chitosan nanoparticles and confirmation of their composition. The FTIR spectra of chitosan nanoparticles often exhibit absorption bands that are slightly shifted around 3,425, 2,924, 1,303, and 890 cm^{-1}. Additionally, there is a chemical bond formed between the phosphate ion from STPP and the amino group from chitosan. The presence of two additional absorption bands at 1,651 and 1,554 cm^{-1} can be attributed to $-CONH_2$ and N$-$H, respectively. These bands suggest growth in both inter- and intra-molecular forces within the nanoparticles (Garavand et al., 2022).

4.7.5 Nuclear Magnetic Resonance Spectroscopy

Nuclear magnetic resonance (NMR) spectroscopy enables the investigation of the structure of carbon nanomaterials (CNPs) and the identification of cross-linking and functionalization processes. The study of nuclear magnetic resonance (NMR) events in nuclei with non-zero spin is achieved by exposing them to a strong magnetic field. This approach is frequently used to examine the interactions or coordination between the ligand and the diamagnetic or antiferromagnetic CNPs' surface (Garavand et al., 2022).

4.7.6 Atomic Force Microscopy

Atomic force microscopy (AFM) is commonly used for the study of the dimensions and three-dimensional surface features of carbon nanoparticles. This is because AFM has a high spatial resolution, allowing for detailed analysis. Additionally, AFM can be used in various environments, including solutions and specific atmospheres. It is also capable of analyzing a wide range of materials, including biological samples, and can achieve atomic resolutions with minimal sample requirements (Kola-Mustapha, 2018; Garavand et al., 2022).

4.7.7 X-Ray Photo Electron Spectroscopy

X-ray photoelectron spectroscopy (XPS) is a method used to accurately determine the chemical composition of a sample's surface. It relies on the photoelectric effect and is performed under extremely low-pressure conditions in a vacuum. The provided information encompasses the elemental composition, electronic structure, oxidation states of the elements, ligand exchange interactions, surface functionalization of CNPs, and core/shell architectures. Furthermore, numerous authors have documented the XPS analysis of drug-loaded CNPs, employing XPS in a depth profile setup to ascertain the drug concentration inside the outer layers of the surface (Korin et al., 2017; Garavand et al., 2022).

4.7.8 Differential Scanning Calorimetry

Differential scanning calorimetry (DSC) is used for analyzing the thermal characteristics of chitosan nanoparticles, such as the glass transition temperature and melting temperatures (Ebert et al., 2017).

4.8 APPLICATION OF CHITOSAN NANOPARTICLES IN GREEN NANOTECHNOLOGY

Chitosan nanoparticles have found various applications in green nanotechnology, which is a field focused on the development of environmentally friendly and sustainable nanomaterials processes (Sathiyabama et al., 2024). Here are several applications of chitosan nanoparticles in green nanotechnology:

4.8.1 DRUG DELIVERY SYSTEMS

Chitosan nanoparticles can be used as transporters for delivering drugs and other bio-active compounds (Wu et al., 2024; Guadarrama-Escobar et al., 2023). Their biocompatibility and ability to encapsulate a wide range of substances make them suitable for targeted and controlled drug release, reducing the use of synthetic and potentially harmful drug delivery systems (Aljebory et al., 2017; Nagpal et al., 2010b). Chitosan nanoparticles are well suited for controlled drug delivery systems (Sachdeva et al., 2023). Chitosan forms colloidal particles and controls bioactive compounds through chemical crosslinking, ionic crosslinking, and ionic complexation. Chemical modification can modify the release characteristics of drugs and serve as a coating agent for liposome formulations (Prabaharan& Mano, 2005; Aljebory et al., 2017).

4.8.2 AGRICULTURE AND CROP PROTECTION

Chitosan nanoparticles have been used in agriculture for crop protection and enhancement (Alenazi et al., 2024). They can encapsulate and release agrochemicals like pesticides, herbicides, and fertilizers in a controlled and sustainable manner (Kashyap et al., 2015).

4.8.3 WATER PURIFICATION

Chitosan nanoparticles have been utilized to remove heavy metals, dyes, and organic pollutants from water. They can efficiently adsorb and remove contaminants, contributing to green and sustainable water purification processes (Abd-Elhakeem et al., 2016).

4.8.4 FOOD PACKAGING

To improve the barrier properties, extend life, and reduce the waste of food, these chitosan nanoparticles can be incorporated into biodegradable food packaging materials (Garavand et al., 2022).

4.8.5 BIO SENSORS

The field of biosensor research and development has attracted attention because of its advantageous features as analytical instruments, such as compact size, mobility, affordability, and ease of use in comparison to traditional laboratory-based techniques (Shukla et al., 2013). Chitosan nanoparticles can be functionalized to produce green sensors that may detect a variety of environmental factors, such as the presence of pathogens, pollutants, or poisons.

4.8.6 BRAIN TARGETING

Chitosan nanoparticles are being used for the purpose of delivering drugs to the brain, specifically for the treatment of disorders in the central nervous system. After coating with polysorbate 80, they have improved brain targeting efficiency via nasal route (Fazil et al., 2012).

4.8.7 Biomedical Applications

Due to their biocompatibility, chitosan nanoparticles are used in tissue engineering (Nagpal et al., 2010b), wound healing, and regenerative medicine (Beulah et al., 2019; Guo et al., 2024). They can also be used for the development of environmentally friendly and durable healthcare devices (Sharaf et al., 2024).

4.9 CHALLENGES AND FUTURE DIRECTIONS

The application of chitosan nanoparticles is restricted by the lack of standardized manufacturing methods and quality control protocols. Biocompatibility and toxicity are vulnerable to variation depending on factors such as size, surface changes, and application environment. Therefore, more extensive research is required to address these complexities. The ecological consequences of chitin extraction, obtained from discarded shellfish remains, must be taken into account. Obtaining regulatory approvals and achieving standardization across several sectors pose a significant challenge. Possible future areas of focus include breakthroughs in drug delivery, tissue engineering, and regenerative medicine within the biomedical field. Additionally, there is potential for the development of customized nanoparticles, as well as the exploration of their multifunctional properties. Other possible fields of interest include environmental remediation, green synthesis, and the establishment of sustainable sourcing practices. Developing through regulatory frameworks is crucial for ensuring the safety, effectiveness, and environmental impact evaluations of products. The promotion and acceptance of environmentally friendly nano technological solutions are essential for encouraging innovation. Collaborative research involving academia, industry, and regulatory authorities is crucial for assuring sustainable development.

4.10 CONCLUSION

In summary, the identification of chitosan as a sustainable and environmentally friendly material has ushered in significant breakthroughs in the realm of nanotechnology applications. Through a comprehensive examination of its attributes and applications, it becomes evident that chitosan presents a promising avenue for addressing both environmental concerns and technological imperatives. Demonstrating biodegradability and biocompatibility, chitosan emerges as an excellent candidate for a range of applications, including drug delivery systems and environmental fields, aligning seamlessly with the global movement toward more sustainable practices. Its adaptability and potential for tailored modifications enhance its functionality, making it a versatile solution across diverse fields. This adaptability, coupled with its eco-friendly characteristics, positions chitosan as a pivotal player in the quest for sustainable and conscientious nano-solutions. Collaboration between the scientific, engineering, and environmental communities becomes paramount to fully unlock the potential of chitosan materials in addressing pressing global challenges. To encapsulate, the discovery of chitosan nanoparticles signifies a milestone in bridging technology and sustainability. By leveraging the distinctive features of chitosan nano-materials, we not only propel the advancement of nanotechnologies but also chart a course toward a more sustainable and ecologically conscious future. This chapter extends an invitation to researchers and industry leaders, urging them to harness the potential of chitosan nano-materials in shaping a more responsible and environmentally friendly era for nanotechnology.

REFERENCES

Abd-Elhakeem, M. A., Ramadan, M. M, & Basaad, F. S. (2016). Removing of heavymetals from water by chitosan nanoparticles. *Journal of Advances in Chemistry, 11*(7), 3765–3771.

Agarwal, M., Agarwal, M. K., Shrivastav, N., Pandey, S., Das, R., & Gaur, P. (2018). Preparation of chitosan nanoparticles and their in-vitro characterization. *International Journal of Life-Sciences Scientific Research, 4*(2), 1713–1720. https://doi.org/10.21276/ijlssr.2018.4.2.17.

Agnihotri, S. A., Mallikarjuna, N. N., & Aminabhavi, T. M. (2004). Recent advances on chitosan-based micro- and nanoparticles in drug delivery. *Journal of Controlled Release, 100*(1), 5–28. https://doi.org/10.1016/j.jconrel.2004.08.010.

Alenazi, M. M., El-Ebidy, A. M., El-Shehaby, O. A., Seleiman, M. F., Aldhuwaib, K. J., Abdel-aziz, H. M. M. (2024). Chitosan and chitosan nanoparticles differentially alleviate salinity stress in Phaseolus vulgaris L. *Plants, 13*(3), 398.

Aljebory, A. M., Alsalman, T. M., Aljebory, A., & Alsalman, T. M. (2017). Chitosan nanoparticles: Review article. *Imperial Journal of Interdisciplinary Research (IJIR), 3*(7), 233–242. https://www.onlinejournal.in/IJIRV3I7/039.pdf.

Alonso, M. J., & Sánchez, A. (2010). The potential of chitosan in ocular drug delivery. *Journal of Pharmacy and Pharmacology, 55*(11), 1451–1463. https://doi.org/10.1211/0022357022476.

Azuma, K., Izumi, R., Osaki, T., Ifuku, S., Morimoto, M., Saimoto, H., Minami, S., & Okamoto, Y. (2015). Chitin, chitosan, and its derivatives for wound healing: Old and new materials. *Journal of Functional Biomaterials, 6*(1). https://doi.org/10.3390/jfb6010104.

Banerjee, T., Mitra, S., Kumar Singh, A., Kumar Sharma, R., & Maitra, A. (2002). Preparation, characterization and biodistribution of ultrafine chitosan nanoparticles. *International Journal of Pharmaceutics, 243*(1–2), 93–105. https://doi.org/10.1016/S0378-5173(02)00267-3.

Berthold, A., Cremer, K., & Kreuter, J. (1996). Preparation and characterization of chitosan microspheres as drug carrier for prednisolone sodium phosphate as model for anti-inflammatory drugs. *Journal of Controlled Release, 39*(1), 17–25. https://doi.org/10.1016/0168-3659(95)00129-8.

Beulah, P., Jinu, U., Ghorbanpour, M., & Venkatachalam, P. (2019). Green engineered chitosan nanoparticles and its biomedical applications-an overview. In *Advances in Phytonanotechnology: From Synthesis to Application*. Elsevier Inc. https://doi.org/10.1016/B978-0-12-815322-2.00015-8.

Borges, O., Tavares, J., de Sousa, A., Borchard, G., Junginger, H. E., & Cordeiro-da-Silva, A. (2007). Evaluation of the immune response following a short oral vaccination schedule with hepatitis B antigen encapsulated into alginate-coated chitosan nanoparticles. *European Journal of Pharmaceutical Sciences, 32*(4–5), 278–290. https://doi.org/10.1016/j.ejps.2007.08.005.

Bugnicourt, L., & Ladavière, C. (2016). Interests of chitosan nanoparticles ionically cross-linked with tripolyphosphate for biomedical applications. *Progress in Polymer Science, 60*, 1–17. https://doi.org/10.1016/j.progpolymsci.2016.06.002.

Calvo, P., Vila-Jato, J. L., & Alonso, M. J. (1997). Evaluation of cationic polymer-coated nanocapsules as ocular drug carriers. *International Journal of Pharmaceutics, 153*(1), 41–50. https://doi.org/10.1016/S0378-5173(97)00083-5.

de Souza, M. L., Oliveira, D. D., Pereira, N. de P., & Soares, D. M. (2018). Nanoemulsions and dermatological diseases: Contributions and therapeutic advances. *International Journal of Dermatology, 57*(8), 894–900. https://doi.org/10.1111/ijd.14028.

Devi, J. (2013). Antihyperlipidemic effect of ambrex, a polyherbal formulation against experimentally induced hypercholesterolemia in rats. *African Journal of Pharmacy and Pharmacology, 7*(25), 1737–1743. https://doi.org/10.5897/ajpp12.1316.

Ebert, D. D., Nobis, S., Lehr, D., Baumeister, H., Riper, H., Auerbach, R. P., Snoek, F., Cuijpers, P., & Berking, M. (2017). The 6-month effectiveness of Internet-based guided self-help for depression in adults with Type 1 and 2 diabetes mellitus. *Diabetic Medicine, 34*(1), 99–107. https://doi.org/10.1111/dme.13173.

Ehrlich, H., Shaala, L. A., Youssef, D. T. A., Zoltowska-Aksamitowska, S., Tsurkan, M., Galli, R., Meissner, H., Wysokowski, M., Petrenko, I., Tabachnick, K. R., Ivanenko, V. N., Bechmann, N., Joseph, Y., & Jesionowski, T. (2018). Discovery of chitin in skeletons of non-verongiid Red Sea demosponges. *PLoS One, 13*(5), 1–18. https://doi.org/10.1371/journal.pone.0195803.

Elmoghayer, M. E., Saleh, N. M., & Abu Hashim, I. I. (2024). Enhanced oral delivery of hesperidin-loaded sulfobutylether-β-cyclodextrin/chitosan nanoparticles for augmenting its hypoglycemic activity: In vitro-in vivo assessment study. *Drug Delivery and Translational Research, 14*(4), 895–917. https://doi.org/10.1007/s13346-023-01440-6.

Fazil, M., Md, S., Haque, S., Kumar, M., Baboota, S., Sahni, J. K., & Ali, J. (2012). Development and evaluation of rivastigmine loaded chitosan nanoparticles for brain targeting. *European Journal of Pharmaceutical Sciences, 47*(1), 6–15. https://doi.org/10.1016/j.ejps.2012.04.013.

Garavand, F., Cacciotti, I., Vahedikia, N., Rehman, A., Tarhan, Ö., Akbari-Alavijeh, S., Shaddel, R., Rashidinejad, A., Nejatian, M., Jafarzadeh, S., Azizi-Lalabadi, M., Khoshnoudi-Nia, S., & Jafari, S. M. (2022). A comprehensive review on the nanocomposites loaded with chitosan nanoparticles for food packaging. *Critical Reviews in Food Science and Nutrition, 62*(5), 1383–1416. https://doi.org/10.1080/10408398.2020.1843133.

Gomes, L. P., Paschoalin, V. M. F., & Del Aguila, E. M. (2017). Chitosan nanoparticles: Production, physicochemical characteristics and nutraceutical applications. *Revista Virtual de Quimica, 9*(1), 387–409.

Grenha, A. (2012). Chitosan nanoparticles: A survey of preparation methods. *Journal of Drug Targeting, 20*(4), 291–300. https://doi.org/10.3109/1061186X.2011.654121.

Guadarrama-Escobar, O. R., Serrano-Castañeda, P., Anguiano-Almazán, E., Vázquez-Durán, A., Peña-Juárez, M. C., Vera-Graziano, R., Morales-Florido, M. I., Rodriguez-Perez, B., Rodriguez-Cruz, I. M., Miranda-Calderón, J. E., & Escobar-Chávez, J. J. (2023). Chitosan nanoparticles as oral drug carriers. *International Journal of Molecular Sciences, 24*(5), 1–17. https://doi.org/10.3390/ijms24054289.

Guo, Y., Qiao, D., Zhao, S., Liu, P., Xie, F., & Zhang, B. (2024). Biofunctional chitosan-biopolymer composites for biomedical applications. *Materials Science and Engineering R: Reports, 159*(February), 100775. https://doi.org/10.1016/j.mser.2024.100775.

Hamdi, M., Hammami, A., Hajji, S., Jridi, M., Nasri, M., & Nasri, R. (2017). Chitin extraction from blue crab (Portunus segnis) and shrimp (Penaeus kerathurus) shells using digestive alkaline proteases from P. segnis viscera. *International Journal of Biological Macromolecules, 101*, 455–463. https://doi.org/10.1016/j.ijbiomac.2017.02.103.

Hameed, A. Z., Raj, S. A., Kandasamy, J., Baghdadi, M. A., & Shahzad, M. A. (2022). Chitosan: A sustainable material for multifarious applications. *Polymers, 14*(12), 1–34. https://doi.org/10.3390/polym14122335.

Hoang, N. H., Thanh, T. Le, Sangpueak, R., Treekoon, J., Saengchan, C., Thepbandit, W., Papathoti, N. K., Kamkaew, A., & Buensanteai, N. (2022). Chitosan nanoparticles-based ionic gelation method: A promising candidate for plant disease management. *Polymers*, *14*(4), 1–28. https://doi.org/10.3390/polym14040662.

Hongkulsup, C., Khutoryanskiy, V. V., & Niranjan, K. (2016). Enzyme assisted extraction of chitin from shrimp shells (Litopenaeus vannamei). *Journal of Chemical Technology and Biotechnology*, *91*(5), 1250–1256. https://doi.org/10.1002/jctb.4714.

Hu, Y., Jiang, X., Ding, Y., Ge, H., Yuan, Y., & Yang, C. (2002). Synthesis and characterization of Chitosan-poly (acrylic acid) nanoparticles. *Biomaterials*, *23*(15), 3193–3201. https://doi.org/10.1016/S0142-9612(02)00071-6.

Iber, B. T., Kasan, N. A., Torsabo, D., & Omuwa, J. W. (2022). A review of various sources of chitin and chitosan in nature. *Journal of Renewable Materials*, *10*(4), 1097–1123. https://doi.org/10.32604/JRM.2022.018142.

Janes, K. A., Fresneau, M. P., Marazuela, A., Fabra, A., & Alonso, M. J. (2001). Chitosan nanoparticles as delivery systems for doxorubicin. *Journal of Controlled Release*, *73*(2–3), 255–267. https://doi.org/10.1016/S0168-3659(01)00294-2.

Jha, R., & Mayanovic, R. A. (2023). A review of the preparation, characterization, and applications of chitosan nanoparticles in nanomedicine. *Nanomaterials*, *13*(8). https://doi.org/10.3390/nano13081302.

Jiang, T., James, R., Kumbar, S. G., & Laurencin, C. T. (2014). Chitosan as a biomaterial. *Natural and Synthetic Biomedical Polymers*, *18*(1), 91–113. https://doi.org/10.1016/b978-0-12-396983-5.00005-3.

Kashyap, P. L., Xiang, X., & Heiden, P. (2015). Chitosan nanoparticle based delivery systems for sustainable agriculture. *International Journal of Biological Macromolecules*, *77*, 36–51. https://doi.org/10.1016/j.ijbiomac.2015.02.039.

Katas, H., & Alpar, H. O. (2006). Development and characterisation of chitosan nanoparticles for siRNA delivery. *Journal of Controlled Release*, *115*(2), 216–225. https://doi.org/10.1016/j.jconrel.2006.07.021.

Kola-Mustapha, A. T. (2018). Microscopy of nanomaterial for drug delivery. In *Characterization and Biology of Nanomaterials for Drug Delivery: Nanoscience and Nanotechnology in Drug Delivery*. Elsevier Inc. https://doi.org/10.1016/B978-0-12-814031-4.00010-6.

Korin, E., Froumin, N., & Cohen, S. (2017). Surface analysis of nanocomplexes by X-ray Photoelectron Spectroscopy (XPS). *ACS Biomaterials Science and Engineering*, *3*(6), 882–889. https://doi.org/10.1021/acsbiomaterials.7b00040.

Kou, S. (Gabriel), Peters, L. M., & Mucalo, M. R. (2021). Chitosan: A review of sources and preparation methods. *International Journal of Biological Macromolecules*, *169*, 85–94. https://doi.org/10.1016/j.ijbiomac.2020.12.005.

Kozma, M., Acharya, B., & Bissessur, R. (2022). Chitin, chitosan, and nanochitin: Extraction, synthesis, and applications. *Polymers*, *14*(19), 3989.

Liu, Y., Hardie, J., Zhang, X., & Rotello, V. M. (2017). Effects of engineered nanoparticles on the innate immune system. *Seminars in Immunology*, *34*(July), 25–32. https://doi.org/10.1016/j.smim.2017.09.011.

Liu, Y., Xing, R., Yang, H., Liu, S., Qin, Y., Li, K., Yu, H., & Li, P. (2020). Chitin extraction from shrimp (Litopenaeus vannamei) shells by successive two-step fermentation with Lactobacillus rhamnoides and Bacillus amyloliquefaciens. *International Journal of Biological Macromolecules*, *148*, 424–433. https://doi.org/10.1016/j.ijbiomac.2020.01.124.

Ma, Z., Yeoh, H. H. I. N., & Lim, L. (2002). Formulation pH modulates the interaction of insulin with chitosan nanoparticles. *Journal of Pharmaceutical Sciences*, *91*(6), 1396–1404.

Magnuson, B. A., Jonaitis, T. S., & Card, J. W. (2011). A brief review of the occurrence, use, and safety of food-related nanomaterials. *Journal of Food Science*, *76*(6), 126–133. https://doi.org/10.1111/j.1750-3841.2011.02170.x.

Mao, H. Q., Roy, K., Troung-Le, V. L., Janes, K. A., Lin, K. Y., Wang, Y., August, J. T., & Leong, K. W. (2001). Chitosan-DNA nanoparticles as gene carriers: Synthesis, characterization and transfection efficiency. *Journal of Controlled Release*, *70*(3), 399–421. https://doi.org/10.1016/S0168-3659(00)00361-8.

Mikušová, V., & Mikuš, P. (2021). Advances in chitosan-based nanoparticles for drug delivery. *International Journal of Molecular Sciences*, *22*(17), 1–93. https://doi.org/10.3390/ijms22179652.

Mitra, S., Gaur, U., Ghosh, P. C., & Maitra, A. N. (2001). Tumour targeted delivery of encapsulated dextran-doxorubicin conjugate using chitosan nanoparticles as carrier. *Journal of Controlled Release*, *74*(1–3), 317–323. https://doi.org/10.1016/S0168-3659(01)00342-X.

Mohammed, M. A., Syeda, J. T. M., Wasan, K. M., & Wasan, E. K. (2017). An overview of chitosan nanoparticles and its application in non-parenteral drug delivery. *Pharmaceutics*, *9*(4). https://doi.org/10.3390/pharmaceutics9040053.

Morais, E. S., Da Costa Lopes, A. M., Freire, M. G., Freire, C. S. R., Coutinho, J. A. P., & Silvestre, A. J. D. (2020). Use of ionic liquids and deep eutectic solvents in polysaccharides dissolution and extraction processes towards sustainable biomass valorization. *Molecules*, *25*(16). https://doi.org/10.3390/molecules25163652.

Nagpal, K., Singh, S. K., & Mishra, D. N. (2010a). Chitosan nanoparticles: A promising system in novel drug delivery. *Chemical and Pharmaceutical Bulletin*, *58*(11), 1423–1430. https://doi.org/10.1248/cpb.58.1423.

Nagpal, K., Singh, S. K., & Mishra, D. N. (2010b). Nagpal. *Chem Pharm Bull*, *58*(11), 1423–1430.

Nasir, A. (2010). Nanotechnology and dermatology: Part I-potential of nanotechnology. *Clinics in Dermatology*, *28*(4), 458–466. https://doi.org/10.1016/j.clindermatol.2009.06.005.

Ochekpe, N. A., Olorunfemi, P. O., & Ngwuluka, N. C. (2009). Nanotechnology and drug delivery part 2: Nanostructures for drug delivery. *Tropical Journal of Pharmaceutical Research*, *8*(3), 275–287. https://doi.org/10.4314/tjpr.v8i3.44547.

Ohya, Y., Shiratani, M., Kobayashi, H., & Ouchi, T. (1994). Release behavior of 5-fluorouracil from chitosan-gel nanospheres immobilizing 5-fluorouracil coated with polysaccharides and their cell specific cytotoxicity. *Journal of Macromolecular Science, Part A, 31*(5), 629–642. https://doi.org/10.1080/10601329409349743.

Paolicelli, P., De La Fuente, M., Sánchez, A., Seijo, B., & Alonso, M. J. (2009). Chitosan nanoparticles for drug delivery to the eye. *Expert Opinion on Drug Delivery*, *6*(3), 239–253. https://doi.org/10.1517/17425240902762818.

Pourshahab, P. S., Gilani, K., Moazeni, E., Eslahi, H., Fazeli, M. R., & Jamalifar, H. (2011). Preparation and characterization of spray dried inhalable powders containing chitosan nanoparticles for pulmonary delivery of isoniazid. *Journal of Microencapsulation*, *28*(7), 605–613. https://doi.org/10.3109/02652048.2011.599437.

Prabaharan, M., & Mano, J. F. (2005). Chitosan-based particles as controlled drug delivery systems. *Drug Delivery: Journal of Delivery and Targeting of Therapeutic Agents*, *12*(1), 41–57. https://doi.org/10.1080/10717540590889781.

Psimadas, D., Georgoulias, P., Valotassiou, V., & Loudos, G. (2012). Molecular nanomedicine towards cancer . *Journal of Pharmaceutical Sciences*, *101*(7), 2271–2280. https://doi.org/10.1002/jps.23146

Rajitha, P., Gopinath, D., Biswas, R., Sabitha, M., & Jayakumar, R. (2016). Chitosan nanoparticles in drug therapy of infectious and inflammatory diseases. *Expert Opinion on Drug Delivery*, *13*(8), 1177–1194. https://doi.org/10.1080/17425247.2016.1178232.

Rampino, A., Borgogna, M., Blasi, P., Bellich, B., & Cesàro, A. (2013). Chitosan nanoparticles: Preparation, size evolution and stability. *International Journal of Pharmaceutics*, *455*(1–2), 219–228. https://doi.org/10.1016/j.ijpharm.2013.07.034.

Rodrigues, S., Dionísio, M., López, C. R., & Grenha, A. (2012). Biocompatibility of chitosan carriers with application in drug delivery. *Journal of Functional Biomaterials*, *3*(3), 615–641. https://doi.org/10.3390/jfb3030615.

Sachdeva, B., Sachdeva, P., Negi, A., Ghosh, S., Han, S., Dewanjee, S., Jha, S. K., Bhaskar, R., Sinha, J. K., Paiva-Santos, A. C., Jha, N. K., & Kesari, K. K. (2023). Chitosan nanoparticles-based cancer drug delivery: Application and challenges. *Marine Drugs*, *21*(4), 1–23. https://doi.org/10.3390/md21040211.

Salar, S., Mehrnejad, F., Sajedi, R. H., & Arough, J. M. (2017). Chitosan nanoparticles-trypsin interactions: Bio-physicochemical and molecular dynamics simulation studies. *International Journal of Biological Macromolecules*, *103*, 902–909. https://doi.org/10.1016/j.ijbiomac.2017.05.140.

Sarmento, B., Ferreira, D., Veiga, F., & Ribeiro, A. (2006). Characterization of insulin-loaded alginate nanoparticles produced by ionotropic pre-gelation through DSC and FTIR studies. *Carbohydrate Polymers*, *66*(1), 1–7. https://doi.org/10.1016/j.carbpol.2006.02.008.

Sathiyabama, M., Boomija, R. V., Muthukumar, S., Gandhi, M., Salma, S., Prinsha, T. K., & Rengasamy, B. (2024). Green synthesis of chitosan nanoparticles using tea extract and its antimicrobial activity against economically important phytopathogens of rice. *Scientific Reports*, *14*(1), 1–10. https://doi.org/10.1038/s41598-024-58066-y.

Sedaghat, F., Yousefzadi, M., Toiserkani, H., & Najafipour, S. (2017). Bioconversion of shrimp waste Penaeus merguiensis using lactic acid fermentation: An alternative procedure for chemical extraction of chitin and chitosan. *International Journal of Biological Macromolecules*, *104*, 883–888. https://doi.org/10.1016/j.ijbiomac.2017.06.099.

Shaala, L. A., Asfour, H. Z., Youssef, D. T. A., ółtowska-Aksamitowska, S. Z., Wysokowski, M., Tsurkan, M., Galli, R., Meissner, H., Petrenko, I., Tabachnick, K., Ivanenko, V. N., Bechmann, N., Muzychka, L. V., Smolii, O. B., Martinović, R., Joseph, Y., Jesionowski, T., & Ehrlich, H. (2019). New source of 3D chitin scaffolds: The red sea demosponge pseudoceratina arabica (pseudoceratinidae, verongiida). *Marine Drugs*, *17*(2). https://doi.org/10.3390/md17020092.

Sharaf, M., Zahra, A. A., Alharbi, M., Mekky, A. E., Shehata, A. M., Alkhudhayri, A., Ali, A. M., Al Suhaimi, E. A., Zakai, S. A., Al Harthi, N., & Liu, C.-G. (2024). Bee chitosan nanoparticles loaded with apitoxin as a novel approach to eradication of common human bacterial, fungal pathogens and treating cancer. *Frontiers in Microbiology*, *15*(March). https://doi.org/10.3389/fmicb.2024.1345478.

Sharifi-Rad, J., Quispe, C., Butnariu, M., Rotariu, L. S., Sytar, O., Sestito, S., Rapposelli, S., Akram, M., Iqbal, M., Krishna, A., Kumar, N. V. A., Braga, S. S., Cardoso, S. M., Jafernik, K., Ekiert, H., Cruz-Martins, N., Szopa, A., Villagran, M., Mardones, L., ... Calina, D. (2021). Chitosan nanoparticles as a promising tool in nanomedicine with particular emphasis on oncological treatment. *Cancer Cell International*, *21*(1), 1–21. https://doi.org/10.1186/s12935-021-02025-4.

Shukla, S. K., Mishra, A. K., Arotiba, O. A., & Mamba, B. B. (2013). Chitosan-based nanomaterials: A state-of-the-art review. *International Journal of Biological Macromolecules*, *59*, 46–58. https://doi.org/10.1016/j.ijbiomac.2013.04.043.

Tan, Y. N., Lee, P. P., & Chen, W. N. (2020). Microbial extraction of chitin from seafood waste using sugars derived from fruit waste-stream. *AMB Express*, *10*(1). https://doi.org/10.1186/s13568-020-0954-7.

Tang, Y. F., Du, Y. M., Hu, X. W., Shi, X. W., & Kennedy, J. F. (2007). Rheological characterisation of a novel thermosensitive chitosan/poly (vinyl alcohol) blend hydrogel. *Carbohydrate Polymers*, *67*(4), 491–499. https://doi.org/10.1016/j.carbpol.2006.06.015.

Tian, X.-X., & Groves, M. J. (2010). Formulation and biological activity of antineoplastic proteoglycans derived from *Mycobacterium vaccae* in chitosan nanoparticles. *Journal of Pharmacy and Pharmacology*, *51*(2), 151–157. https://doi.org/10.1211/0022357991772268.

Wang, J. J., Zeng, Z. W., Xiao, R. Z., Xie, T., Zhou, G. L., Zhan, X. R., & Wang, S. L. (2011). Recent advances of chitosan nanoparticles as drug carriers. *International Journal of Nanomedicine*, 6, 765–774. https://doi.org/10.2147/ijn.s17296.

Wijayadi, L. J., & Rusli, T. R. (2019). Characterized and synthesis of chitosan nanoparticle as nanocarrier system technology. *IOP Conference Series: Materials Science and Engineering*, *508*(1). https://doi.org/10.1088/1757-899X/508/1/012143.

Wu, T., He, Y., Ding, L., Ding, F., & Tan, F. (2024). Preparation and characterization of magnetic ferrite-chitosan nanoparticles delivery for DOX. *Inorganica Chimica Acta*, *559*(September 2023), 121791. https://doi.org/10.1016/j.ica.2023.121791.

Yang, Z., Peng, H., Wang, W., & Liu, T. (2010). Crystallization behavior of poly (ε-caprolactone)/layered double hydroxide nanocomposites. *Journal of Applied Polymer Science*, *116*(5), 2658–2667. https://doi.org/10.1002/app.31787

Yao, K., Li, J., Yao, F., & Yin, Y. (2012). *Chitosan-Based Hydrogels*. Taylor & Francis Group, LLC, Boca Raton. https://doi.org/10.1201/b11048

Youssef, A. M., El-Nahrawy, A. M., & Abou Hammad, A. B. (2017). Sol-gel synthesis and characterizations of hybrid chitosan-PEG/calcium silicate nanocomposite modified with ZnO-NPs and (E102) for optical and antibacterial applications. *International Journal of Biological Macromolecules*, *97*, 561–567. https://doi.org/10.1016/j.ijbiomac.2017.01.059.

Yu, D., Yu, Z., Zhao, W., Regenstein, J. M., & Xia, W. (2022). Advances in the application of chitosan as a sustainable bioactive material in food preservation. *Critical Reviews in Food Science and Nutrition*, *62*(14), 3782–3797. https://doi.org/10.1080/10408398.2020.1869920.

5 Eco-materials
Opportunity, Applications, and Challenges

Yogesh Kumar Sonia and Sapna Meena

5.1 INTRODUCTION

This chapter may discuss eco-materials that are environmentally friendly and sustainable, such as recycled materials, biodegradable materials, and materials with a lower carbon footprint. Along with covering their qualities, production methods, and renewable energy, it might also go into detail on different renewable energy sources, such as solar, wind, hydro, and geothermal energy. [1–3]. It might discuss the technology, efficiency, and environmental benefits of these energy sources, Green building materials are those that are sustainable, energy-efficient, and environmentally friendly [4]. This chapter could cover materials like low VOC paints, recycled insulation, energy-efficient windows, and sustainable construction practices; a wide range of subjects related to sustainability, such as environmental conservation, waste reduction, water and energy efficiency; and sustainable practices in various industries. Nanotechnology involves manipulating materials at the nanoscale. A chapter on this subject might discuss the principles of nanotechnology, its applications in various fields (e.g., medicine, electronics, materials science), and potential environmental and safety consideration applications. However, the journey toward eco-material adoption is not without its challenges [5–7]. This chapter addresses the obstacles and limitations faced by eco-materials, such as cost considerations, performance trade-offs, and market acceptance. It discusses the need for innovation and collaborative efforts to overcome these challenges, while also providing insights into ongoing research and development efforts aimed at expanding the scope and capabilities of eco-materials [8]. Nanoscience is the study of structures and molecules on a scale of nanometers between 1 and 100 nm and the technique that uses it in practical applications such as devices. Nanotechnology has great promise at the atomic level to change many parts of medical treatment, such as diagnosis, regenerative medicine, development of vaccines, and medication delivery. Nanofibers have been wound dressing, surgical textiles and implants, tissue technology, and artificial organ components. Nanotechnology is used to develop smaller and more powerful electronic devices, lasers, and medical diagnoses [9,10]. The strategy plans of many countries, and the sustainable development in relation to parameters such as economic, social, and industrial. DNA technology has contributed to healthcare through the production of pharmaceutical proteins and gene therapy. The future could also enable objects to harvest energy from their environment. The aim of the healthcare providers is the delivery of safe, effective care for patients and clients. Food products are tested for infections and poisons, and their barrier qualities are strengthened. [11]. There are active and passive nano assemblies, general nano systems, and small molecular nano systems. Rural healthcare and hospitals provide new technology as they experience population loss. They promote health and wellness, including incentives or rewards related to health and wellness. The right amount maintains good health improves physical and mental health and protects against various heart disease sleep. Fruits and vegetables are high in nutrients and boost your immune system. Bone and muscle increase through regular exercise and physical activity. Sugar has positive health effects containing empty calories, weight gain, and elevated blood sugar levels. Tissue growth and regenerative medicine solubility [12]. A good lifestyle for brushing your teeth is to have a few drinks every night. In this context, we are discussing the various applications of eco-friendly nanomaterials, especially for health systems.

5.2 OPPORTUNITIES

Eco-materials, also known as sustainable or green materials, offer a range of opportunities in various industries due to their reduced environmental impact and potential for improving sustainability.

5.2.1 ENVIRONMENTAL BENEFITS

Eco-materials are designed to reduce the environmental footprint of production, consumption, and disposal. They can help reduce pollution, resource depletion, and greenhouse gas emissions, contributing to a cleaner and more sustainable planet. Nowadays, there is increased thought of the design concept and material choice concurrently at the initial stage of product development. In this context, Eco-design tools are essential to diminish the environmental effect due to the product's materials and related processes. Inspiration approaches have emerged as a driving force in innovation processes [13,14].

5.2.2 COST SAVINGS

While the initial cost of some eco-materials may be higher, they often lead to long-term cost savings. For example, energy-efficient materials can reduce operational expenses by lowering energy consumption and maintenance costs. In this context, agrochemical industries have many more subproducts, leading to an enormous amount of waste. However, these residues, of little or no value, can be measured as a renewable source of raw materials, where their use has a dual advantage: decreasing the pollution they produce and changing them into value-added materials, both solid and liquid [15,16].

5.2.3 MARKET DEMAND

Consumer and regulatory demand for eco-friendly products is increasing. Using eco-materials can make your products more appealing to environmentally conscious consumers and help you comply with regulations aimed at reducing environmental impact. The concept and new categorization of environmentally conscious materials, eco-materials, are proposed. Advanced steps ranging from high eco-efficiency of products to consumer-oriented and regional community adaptation are introduced. An additional new concept, the robust design of materials, is also proposed as part of a new range of activities in eco-materials [17,18].

5.2.4 Innovation and Product Development

Developing and using eco-materials can drive innovation in various industries. This includes the development of new materials and technologies that have a smaller environmental footprint and may offer novel properties or features.

5.2.5 Resource Efficiency

Eco-materials are often designed to be more resource-efficient, which can lead to reduced waste, less resource extraction, and more sustainable supply chains. This is particularly relevant in industries like construction and manufacturing [19].

5.2.6 Health and Safety

Many eco-materials are less harmful to human health and safety. This is especially significant in industries where exposure to hazardous materials is a concern. Concerned with material designs have been deemed as one of the effective approaches to address the above-mentioned challenges and promote the growth of advanced lithium-ion (Li) batteries for practical application. Naturally, biomass materials always exist in the form of biological macromolecules or carbohydrate compounds in plant or animal cells, which endow themselves with diverse microstructures, complex compositions, and abundant functional groups. These unique physicochemical properties offer an emerging opportunity to be compatible with high-energy advanced lithium-ion batteries [20,21].

5.2.7 Sustainable Architecture and Construction

Eco-materials are increasingly used in architecture and construction. Green building materials can improve energy efficiency, reduce construction waste, and provide healthier indoor environments. However, broadly speaking, any material that exhibits environmental attributes, such as low carbon emissions, minimal embodied energy, and recyclability, can be classified as an eco-material. A building material achieves this classification when it undergoes a comprehensive evaluation of its life cycle through a life cycle assessment (LCA) and formally demonstrates sustainability [22,23].

5.2.8 Renewable Energy

The renewable energy sector relies heavily on eco-materials. Solar panels, wind turbines, and energy storage solutions, for example, all require materials that are environmentally friendly and sustainable. The sustainable use of energy and natural resources is an essential component of resilient and modern societies. The construction industry plays a key role in this endeavor, since it accounts for over 30% of natural resource extraction and contributes to 25% of solid waste generation [24]. In addition, the construction sector is a major consumer, overwhelming around 40% of the world's energy supply and 12% of the world's water resources. Due to these negative impacts, the building industry and researchers in this field are increasingly challenged to find ways to reduce such impacts, and there is an increasing research focus on the exploration of sustainable, environmentally approachable building materials, mentioned here as eco-materials [25].

5.2.9 Agriculture and Food

Eco-materials are used in sustainable agriculture to reduce the environmental impact of farming practices. Biodegradable packaging materials and organic farming materials are examples of such applications. In the last 50 years, agricultural activities have developed more intensely and the use of fertilizers and pesticides has given rise to disparity between productivity and environmental factors, and soil requiring more organic matter. In these circumstances, agricultural wastes could be integrated by clean technologies or be returned as an organic substrate to the soil as valuable materials, with minimum risk to environmental factors and ecosystems [26]. Moreover, these materials could be used as animal food, combustion raw materials, or disposed of in landfills, but with high implications on environmental issues. Due to their microbial decomposition, agri-food wastes could be associated with some potential risks to the environment, and their treatment is compulsory [27].

5.2.10 Transportation

Lightweight and fuel-efficient eco-materials are crucial in the transportation industry. These materials can help reduce fuel consumption and emissions in vehicles and aircraft.

5.2.11 Waste Management

Eco-materials are also relevant in waste management, as they can be used to create eco-friendly products and packaging that are easier to recycle or compost. However, other than being derived from non-renewable fossil fuel feedstock, they presently possess linear cradle-to-grave life cycles, with only a small minority being recycled [28]. With the demand for plastics probable to remain unabated for the foreseeable future, a more sustainable circular plastics economy is needed to reduce material wastage and maximize resource productivity. In this context, we give a broad conceptual overview of a possible future circular plastics economy, with a strong emphasis on recycling technologies, adoption of eco-design principles to guide plastic life cycles in the future, and the use of renewable sources (e.g., biomass) as feedstock for plastic production instead of fossil fuels [29].

5.2.12 Circular Economy

Eco-materials support the transition toward a circular economy by promoting the reuse, recycling, and repurposing of materials and products.

5.2.13 CORPORATE RESPONSIBILITY

Many companies are adopting eco-materials as part of their corporate responsibility initiatives, demonstrating their commitment to environmental sustainability [30].

5.2.14 MORPHOLOGICAL EFFECT

In synthesized nanomaterials, controlling morphology is crucial for achieving desired characteristics and functionalities. Morphology refers to the shape and structure of nanomaterials, which can significantly influence their properties and applications.

Physical and Chemical Properties: Morphology influences various physical and chemical properties such as surface area, reactivity, optical, electrical, and magnetic properties. For example, nanoparticles with different shapes may exhibit distinct plasmonic or catalytic properties due to variations in their surface structures [31].

Functionality and Performance: Tailoring the morphology allows for fine-tuning the functionality and performance of nanomaterials for specific applications. For instance, nanomaterials with controlled morphologies can be designed to enhance sensitivity and selectivity in sensors or improve efficiency in catalysts [32].

Overall, the role of morphologies in synthesized nanomaterials is multifaceted, encompassing fundamental properties, functional performance, assembly behavior, mechanical characteristics, biological interactions, and synthesis scalability. By precisely engineering the morphology of nanomaterials, researchers can unlock their full potential for a wide range of applications across various fields.

Incorporating eco-materials into your products, processes, or projects can lead to numerous benefits, including reduced environmental impact, cost savings, and increased market competitiveness. As sustainability continues to be a priority for consumers and regulators, the opportunities associated with eco-materials are likely to grow.

5.3 APPLICATIONS

Eco-materials, also known as environmentally friendly or sustainable materials, have gained increasing prominence in various fields as the world seeks more sustainable solutions to address pressing environmental challenges. These materials are designed and utilized with a primary focus on reducing environmental impact, conserving natural resources, and enhancing overall sustainability. In this introduction, we will explore the applications of eco-materials in several key domains, including renewable energy, energy storage materials, green building materials, environmental sustainability, and nanotechnology.

5.3.1 RENEWABLE ENERGY

The shift toward renewable energy sources, such as solar, wind, and hydropower, is a critical response to global climate change and the depletion of finite fossil fuels. Eco-materials play a pivotal role in this transition. They are employed in the manufacturing of photovoltaic solar panels, wind turbine components, and materials for hydropower systems. These materials enhance the efficiency and sustainability of energy production, contributing to a cleaner and greener energy landscape [33].

5.3.2 Energy Storage Materials

Efficient energy storage solutions are imperative for harnessing the intermittent nature of renewable energy sources. Eco-materials find applications in the development of sustainable energy storage technologies, including advanced battery systems, supercapacitors, and hydrogen storage materials. Contextually, supercapacitors are classified into electric double-layer capacitors (EDLCs), pseudocapacitors, and hybrid capacitors based on the electrode materials used in the positive and negative electrodes of the device. These types of supercapacitors mostly involve two distinctive charge storage mechanisms, i.e., Faradaic (charge transfer between the electrodes and electrolytes involving redox reactions) and non-Faradaic (electrostatic mechanism) types [34]. Based on the configuration, supercapacitors are classified into symmetric (where two identical EDLCs or pseudocapacitive electrodes are utilized), asymmetric (two kinds of electrodes with similar charge storage mechanism), and hybrid (two kinds of electrodes with dissimilar charge storage mechanism) supercapacitors. Among these, asymmetric and hybrid supercapacitors are best suitable to the present technological scenario, due to their high deliverable energy density at elevated power density conditions and excellent long-term service stability [34]. These materials improve the energy density, charge-discharge efficiency, and overall environmental impact of energy storage solutions [34,35].

Energy density is a crucial property of the evolution of the performance of energy storage devices so we have discussed detail about Energy density. Energy density refers to the amount of energy stored in a given volume or mass of a substance. In simpler terms, it tells us how much energy we can extract or store from a certain amount of material. Higher energy density means more energy stored per unit volume or mass, which is desirable for many applications as it allows for longer operating times, increased power output, or reduced size and weight of energy storage devices. Understanding energy density is essential for designing efficient and compact energy storage systems, like batteries, capacitors, or fuel cells. It is a crucial concept in various fields such as materials science, nanomaterials, physics, chemistry, and engineering, particularly in the chapter on energy storage and conversion technologies. Assessing the performance of fuels and other energy sources also relies on understanding energy density [36].

These examples demonstrate how nanomaterials play a crucial role in enhancing the energy density of various energy storage and conversion devices, paving the way for more efficient and sustainable energy technologies.

Energy density comes in different types depending on the form of energy being stored or converted.

Gravimetric energy density, for example, refers to the amount of energy stored per unit mass (e.g., watt-hours per kilogram, or joules per gram). It characterizes batteries, where a higher gravimetric energy density translates to longer runtime for portable electronic devices or electric vehicles without significant weight addition [37].

Volumetric energy density measures the energy stored per unit volume (e.g., watt-hours per liter, or joules per cubic centimeter). This is crucial for applications like mobile devices or electric vehicles, where maximizing energy storage within limited space is vital [38].

Now, let's delve into some examples of nanomaterials utilized to enhance energy density:

Nanostructured Electrodes for Batteries: Nanoparticles, nanowires, or nanostructured thin films are employed in battery electrodes to increase surface area and improve ion/electron transport kinetics. This enhances the battery's energy density, allowing more active material participation in electrochemical reactions, resulting in higher capacity and improved performance [39].

Nanostructured Catalysts for Fuel Cells: Fuel cells directly convert chemical energy into electrical energy through electrochemical reactions. Nanostructured catalysts, like platinum nanoparticles supported on carbon nanotubes or graphene, significantly enhance catalytic activity and surface area available for reactions, leading to higher power density and overall energy efficiency [40].

Nanocomposites for Supercapacitors: Supercapacitors, also known as ultracapacitors, store energy through ion adsorption at the electrode-electrolyte interface. Incorporating nanomaterials such as carbon nanotubes, graphene, or metal oxides into electrode materials increases surface area and improves charge storage capacity, resulting in higher energy density and faster charging/discharging rates compared to traditional capacitors [41].

Nanoporous Materials for Hydrogen Storage: Hydrogen boasts high energy density by weight but low density by volume, presenting challenges in storage and transportation. Nanoporous materials like metal-organic frameworks (MOFs) or nanoporous carbons offer high surface area and tunable pore structures, efficiently adsorbing and storing hydrogen molecules, thus improving the volumetric energy density of hydrogen storage systems [42,43]. These examples underscore how nanomaterials play a crucial role in enhancing the energy density of various energy storage and conversion devices, thereby paving the way for more efficient and sustainable energy technologies.

5.3.2.1 Ecomaterials Used for CO_2 Reduction

Eco-materials, also known as sustainable materials or green materials, can play a significant role in CO_2 reduction. These materials are designed and manufactured with environmental sustainability in mind, often aiming to minimize resource consumption, reduce pollution, and lower carbon emissions throughout their lifecycle.

Here's how eco-materials contribute to CO_2 reduction:

Renewable Resources: Eco-materials are often made from renewable resources such as bamboo, recycled plastics, or reclaimed wood. Using renewable resources reduces the dependence on fossil fuels for manufacturing processes, which in turn lowers CO_2 emissions [44].

Energy Efficiency: Eco-materials are often designed to be more energy-efficient during the manufacturing, transportation, and use phases. By requiring less energy to produce and transport, they contribute to lower overall carbon emissions [45].

Carbon Capture and Storage: Some eco-materials, such as certain types of concrete, are engineered to absorb and store carbon dioxide during their lifecycle. This carbon capture and storage capability can help offset CO_2 emissions from other sources [46].

Durability and Longevity: Many eco-materials are designed to be durable and long-lasting, reducing the need for frequent replacements or maintenance. This longevity reduces the overall environmental impact associated with manufacturing and disposing of materials, including CO_2 emissions [47].

Recyclability and Biodegradability: Eco-materials are often recyclable or biodegradable at the end of their lifecycle. Recycling materials reduces the energy and emissions associated with producing new materials from virgin resources. Biodegradable materials break down naturally, avoiding the accumulation of waste in landfills, which can release methane, a potent greenhouse gas [47].

Low Emissions in Use: Some eco-materials, such as low VOC (Volatile Organic Compound) paints or energy-efficient building materials, contribute to lower emissions during their use phase. This can lead to reduced energy consumption for heating, cooling, or maintenance, indirectly lowering CO_2 emissions [47].

Overall, the use of eco-materials can contribute to CO_2 reduction by addressing various aspects of material production, usage, and disposal, while promoting environmental sustainability and resource efficiency.

5.3.2.2　Nanomaterials Used for H_2 Storage

Nanoparticles used for hydrogen (H_2) storage offer specific advantages and functionalities, contributing to the advancement of clean energy solutions. Here's a breakdown of their key characteristics and applications:

Material Composition: Nanoparticles for H_2 storage typically comprise materials with high surface area-to-volume ratios. Examples include metal hydrides like magnesium hydride, complex metal alloys such as palladium alloys, and porous carbon-based materials like carbon nanotubes and graphene [48].

Surface Modification: Surface modifications play a crucial role in enhancing H_2 adsorption and desorption kinetics. This involves techniques such as coating nanoparticles with catalysts such as platinum or functionalizing their surfaces with reactive groups to improve hydrogen interaction [49].

Nanostructuring: Nanostructuring techniques, including size and shape control, are employed to optimize nanoparticle properties. By manipulating surface area and pore structure, nanoparticles can offer increased H_2 storage capacity. For example, nanoparticles with hierarchical structures or high aspect ratios provide enhanced accessibility for hydrogen molecules [50].

Catalytic Effects: Certain nanoparticles exhibit catalytic effects that facilitate hydrogen dissociation and diffusion within the storage material. These catalysts promote reversible hydrogenation/dehydrogenation reactions, thereby improving the efficiency and kinetics of H_2 uptake/release [51].

Thermal Management: Engineered nanoparticles also improve thermal conductivity and heat dissipation during hydrogen absorption and desorption processes. This mitigates temperature-related issues like overheating or thermal instability, ensuring safe and efficient H_2 storage [52].

Through these specific characteristics and design considerations, nanoparticles play a vital role in the development of efficient and practical solutions for hydrogen storage, driving forward the utilization of hydrogen as a clean energy carrier.

5.3.3 GREEN BUILDING MATERIALS

The construction and operation of buildings account for a substantial portion of energy consumption and carbon emissions. Eco-materials are integrated into the construction industry to create green building materials [53]. These materials include recycled low-impact, and energy-efficient products that enhance the sustainability and energy efficiency of buildings. They contribute to reduced energy consumption, better indoor air quality, and long-term environmental sustainability in the built environment [54].

5.3.4 ENVIRONMENTAL SUSTAINABILITY

Eco-materials are integral to broader environmental sustainability efforts. They play a role in waste reduction, pollution prevention, and resource conservation. Through recycling, upcycling, and utilizing renewable resources, eco-materials contribute to the reduction of the environmental footprint across industries [55]. In this chapter, we discuss details uses of eco-materials in the field of various environmental sustainability.

Material Selection: Eco-materials are carefully chosen to minimize environmental impact. They often include renewable, recycled, or reclaimed resources, reducing the demand for virgin materials and conserving natural resources.

Resource Efficiency: Eco-materials aim to maximize resource efficiency throughout their lifecycle. This involves reducing raw material consumption, energy usage, and waste generation during the manufacturing, distribution, use, and disposal phases.

Low Carbon Footprint: Eco-materials are designed to have a low carbon footprint, meaning they emit minimal greenhouse gases (GHGs) during production and use. This helps mitigate climate change and reduce overall environmental impact.

Pollution Reduction: Eco-materials are formulated to minimize pollution and harmful emissions. This includes reducing air and water pollution, as well as minimizing waste generation and hazardous material usage.

Biodegradability and Recyclability: Eco-materials are often biodegradable or recyclable at the end of their lifecycle. Biodegradable materials break down naturally, reducing landfill waste and preventing environmental pollution. Recyclable materials can be reused or repurposed, conserving resources and reducing the need for new material production.

Energy Efficiency: Eco-materials prioritize energy efficiency in their manufacturing processes. This involves using renewable energy sources, optimizing production techniques, and reducing energy consumption to minimize environmental impact and reliance on fossil fuels.

Longevity and Durability: Eco-materials are engineered to be durable and long-lasting, reducing the need for frequent replacements and extending product lifespan. This helps conserve resources, reduce waste, and lower environmental impact over time.

Sustainable Practices: Eco-materials are often produced using sustainable practices that prioritize environmental, social, and economic responsibility. This may include ethical sourcing, fair labor practices, and adherence to environmental regulations and certifications

5.3.5 NANOTECHNOLOGY AND THEIR APPLICATIONS

Nanotechnology, which deals with materials and devices at the nanoscale, is a frontier where eco-materials exhibit significant promise. These materials are engineered at the molecular and atomic level to achieve unique properties. Nanoscale eco-materials find applications in drug delivery systems, water purification, pollutant removal, and advanced diagnostics, contributing to both environmental sustainability and improved healthcare. Furthermore, we are discussing in detail applications in the area of nanomedicines [56].

5.3.5.1 Nanotechnology in Cosmetics and Rejuvenation

Cosmetic Manufacturers use Nanoscale-size ingredients to provide better UV Protection, Deeper skin penetration, long-lasting effects, increased color, and much more. In the cosmetic industry, nanoparticles are present in shampoos, kinds of toothpaste, anti-wrinkle creams, anti-cellulite creams, whitening skin, moisturizers, face powders, sunscreens, perfumes, etc. [57]. Nanoparticles are used as UV filters. Sunscreens contain titanium dioxide and zinc oxide nanoparticles as inorganic UV filters to protect the skin from damaging UV light using scattering, absorption, and reflection. Nanoparticles are increasingly used in cosmetics due to their unique properties and potential benefits for skincare and beauty products. Titanium dioxide (TiO_2) nanoparticles in sunscreen provide a prime example of this application: such as titanium dioxide (TiO_2) is commonly used in sunscreens for its ability to scatter and absorb ultraviolet (UV) radiation. When reduced to nanoscale, TiO_2 particles (typically less than 100 nanometers in diameter) exhibit enhanced optical properties, such as improved UV protection and transparency compared to larger particles. Titanium dioxide nanoparticles offer broad-spectrum UV protection by effectively blocking both UVA and UVB rays. Their small size allows for uniform dispersion in sunscreen formulations, ensuring even coverage and consistent protection across the skin surface [58].

Nanoemulsions are also used in hair care, lip care, and skin care products. It is composed of squalene (sebum lipid) and diphencyprone (DPCP) for alopecia areata treatment that induce hair regrowth through a local immune response [58].

5.3.5.2 Composite Materials

Nanoscale additives in polymer composite materials are being used to make cars, airplanes, and spacecraft lighter. For example, carbon nanotube sheets are being used in next-generation air vehicles in the hopes of gaining high fuel efficiency [55,56].

5.3.5.3 Infrastructure Sensing

Nano-scale sensors are being used in bridges, tunnels, rails, and buildings. They provide continuous monitoring of its structural integrity and performance. The sensors can be embedded into the construction materials to offer continuous monitoring [58].

5.3.5.4 Automotive Technology

Nanoengineered materials in the automotive industry include high power rechargeable battery systems, thermoelectric materials, tires with lower rolling resistance, and fuel additives for cleaner exhaust and extended range [34].

5.3.6 Medicinal Applications

Healthcare is the improvement of the prevention, diagnosis, treatment, disease, illness, injury, and other physical and mental impairments in people. It is delivered by medicine, dentistry, pharmacy, midwives, audiology, psychologists, physical therapists, and other health professionals [14]. It includes work done in providing primary care, secondary care, and tertiary care, which are organizations established to meet good health. According to the World Health Organization (WHO), a well-functioning healthcare system requires a financing mechanism, a well-trained and adequately paid workforce, reliable information on which to base decisions and policies, and well-maintained health [59].

5.3.6.1 Gene Therapy

Gene therapy is a technique that works by adding new copies of a gene that is broken, or by replacing a defective or missing gene in a patient's cells with a healthy version of that gene. There are three types of gene therapy: ex vivo, in vivo, and in situ. In ex vivo gene therapy, the target cells are removed from the patient's body, engineered either by the addition of the therapeutic gene or by other genetic manipulations that allow correction of the phenotype of the disease. In the future, genetic treatment, or cure certain inherited disorders, such as cystic fibrosis, alpha-1 antitrypsin deficiency, hemophobia, beta thalassemia, and sickle cell disease [17]. They are also used to treat cancers or infections, including HIV. Gene therapy promises to treat a such as cancer, cystic fibrosis, heart disease, diabetes, hemophobia, and AIDS. life-saving for some people specific medical expenses and can cause side effects [60]. Gene therapy is a new generation gene that is delivered to a targeted tissue in the body to produce a missing or nonfunctioning protein. DNA is the code that controls much of your body's function, from making you grow taller to regulating your proper work for body systems. Genetic engineering disorders in humans are treated by replacing the defective gene with a functioning one. Important drugs, vaccines, and other products have been harvested from organisms developed that aid food security by increasing yield, nutritional value, and tolerance to environmental stresses [59,60].

5.3.6.2 Drug Delivery

Using various excipients, drug carriers, and medical devices in the formulation process allows for the pharmacokinetics and specificity of drug delivery. Medical methods for drug delivery by swallowing, inhalation, absorption through the skin, or by injection. The drug is to desired body site for drug release and absorption, of the subsequent transport of the active ingredients across the biological membranes to the action [59]. The field of research to develop new technologies and approaches to transport drugs through the body. The regarded dose right time, right drug, right dose, as standard for safety. Drug delivery systems provide enhanced efficacy and reduced toxicity for long-circulating macromolecules such as liposomes and can exploit the 'enhanced permeability and retention' effect for tumor vessels [61].

5.3.6.3 Cancer Therapy

Cancer is a disease driven by inherently nanostructural problems. Treatments include surgery, chemotherapy, and radiation because they target the use of the human body. Surgery is an operation to remove cancer, while chemotherapy, radiation, targeted therapy, and immunotherapy are among the different types of treatment options. These treatments utilize high-energy rays to kill cancer cells and hormones that block cells from growing. They are particularly effective for thyroid, prostate, and testicular cancer detection and treatment. Despite many considering it impossible to live long with cancer, recognizing common signs and symptoms can prompt individuals to adopt healthier lifestyles, which might include consuming fruits, vegetables, minerals, vitamins, plant-based foods, grains, beans, and whole foods, as well as prioritizing sleep, maintaining a relaxed mind, engaging in exercise, and maintaining good health [62]. Wilms tumor, also known as nephroblastoma, is a childhood cancer that affects both kidneys and is categorized into stages. Although the term is not commonly used, any type of cancer, including skin, lung, breast, prostate, colorectal, kidney, heart, and bladder cancer, can be diagnosed. Among these, prostate cancer and lung cancer are more prevalent. Treatments for these cancers often involve surgery, such as mammaplasty, tissue expansion, lymph node dissection, lumpectomy, and mastectomy. Medical procedures such as teletherapy and radiation therapy are also employed [63]. Meditation plays a role as an estrogen modulator and can complement chemotherapy, hormones, and bone health treatments to support bones, tissues, and overall bone strength and development [64,65].

5.3.6.4 Genetic Disorder

Genes are blocks of heredity from parents to child, a change in a gene for making a protein. Down syndrome, fragile X syndrome, cystic fibrosis, Huntington's disease, turner syndrome, and sickle cell anemia, are caused due to mutations in the gene responsible for making the hunting protein [66]. Category of disease that certain types of birth defect, chronic diseases, development problems, and sensory deficits that are inherited from one or both parents. There are four types of genetic disorders include, single gene inheritance, multifactorial inheritance, chromosome abnormalities, and mitochondrial inheritance [67]. Children who have parents with asthma are more likely to research has shown that both genetics and environmental factors are involved. A person may carry more than one gene or group of genes that increase or reduce the risk of Alzheimer's, which include gene TP53, TNF, EGFR, VEGFA, APOE, I16, TGFB1, MTHFR, ESR1, AKT1. Environmental variables that can cause mutations include radiation, chemicals, and infectious pathogens. The study of genes and traits of certain qualities or traits are passed from parents to offspring as a result of changes in DNA sequence. Genetic factors play some role in high blood pressure, and stroke, including sickle cell disease, a sign of a family's hypercholesterolemia, and a genetic disorder that causes high cholesterol [67,68].

5.3.6.5 Ocular Disease

The leading causes of blindness are primarily age-related macular degeneration, cataracts, diabetic retinopathy, and glaucoma. Vision loss usually starts in childhood but for some people treatment for stargate disease can help people make the most of their remaining vision. Glaucoma disease is caused by damage to the nerve, and the major risk factor is eye pressure. People having color blindness see color differently than most people but special glasses and contact lenses help them to see the difference between colors and deficiency and don't have problems with activities. Age-related macular degeneration (AMD) in eye disease can blur your central and cause damage to the macular control and sharp, straight-ahead vision. As you age, a cataract region in the lens is more likely to cause blurry, foggy, and less colorful vision. Normal changes following surgery for another eye condition or after an eye injury [59,67,68]. Ocular disease is a very problem in every people, the thing so driving, reading and other see work all problem its. Infection, allergy, vitamin, deficiency, chemical irritants, genetics, smoking, etc are some of the common problems eye disorders include amblyopia and strabismus. Eye pain is stopped adjust the lighting, breaks, of your space, your computer setup change, eye exercises, palming your eyes, warm compress, etc can help us. Surgery, the risk is negligible, the time following the procedure, be blurry, may experience glare or other symptoms your eye [68].

5.3.6.6 Cardiovascular Disease

Aortic disease, peripheral arterial disease, coronary heart disease, and stroke are examples of heart ailments. Reduce low-density lipoprotein cholesterol, enhance blood flow, and perform procedures like valve replacement or grafting while treating cardiovascular disease. Choose healthy food, drinks, meals, snacks, continued physical activity, don't smoke, vegetables, etc. [69]. The most common symptom is a weakness of the face, arm, or leg, most often the side of the body, and confusion, seeing with one or both eyes, difficulty speaking or understanding. The blood supply to the heart muscle occurs with the coronary disease the heart to "cry out in pain" and work harder. To treat patients with blocked or clogged coronary cardiologists threads a balloon-tipped the narrowed or blocked artery and then inflate the balloon to open the vessel [70]. Damage to the heart valves and heart muscle from the primary care level will ultimately result in reduced coronary heart disease. An electrocardiogram (ECG) is a test that records the electrical happening in different areas of the heart and helps identify any problems that are painless and take around [69,71].

5.3.6.7 Brain and CNS Disorder

Functional disorders include headache, epilepsy, dizziness, multiple sclerosis, amyotrophic lateral sclerosis, and Alzheimer's disease. Anxiety is a feeling of fear, worry, and unease that causes physical changes, mental health issues, and anxiety disorders. One type brought on by abuse is CNS depression. Medications used to manage pain, anxiety, sleep difficulties, and stress include sedatives and opioids. [72]. The benefits of Zincovit tablet include recovery from weakness and the absorption of iron leading to RBC formation. Instead of tablets, fresh fruit juice, milkshake, green tea, or smoothies for foods that prevent fatigue can also be taken. The central nervous system (CNS) from substance overdoses, and poisoning, slows down too much, and it can quickly become life-threatening. A group of CNS disorders are epilepsy, stroke, infectious or immunological, migraine, and traumatic brain injury [73].

5.3.6.8 Other Applications

Eco-materials, also known as eco-friendly materials or sustainable materials, are increasingly being explored and utilized in various medicinal applications due to their environmentally friendly nature and potential benefits. Below are some types of eco-materials commonly used in medicinal applications [74–77]:

Biodegradable Polymers: Suitable for drug delivery systems, wound dressings, and tissue engineering scaffolds. These polymers degrade naturally over time into non-toxic byproducts.

Natural Fibers: Renewable and biodegradable materials like cotton, hemp, and bamboo fibers find applications in wound dressings, surgical textiles, and drug delivery systems.

Biocompatible Metals: Biocompatible metals such as titanium and stainless steel, which interact safely with biological systems, are used in medical implants, prosthetics, and surgical instruments.

Biodegradable Ceramics: Used in bone grafts, dental implants, and tissue engineering. Ceramics derived from natural sources or designed to degrade inside the body offer viable solutions.

Plant-Derived Extracts: Incorporating extracts from medicinal plants into formulations for topical treatments, wound healing, and skincare products enhances their efficacy.

Recycled Materials: Repurposed plastics and metals in medical device components, packaging materials, and equipment reduce waste and environmental impact.

Alginate and Chitosan: Biopolymers derived from seaweed and crustacean shells are used in wound dressings, drug delivery systems, and tissue engineering owing to their biocompatibility and biodegradability.

Bioactive Glasses: Glasses containing ions like calcium, phosphate, and silica stimulate tissue growth and repair, making them suitable for bone regeneration, wound healing, and drug delivery.

Green Synthesized Nanomaterials: Investigated for targeted drug delivery, imaging, and diagnostic applications. Nanoparticles synthesized using eco-friendly methods offer promising avenues.

Natural Antimicrobial Agents: Essential oils, plant extracts, and other natural antimicrobial agents prevent infections and promote healing in wound care products, disinfectants, and antimicrobial coatings.

These eco-materials provide sustainable alternatives to conventional materials in medicinal applications, minimizing environmental impact and fostering eco-friendly practices (Figure 5.1).

In summary, eco-materials have emerged as a vital component in the pursuit of a more sustainable and eco-conscious future. Their applications in renewable energy, energy storage, green building materials, environmental sustainability, and nanotechnology represent innovative solutions to the urgent challenges of our time, with the potential to reshape industries and promote a greener and more environmentally responsible world.

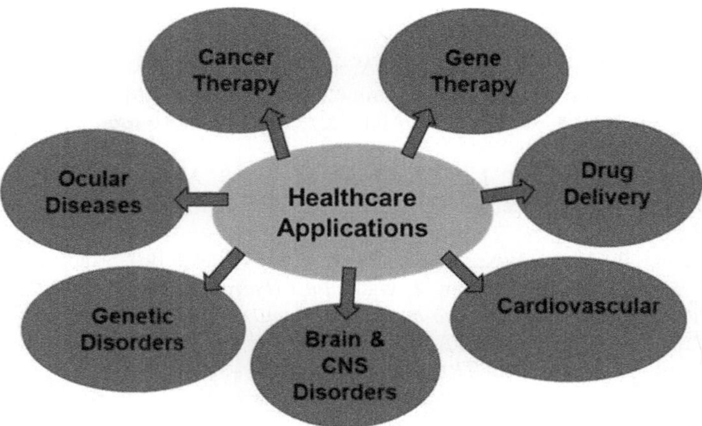

FIGURE 5.1 Schematic arrangement of applications of numerous nanosystems in related healthcare areas.

5.4 CONCLUSION

Eco-materials are a promising avenue in the pursuit of sustainable and environmentally responsible solutions. These materials offer a wide scope of applications, ranging from construction and packaging to electronics and transportation. However, they come with various challenges, such as sourcing renewable resources, ensuring durability, and addressing cost concerns. The successful integration of eco-materials into various industries requires a multi-faceted approach that combines innovation, regulation, and consumer awareness. In conclusion, eco-materials hold great potential in reducing the environmental impact of our industrial processes, but their widespread adoption necessitates continued research and concerted efforts to overcome challenges. Nanomedicine, an emerging field at the intersection of nanotechnology and medicine, harnesses eco-materials for various purposes. These eco-materials find applications in nanomedicine due to their biocompatibility and reduced environmental impact. Key uses of eco-materials in nanomedicine include drug delivery, imaging, tissue engineering, cancer treatment, diagnosis, energy storage, sustainability, and environmental remediation. Eco-materials in nanomedicine offer the advantage of reduced environmental impact and improved biocompatibility, making them a sustainable choice for medical applications. However, ongoing research and development are essential to enhance their efficiency and safety in healthcare.

ACKNOWLEDGEMENTS

Yogesh Kumar Sonia appreciatively acknowledges, the University Grants Commission (UGC) for providing financial support. MNIT Jaipur is acknowledged for providing the laboratory space. Sapna Meena acknowledge department of chemistry, government girls college nagaur, Rajasthan.

REFERENCES

1. R. Roy, R. Ashmika, *IGI Global, ISBN*: 9798369312148 (2024) 288–314.
2. M. Awad, A. Said, M. H. Saad, A. Farouk, M. M. Mahmoud, M. S. Alshammari, M. L. Alghaythi, S. H. A. Aleem, A. Y. Abdelaziz, A. I. Omar, *Alexandria Eng J.* 87 (2024) 213–239.
3. Q. Hassan, P. Viktor, T. J. Al-Musawi, B. M. Ali, S. Algburi, H. M. Alzoubi, A. K. Al-Jiboory, A. Z. Sameen, H. M. Salman, M. Jaszczur, *Renew Energy Focus.* 48 (2024) 100545.
4. J. D. Hunt, B. Zakeri, A. Nascimento, M. A. V. de Freitas, F. do Carmo Amorim, F. Guo, G. J. Witkamp, B. Van Ruijven, Y. Wada, *Int J Hydrogen Energy.* 53 (2024) 875–884.
5. P. Cicconi, *Eco-design and Eco-materials: An interactive and collaborative approach.* Sustain. Mater. Technol.(2020) 135. 10.1016/j.susmat.2019.e00135
6. S. Buzuku, I. Shnai, *J Eur TRIZ Assoc.* 4 (2018) 20–31.
7. Y. Y. Lee, B. Sriram, S. F. Wang, M. M. Stanley, W. C. Lin, S. Kogularasu, G. P. Chang-Chien, M. George, *Appl Surf Sci Adv.* 20 (2024) 100584.
8. B. Rabeie, N. M. Mahmoodi, *J Colloid Interface Sci.* 654 (2024) 495–522.
9. M. Massoudinejad, N. Amanidaz, R. M. Santos, R. Bakhshoodeh, *J Environ Health Sci Eng.* 17 (2019) 1227–1242.
10. B. Aaliya, K. V. Sunooj, M. Lackner, *Int J Biobased Plastics.* 3 (2021) 40–84.
11. J. M. Danatzko, H. Sezen, Q. Chen, *J Green Build.* 8 (2013) 120–135.
12. P. O. Akadiri, P. O. Olomolaiye, *Eng Construct Architect Manage.* 19 (2012) 666–687.
13. P. Cicconi, *Sustai Mater Technol.* 23 (2020) e00135.
14. H. Nguyen-Yamamoto, Eco-Design. *3rd International Symposium on Environmentally Conscious Design and Inverse Manufacturing, IEEE*, Tokyo, Japan. (2003) 706–709.
15. M. A. Martin-Luengo, M. Yates, E. Diaz, E Saez Rojo, L. Gonzalez Gil, *Appl Cat B Environ.* 106 (2011) 488–493.
16. C.-F. Dobrescu, *Int Multidiscip Sci GeoConf: SGEM.* 18 (2018) 115–120.
17. U. Osamu, Y. Shinohara, K. Halada, *Mater Trans.* 55 (2014) 745–749.
18. T. G. Gutowski, *Mater Trans.* 43 (2002) 359–363.
19. L. M. Torres-Martínez, O. V. Kharissova, B. I. Kharisov, *Handbook of Ecomaterials Springer* (2018) 2709–2730.
20. J. Liu, Y. Hong, T. Xinyong, L. Yeru, J. Y. Seung, J.-Q. Huang, T.-Q. Yuan, M.-M. Titirici, Q. Zhang, *Eco Mat.* 2 (2020) 12019.
21. A. S. Majdi, L. María, Jalón, E. Puertas, J. Chiachío, *Appl Sci.* 13 (2023) 12310.
22. V. Echarri, C. A. Brebbia, , *Eco-Architecture VI: Harmonisation between Architecture and Nature* WIT Press, 161 (2016) 169.
23. J. Jovanović, S. Stevović. e "New Functional Materials and High Technology", Institute of Solution Chemistry of Russian Academy of Sciences (ISC-RAS), 2015.-250 ISBN 978-5-905364-10-5. (2015) 262.
24. F. Antoniou, F. Demertzidou, P. Mentzelou, D. Konstantinidis, *IOP Conference Series: Earth and Environmental Science*, vol. 1123, no. 1, p. 012033. IOP Publishing, 2022.
25. S.-P. Jorge, L. F. Ignacio, B. R. Jesús, G. Xavier, *J Cleaner Prod.* 137 (2016) 606–616.
26. E. Matei, M. Râpă, A. M. Predescu, A. A. Turcanu, R. Vidu, C. Predescu, C. Bobirica, L. Bobirica, C. Orbeci, *Materials.* 14 (2021) 4581.
27. D. Merino, C. Casalongué, V. A. Alvarez, *Handbook of Ecomaterials*; Torres-Martínez, LM, Kharissova, OV, Kharisov, BI, Eds, Springer Nature 7 (2018) 2709–2730.
28. J. Y. Lim, T. N. B. Truong, J. Y. Teo, C. G. Wang, Z. Li, *Circularity of Plastics* ,Elsevier, (2023) 1–34.
29. Z. Nie, *Green and Sustainable Manufacturing of Advanced Materia, ISBN: 9780124114975 Elsevier.* (2016) 31–76.
30. S. M. Rahman, A. N. Khondaker, *Renew Sustain Energy Rev.* 16 (2012) 2446–2460.

31. E. Mirzadeh, K. Akhbari, *CrystEngComm.* 39 (2016) 7410–7424.

32. L. Tian, Y. T. Han, J. Jagadese, J. Vittal, *Crystal Growth Design.* 8(2) (2008) 734–738.

33. R. Pacheco, J. Ordóñez, G. Martínez, *Renew Sustain Energy Rev.* 16 (2012) 3559–3573.

34. R. Braga, D. M. Fernandes, A. Adán-Más, T. M. Silva, M. F. Montemor, *Batteries.* 9 (2023) 168.

35. S. B. Sadineni, S. Madala, R. F. Boehm, *Renew Sustain Energy Rev.* 15 (2011) 3617–3631.

36. Y. K. Sonia, M. K. Paliwal, S. K. Meher, *Sustain Energy Fuels.* 5 (2021) 973–985.

37. A. Nomura, K. Ito, Y. W. Denis, Y. Kubo, *J Power Sources.* 592 (2024) 233924.

38. W. Wang, Z. Shen, J. Zhang, L. Pan, C. Shi, X. Zhang, J. J. Zou, *Fuel.* 355 (2024) 129380.

39. Y. K. Sonia, S. Srivastav, S. K. Meher, S. K., *Langmuir.* 26 (2023) 9111–9129.

40. Y. K. Sonia, S. K. Meher, *Energy Fuels.* 37 (2023) 4010–4025.

41. Y. K. Sonia, S. K. Meher, *Sustain Energy Fuels.* 7 (2023) 2895–2909.

42. Y. K. Sonia, S. K. Meher, *ACS Appl Energy Mater.* 5 (2022) 13672–13691.

43. K. Mondal, S. J. Malode, N. P. Shetti, S. A. Alqarni, S. Pandiaraj, A. Alodhayb, *J Energy Storage.* 76 (2024) 109719.

44. I. Elfaleh, F. Abbassi, M. Habibi, F. Ahmad, M. Guedri, M. Nasri, C. Garnier, *Development.* 1 (2023) 3.

45. M. Imani, M. Donn, Z. Balador, Handbook of Ecomaterials , Springer. (2018) 1–24.

46. D. S. Vijayan, P. Devarajan, A. Sivasuriyan, A. Stefańska, E. Koda, A. Jakimiuk, M. D. Vaverková, J. Winkler, C. C. Duarte, N. D. Corticos, *Sustainability.* 15 (2023) 6751.

47. S. I. Plasynski, J. T. Litynski, H. G. McIlvried, R. D. Srivastava, *Crit Rev Plant Sci.* 28 (2009) 123–138.

48. W. Zhao, V. Fierro, C. Zlotea, M. T. Izquierdo, C. Chevalier-César, M. Latroche, A. Celzard, *Int J Hydrogen Energy.* 37 (2012) 5072–5080.

49. M. N. Sepehr, V. Sivasankar, M. Zarrabi, M. S. Kumar, *Chem Eng J.* 228 (2013) 192–204.

50. Y. Xia, Z. Yang, Y. Zhu, *J Mater Chem A.* 1 (2013) 9365–9381.

51. M. J. Zhou, Y. Miao, Y. Gu, Y. Xie, *Adv Mater.* 24 (2024) 2311355.

52. K. Wang, W. Chen, L. Li, *Renew Energy.* 187 (2022) 1118–1129.

53. K. Golić, V. Kosorić, A. K. Furundžić, *Renew Sustain Energy Rev.* 15 (2011) 1533–1544.

54. L. B. Robichaud, V. S. Anantatmula, *J Manage Eng.* 27 (2010) 48–57.

55. J. Sarkis, L. M. Meade, A. R. Presley, *J Cleaner Prod.* 31 (2012) 40–53.

56. A. Mwasha, R. G. Williams, J. Iwaro, *Energy Build.* 43 (2011) 2108–2117.

57. M. A. Green, K. Emery, Y. Hishikawa, W. Warta, *Prog Photovolt: Res Appl.* 23 (2015) 1–9.

58. T. G. Smijs, S. Pavel, *Nanotech Sci Appl.* (2011) 95–112.

59. M. Armand, J. M. Tarascon, *Nature.* 451 (2008) 652–657.

60. C. J. Kibert, Sustainable Construction: Green Building Design and Delivery (2008). John Wiley & Sons, ISBN: 9780470904459.

61. J. A. Hubbell, A. Chilkoti, *Science.* 337 (2012) 303–305.

62. Z. Cheng, M. Li, R. Dey, Y. Chen, *J Hematol Oncol.* 14 (2021) 1–27.

63. K. H. Bae, H. J. Chung, T. G. Park, *Mol Cells.* 31 (2011) 295–302.

64. W. Qiao, B. Wang, Y. Wang, L. Yang, Y. Zhang, P. Shao, *J Nanomater.* 10 (2010) 1–6.

65. O. C. Farokhzad, R. Langer, *Adv Drug Delivery Rev.* 58 (2006) 1456–1459.

66. A. A. Aljabali, M. A. Obeid, H. A. Amawi, M. M. Rezigue, Y. Hamzat, S. Satija, M. M. Tambuwala, *Appl Nanomater Human Health.* 18 (2020) 125–146.

67. X. Zhu, J. Li, H. He, M. Huang, X. Zhang, S. Wang, *Biosens Bioelectron.* 74 (2015) 113–133.

68. M. Ebrahimi, M. Asadi, O. Akhavan, *ACS Biomater Sci Eng.* 8 (2021) 54–81.

69. W. Jiang, D. Rutherford, T. Vuong, H. Liu, *Bioactive Mater.* 2 (2017) 185–198.

70. H. Chopra, S. Bibi, A. K. Mishra, V. Tirth, S. V. Yerramsetty, S.V. Murali, S. U. Ahmad, Y. K. Mohanta, M. S. Attia, A. Algahtani, F. Islam, *J Nanomater.* 22 (2022) 1–25.
71. A Sharma, P. Maheshwari, R. Tekade, R. K. Tekade, *Current Pharm Des.* 21 (2015) 4465–4478.
72. B. Conklin, B. M. Conley, Y. Hou, M. Chen, K. B. Lee, *Adv Drug Delivery Rev.* 192 (2023) 114636.
73. D. Furtado, M. Björnmalm, S. Ayton, A. I. Bush, K. Kempe, F. Caruso, *Adv Mater.* 30 (2018) 1801362.
74. I. Elfaleh, F. Abbassi, M. Habibi, F. Ahmad, M. Guedri, M. Nasri, C. Garnier, *Results Eng.* (2023) 101271.
75. S. B. Jaffri, K. S. Ahmad, K. H. Thebo, F. Rehman, *Rev Inorg Chem.* 41 (2021) 131–150.
76. K. Y. Song, S. W. Kim, D. C. Nguyen, J. Y. Park, T. T. Luu, D. Choi, J. M. Baik, S. An, *Eco Mat.* 23 (2023) 12357.
77. P. P. Das, V. Chaudhary, *Cleaner Eng Tech.* 4 (2021) 100182.

Part II

Green Energy Applications

6 Green Synthesis
of Hydroxyapatite
Nanoparticles

*Sustainable Approaches for
Biomedical Advancements*

*Prabha Gurawalia, Sudip Majumder,
Chandra Mohan Srivastava, Sonika Charak,
and Manish Shandilya*

6.1 INTRODUCTION

The term "eco-material" refers to a broad category of materials manufactured, used, and designed with a mindset toward ecological sustainability. "Eco-material" usually describes materials that are sustainable or beneficial to the environment at every stage of their lifecycle, from production to disposal. These materials are made with resource conservation and a minimal impact on the environment in mind. Eco-materials can be applied to a range of industries, such as consumer goods, manufacturing, packaging, and building.

As technology progresses and environmental consciousness increases, the field of eco-materials is continually expanding. To lessen the negative effects of human activity on the environment, these materials are frequently employed in environmental friendly product design, sustainable building, and numerous industrial applications. The environmental effects of conventional materials are a subject of growing concern, particularly concerning their manufacturing, use, and disposal. Eco-materials are designed to address this issue. Here are some essential traits and features of eco-materials: sustainability, low environmental impact, recyclability and biodegradability, energy efficiency, durability, local sourcing, waste reduction, certifications, and standards, natural or bio-based materials. Even though hydroxyapatite (HAp) exhibits several environmentally friendly qualities, it is crucial to remember that the sustainability of any material also depends on factors like the energy and resources needed for its extraction, processing, and transportation. To fully assess the environmental impact of hydroxyapatite on the environment, it is crucial to consider the complete cycle of the product or application in which it is employed [1]. HAp is a type of naturally occurring calcium apatite, with the chemical formula $Ca_{10}(PO_4)_6(OH)_2$. It is the primary mineral in vertebrates' bone and teeth enamel. While HAp is most frequently associated with its use in medical and dental applications, its potential as an eco-material is an intriguing and developing field of study. Here are some reasons why HAp nanoparticles (NPs) can be considered as eco-materials:

Biocompatibility: HAp is biocompatible, meaning that the body tolerates it well. As it integrates well with living tissue, it can be employed in many medical applications, including orthopedic and dental implants. This lessens the need for substances that could be harmful to health.

Natural Origin: Geological deposits are a rich source of HAp, which occurs naturally. This lessens the negative effects on the environment caused by the mining and processing of synthetic materials. HAp is slowly absorbed by the body over time, minimizing waste generation and adverse environmental effects. This is crucial in the context of medical implants.

Low Toxicity: HAp does not emit any poisonous compounds or chemicals into the environment. It is therefore a safer option than some other synthetic materials.

Sustainable Production: HAp can be synthesized using ecofriendly methods like wet chemical precipitation, which can be planned to minimize waste and energy usage [2].

Regenerative Properties: HAp stimulates tissue regeneration and repair, making it valuable for regenerative medicine.

Recyclability: Materials made of HAp can often be recycled, promoting a more circular economy, and reducing the demand for raw materials.

Mineral Composition: As HAp is a mineral, it is composed of substances widely distributed in the Earth's crust. When sourced without a substantial negative influence on the environment, this composition contributes to eco-friendliness [3].

Biogenic Production: Scientists are exploring methods to produce HAp through biogenic processes, such as those initiated by microorganisms. This could potentially make the material production process even more environmentally friendly [4].

Catalytic Properties: The catalytic potential of HAp has been studied, especially concerning environmental applications. It could play a role in initiatives aimed at reducing pollution or enhancing environmentally friendly chemical processes [5].

The widespread usage of HAp as an eco-material depends on several aspects, including the scalability of production processes, cost-effectiveness, and specialized applications. It is vital to keep in mind that HAp may have some potential eco-friendly qualities. Furthermore, research on eco-materials is ongoing, and future advancements may further improve the environmental profile of HAp-based materials. The green synthesis of HAp NPs is a sustainable method that aims to generate HAp NPs or powders with little negative influence on the environment, reduced chemical waste, and the use of eco-friendly reagents. Green synthesis, which combines nanotechnology and environmental principles, is the process of synthesizing nanoparticles from plant and animal sources. In this chapter, we explored different conventional and green synthesis methods, delving into their diverse applications, and addressing the challenges associated with green synthesis approaches.

6.2 CONVENTIONAL METHODS FOR SYNTHESIZING HYDROXYAPATITE NANOPARTICLES

Chemical procedures are commonly used in traditional methods for producing HAp NPs, which have been employed for many years in a variety of industrial and scientific applications. Here is a summary of some of the popular traditional techniques for HAp synthesis:

6.2.1 PRECIPITATION METHOD

In this process, aqueous solutions containing calcium and phosphate sources are combined. While sources of phosphate can be ammonium phosphate, phosphoric acid, or other phosphate salts, sources of calcium are frequently obtained from calcium nitrate, calcium chloride, or calcium hydroxide. Usually, ammonium hydroxide or sodium hydroxide is added to the solution to alter the pH to a slightly alkaline level. HAp NPs detach from the solution, precipitate out, and are subsequently given additional processing, like washing and drying.

6.2.2 HYDROTHERMAL METHOD

In this process, calcium and phosphate sources react in a high-temperature, high-pressure aqueous environment. The production of HAp crystals is facilitated by the high temperature and pressure conditions. Specific particle sizes and shapes of well-crystalline HAp NPs can be produced through hydrothermal synthesis.

6.2.3 SOL-GEL METHOD

The sol-gel method involves hydrolyzing and condensing a solution containing phosphate esters and metal alkoxides (for example, calcium alkoxide). The gel is then heated and dried to create a solid that may be ground into a powder. The HAp NPs content and morphology can be precisely controlled by sol-gel synthesis.

6.2.4 MECHANICAL MIXING

Calcium and phosphate sources are ground up during mechanical mixing to produce a homogenous mixture. When the resultant mixture is heated, a solid-state reaction occurs, and HAp NPs are produced. Although this procedure is rather straightforward, high temperatures could be needed to speed up the process.

6.2.5 COPRECIPITATION METHOD

Coprecipitation is the simultaneous addition of calcium and phosphate sources to an aqueous solution and their subsequent co-precipitation. Control of the pH, temperature, and mixing conditions are crucial to ensure the proper stoichiometric of HAp NPs.

6.2.6 SOLID-STATE REACTION

In a solid-state reaction, calcium and phosphate sources are combined, and then, the mixture is heated at a high temperature, so that chemical reactions can occur. This process is frequently used to produce HAp NPs from calcium and phosphate minerals that are found in nature.

6.2.7 ELECTRODEPOSITION

HAp NPs are deposited onto an electrode using electrodeposition, which includes electrolyzing a solution containing calcium and phosphate ions. It is a technique that is frequently used to evenly coat substrates with HAp NPs for a variety of purposes.

6.2.8 SPRAY PYROLYSIS

This technique involves spraying tiny droplets of a solution containing calcium and phosphate precursors into a hot furnace. As the droplets break down, HAp NPs are generated and accumulated into a powder.

The preferred HAp NPs features, such as particle size, crystal structure, purity, and morphology, as well as the application, determine the synthesis process to be used. Numerous environmental and sustainability issues may arise from using traditional ways of manufacturing HAp NPs. These difficulties are mostly caused by the usage of chemicals, energy, waste production, resource utilization, toxic byproducts, transportation, and distribution. Although conventional methods have been widely employed, research into more sustainable and environmentally friendly approaches, including green synthesis, which seeks to lessen the impact of HAp NPs production on the environment, is still ongoing [6] (Figure 6.1).

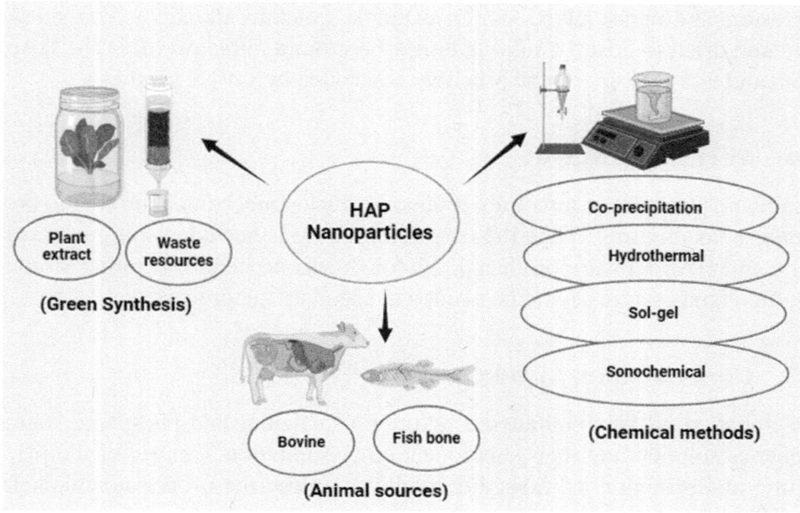

FIGURE 6.1 Different methods for synthesis of hydroxyapatite nanomaterial.

TABLE 6.1

Advantages and Disadvantages of Synthesis Methods for Hydroxyapatite Nanomaterial

Methods	Advantages	Disadvantages
Precipitation method	• It allows precise control over the composition of the synthesized HAp. • Used to synthesize various forms of HAp, including nanoparticles, microspheres, and coatings. • It is cost-effective compared to other synthesis methods because it involves simple and inexpensive reagents.	• Achieving uniform particle size distribution leading to variations in properties such as surface area and reactivity. • Its reaction is sensitive to processing conditions like pH, temperature, and reaction time, requiring careful optimization to obtain desired properties. • Some precipitation reagents used in the synthesis process may be environmentally hazardous.
Hydrothermal method	• It allows for precise control over the morphology and particle size of HAp crystals. • It promotes the growth of well crystallized HAp crystals. • It promotes the formation of highly pure HAp with minimal impurities which leading to improved material properties.	• Maintaining high pressure and temperature reactors can consume high energy. • It requires prolonged reaction times ranging from hours to days. • High energy consumption and specialized equipment associated with this method may contribute to environmental impact.
Sol-gel method	• Occurs at low temperature, reducing energy consumption. • It is adapted to incorporate modifiers or organic molecules into the HAp structure, imparting desired functionalities such as drug delivery capabilities.	• It involves multiple synthesis steps, including precursor preparation, gel formation, drying and calcination, which can increase the complexity. • Selection of appropriate precursors and controlling the reaction stability and reactivity can be challenging, impacting on yield and quality. • It requires specialized equipment, such a rotary evaporators, drying ovens and furnaces, for precise control over size.
Mechanical mixing	• It can be used for large-scale production of HAp materials to meet industrial demands. • Proper mixing techniques can promote Uniform distribution of reactants and enhance homogeneity.	• It may offer limited control over reaction parameters such as temperature and pressure which can influence the kinetics and outcome of the synthesis process. • It produces HAp materials with lower crystallinity compared to other synthesis methods, which can affect mechanical strength

(Continued)

TABLE 6.1 (*Continued*)
Advantages and Disadvantages of Synthesis Methods for Hydroxyapatite Nanomaterial

Methods	Advantages	Disadvantages
Coprecipitation method	• It yields a high quantity of HAp NPs or powders in a single synthesis batch. • Used for precise control over the stoichiometry of the synthesized NPs. • It produces HAp with high crystallinity and purity.	• Its reactions are sensitive to changes in reaction conditions to achieve desired material properties and minimize impurities. • It may result in the formation of by-products. • Uncontrolled agglomeration of HAp NPs may occur during coprecipitation.
Solid-state reaction method	• Its reaction allows for precise control over the stoichiometry of the synthesized HAp. • This method is relatively simple and versatile approach for HAp synthesis. • Its reactions typically exhibit high temperature stability, making it suitable for applications requiring heat treatment.	• Its reactions may yield HAp NPs with a wide range of particle sizes and morphologies, depending on the synthesis conditions. • Achieving high quality HAp via solid state reaction may require specialized equipment such as high temperature furnaces, ball mills.
Electrodeposition method	• It allows HAp deposition with excellent adhesion and conformity to the substrate surface. • It can be selectively applied to specific areas of a substrate, allowing for localized coating of HAp onto targeted regions.	• It typically operates at relatively slow deposition rates, leading to extended coating times and limited throughput compared to other synthesis methods. • The composition of the electrolyte solution used in electrodeposition may influence the properties of the synthesize coating. • Electrodeposition may exhibit limited electrochemical stability under certain conditions, particularly in corrosive or acidic environments.
Spray pyrolysis method	• It allows the deposition of uniform HAp coating onto the substrates with complex shapes and surfaces, ensuring consistent material properties. • It can be used for large scale production of HAp coatings, making it suitable for industrial application.	• The suitability of substrates for spray pyrolysis deposition may be limited by their thermal stability and surface roughness. • For preparation of precursor solutions for spray pyrolysis deposition may involve additional steps such as solubilization, filtration and degassing to ensure uniform coating quality. • It may require a controlled atmosphere environment to prevent oxidation or unwanted reactions during deposition, adding complexity and cost to the synthesis setup.

(Continued)

TABLE 6.1 (*Continued*)

Advantages and Disadvantages of Synthesis Methods for Hydroxyapatite Nanomaterial

Methods	Advantages	Disadvantages
Biological synthesis method	• It may utilize biological organisms or biomolecules that inherently possess functional groups. • It mimics natural biomineralization processes found in living organisms, resulting in HAp materials with hierarchical structures and properties that closely resembles those found in biological tissues like bone and teeth.	• These methods may involve complex biological processes and interactions between organisms and precursor materials. • It may require specialized equipment and expertise in biotechnology, biochemistry, increase the complexity and cost of the synthesis process compared to conventional methods. • It may offer limited control over the properties.
Ionic liquids method	• Ionic liquids are often considered greener alternatives to traditional organic solvents, as they are non-volatile, non-flammable, and recyclable. • Use of ionic liquids in hydroxyapatite synthesis allows for the incorporation of various additives, dopants, or modifiers to impart desired functionalities. • It can dissolve a wide range of precursor materials and facilitate homogeneous mixing and reaction.	• Ionic liquids can exhibit high viscosity, which may hinder mass transfer and reaction kinetics. • Some ionic liquids may exhibit toxicity or biocompatibility issues, particularly if residual amounts are present in the synthesized hydroxyapatite. • The extraction and purification of hydroxyapatite synthesized in ionic liquids may pose challenges due to the unique properties of these solvents.
Natural precursors method	• Hydroxyapatite synthesized from it often exhibits excellent biocompatibility, making it suitable for various biomedical. • The use of natural precursors such as bone, eggshell, or marine organisms for HAp synthesis reduces reliance on synthetic chemicals and minimizes environmental impact.	• It may pose biological risks such as allergenicity, immunogenicity or transmission of infectious agents, particularly if proper sterilization and purification procedures are not followed. • It may be subject to regulatory requirements and safety standards governing the use of biological materials in medical devices and pharmaceuticals.
Microwave assisted synthesis method	• It provides rapid and uniform heating throughout the reaction mixture, promoting homogeneous nucleation and crystallization of HAp. • It occurs at significantly faster reaction rates compared to conventional heating methods. • It typically requires fewer chemicals and solvents compared to conventional methods, reducing waste generation and environmental impact.	• Uneven distribution of microwaves or localized hot spots within the reaction mixture may occur. • Optimizing reaction parameters for microwave assisted synthesis may require extensive experimentation and parameter tuning. • It requires careful handling of microwave radiation to prevent exposure and ensure operator safety, necessitating appropriate training.

(Continued)

TABLE 6.1 (*Continued*)

Advantages and Disadvantages of Synthesis Methods for Hydroxyapatite Nanomaterial

Methods	Advantages	Disadvantages
Ultrasound assisted synthesis method	• Ultrasound can influence the morphology and shape of HAp crystals by controlling the intensity and frequency of ultrasound waves. • Ultrasound waves can facilitate mass transfer and accelerate reaction kinetics by introducing cavitation, microstreaming, and acoustic streaming.	• It induced cavitation can generate gas bubbles in the reaction mixture. • Prolonged exposure to high intensity ultrasound waves may lead to degradation or modification of HAp. • Ultrasound equipment for synthesis can be complex and expensive, requiring specialized ultrasonic reactors, transducers, and control systems.
Green precursors method	• Renewable resources such as plants, algae, agricultural waste, making them readily available and cost effective. • It may contain naturally occurring organic compounds, minerals, or trace elements that contribution chemical complexity.	• It may pose biological risks such as allergenicity, immunogenicity, or transmission of infectious agents. • It may contain contaminants such as heavy metals, pesticides, or microbial pathogens, which can affect the purity and safety.
Supercritical fluids method	• It can serve as environment friendly and solvent free reaction media for HAp synthesis. • The absence of solvents and the inert nature of supercritical fluids minimize the risk of contamination and impurity incorporation during HAp synthesis.	• Not all precursor materials are readily soluble in supercritical fluids, limiting the range of materials and synthesis strategies that can be employed using this method. • Maintaining supercritical conditions typically requires significant energy input, particularly for heating and pressurization.
Waste materials method	• Utilizing waste materials for synthesis offers a sustainable environmentally friendly approach. • Waste materials can be reused and recycled for HAp synthesis, contributing to the efficient utilization of resources. • It allows for the synthesis of HAp with tailored properties such as particle size, surface chemistry, and drug loading capacity, suited to specific applications.	• The use of waste materials for hydroxyapatite synthesis may pose biological risks such as allergenicity, immunogenicity. • Extracting HAp from waste materials may involve complex processing steps such as extraction, purification, and drying. • Waste materials may exhibit variability in composition and quality depending on factors such as source, processing methods, and environmental conditions.

(Continued)

TABLE 6.1 (*Continued*)

Advantages and Disadvantages of Synthesis Methods for Hydroxyapatite Nanomaterial

Methods	Advantages	Disadvantages
Green surfactants method	• Green surfactants are often biodegradable, allowing for their easy removal from synthesized HAp NPs reducing the risk of residual surfactant contamination in the final product.	• The availability of green surfactants suitable for hydroxyapatite synthesis may be limited compared to conventional surfactants. • Optimizing synthesis parameters for green surfactant-assisted hydroxyapatite synthesis may require extensive experimentation.
In situ precipitation method	• In situ precipitation typically yields high product yields with minimal waste generation. • The entire reaction takes place in a single vessel without the need for additional purification or separation steps.	• In situ precipitation may result in hydroxyapatite products with impurities or by-products, particularly if proper purification. • Washing steps are not employed, which can affect material properties and performance.
Biogenic synthesis methods	• Biogenic synthesis methods can utilize a wide range of biological organisms, including bacteria, fungi, algae, and plants. • Biogenic synthesis methods utilize biological organisms or biomolecules as catalysts or templates for hydroxyapatite formation, reducing the need for harsh chemicals.	• The use of biological organisms or biomolecules for hydroxyapatite synthesis may pose biological risks such as allergenicity or immunogenicity. • Biogenic synthesis methods can be complex and require specialized knowledge in microbiology, biochemistry, or biotechnology, as well as careful optimization of growth conditions.

6.3 APPLICATIONS OF HAP NPs SYNTHESIZED BY CONVENTIONAL METHOD

Due to its special qualities and biocompatibility, HAp NPs have a wide range of uses. The following are some significant uses for HAp NPs:

6.3.1 BIOMEDICAL IMPLANTS

To boost biocompatibility and encourage bone tissue formation, HAp NPs are frequently employed as coatings on orthopedic and dental implants. This improves the integration of the implant with the surrounding tissue [7].

6.3.2 DRUG DELIVERY SYSTEMS

HAp NPs can act as carriers for drugs in medication delivery systems. They are useful in controlled release systems because of their high surface area and capacity to encapsulate a variety of medicines, allowing for the prolonged release of pharmaceuticals [8].

6.3.3 TISSUE ENGINEERING

HAp NPs are utilized in tissue engineering to build scaffolds for cell growth and tissue regeneration. They offer an environment that is bioactive and encourages cell adhesion, growth, and differentiation [9].

6.3.4 APPLICATIONS IN DENTISTRY

HAp NPs are added to dental cements and composites to enhance their mechanical and biocompatibility features. They can also help to remineralize tooth enamel, preventing tooth decay [10].

6.3.5 CANCER TREATMENT

By modifying HAp NPs, anticancer medications or imaging agents can be embedded in them. To increase therapeutic effectiveness and lessen negative effects, they are employed in targeted drug delivery [11].

6.3.6 WASTEWATER TREATMENT

The potential for using HAp NPs to remove heavy metal ions from tainted water has been explored. They function well as adsorbents for contaminants because of their large surface area and ion-exchange characteristics [12].

6.3.7 IMAGING AND DIAGNOSIS

Surface-modified HAp NPs can be used in medical imaging procedures including computed tomography (CT) and magnetic resonance imaging (MRI). For more precise diagnosis, they might be designed to target certain tissues or cells [11].

6.3.8 Cosmetics and Sunscreen

Due to their capacity to scatter and absorb damaging UV rays, HAp NPs are utilized in cosmetics and sunscreens. They offer defense against skin aging caused by UV radiation [13].

6.3.9 Bone Cement

In orthopedic surgery, bone cement is used, and HAp NPs are used to improve their mechanical and bioactive qualities. These cements aid in holding implants and artificial joints in place [14].

6.3.10 Antimicrobial Coatings

To inhibit microbial adhesion and lower the risk of infection, HAp NPs can be added to coatings for medical devices including catheters and surgical tools [15].

6.3.11 Food Additives

HAp NPs are a calcium source that can be utilized in the food business to increase the bioavailability of calcium in items like dairy and calcium-fortified foods [16].

6.3.12 Environmental Remediation

HAp NPs have been investigated for their potential in adsorbing and removing different environmental pollutants, such as heavy metals and organic pollutants, from soil and water. The versatility of HAp NPs in various fields makes them a valuable material with diverse applications, particularly in biomedicine and environmental science. Researchers continue to explore new uses and modifications of HAp NPs to further expand their potential applications [17].

Conventional methods for synthesizing HAp NPs present numerous challenges, including environmental impact, resource depletion, energy intensity, health and safety risks, biocompatibility concerns, chemical by-products, waste generation, economic considerations, and long-term sustainability. Consequently, researchers are increasingly turning to various green synthesis approaches. In this chapter, we aim to offer an overview of these green synthesis methods, exploring their diverse applications across multiple domains, and discussing the challenges associated with their synthesis.

6.4 GREEN SYNTHESIS APPROACHES OF HAP NPS SYNTHESIS

By using green and sustainable processes, green synthesis options for HAp NPs aim to lower the environmental impact of HAp production. These techniques frequently put an emphasis on lowering waste, conserving energy, and using fewer dangerous chemicals. HAp NPs synthesis techniques that are green include the following:

6.4.1   NATURAL PRECURSORS

In this process eggshells, bones, or seashells as starting materials that are naturally high in calcium and phosphorus. There is a low demand for chemical extraction and processing because these sources are plentiful and renewable.

6.4.2   BIOLOGICAL SYNTHESIS

This approach involves promoting HAp NPs production using microbes or other biological components. The crystallization of HAp NPs can be facilitated by microbes or natural extracts acting as templates. Eco-friendly biological synthesis can take place in aqueous solutions and at moderate temperatures.

6.4.3   IONIC LIQUIDS

Traditional organic solvents can be replaced with ionic liquids since they are non-volatile and non-flammable. They can be applied in precipitation or sol-gel processes and are thought to be more environmentally friendly.

6.4.4   MICROWAVE-ASSISTED SYNTHESIS

The synthesis of HAp NPs can be sped up using a low-energy technique called microwave heating. It becomes more environmentally friendly since it uses less energy and takes less time to react.

6.4.5   ULTRASOUND-ASSISTED SYNTHESIS

The nucleation and development of HAp NPs crystals can be aided by ultrasonic vibrations. Energy is conserved by shortening the reaction time and enabling synthesis at lower temperatures.

6.4.6   GREEN PRECURSORS

Using ecologically friendly or biodegradable calcium and phosphate sources as precursors. Calcium nitrate, for instance, can be substituted by calcium lactate or calcium citrate. Similarly, sources of phosphate that are fit for human consumption or plant extracts can be used.

6.4.7   SUPERCRITICAL FLUIDS

HAp NPs with controlled size and morphology can be produced via supercritical CO_2, which is often used as a green solvent. Low pressures and temperatures are used during the process.

6.4.8 Waste Materials

A sustainable calcium source for HAp NPs production can be obtained by recycling waste items such as broken fish bones or eggshells.

6.4.9 Green Surfactants

Stabilizing HAp NPs with eco-friendly surfactants, such as plant-based or biodegradable substitutes.

6.4.10 *In Situ* Precipitation

Synthesizing approaches allow HAp NPs to precipitate in the same location where they will be consumed, so lowering transportation costs and limiting environmental effects.

Green synthesis techniques not only limit the negative effects of HAp NPs synthesis on the environment, but they also frequently provide products that are more biocompatible and suited for a variety of biomedical and environmental applications. These techniques are crucial for addressing environmental and health issues brought on by conventional synthesis approaches and are in line with the concepts of sustainable chemistry [18]. Biogenic synthesis, a subset of green synthesis approaches, distinctly relies on biological entities to facilitate the formation of nanoparticles. In contrast, green synthesis is a more comprehensive term, encompassing a wider array of environmentally friendly methods. This includes approaches utilizing biological components alongside other eco-friendly chemical methods. In this context, our discussion has delved into various biogenic synthesis approaches for HAp NPs, exploring their advantages and presenting diverse examples of biogenically synthesized HAp NPs.

6.5 BIOGENIC SYNTHESIS OF HAP NPs

Natural resources like eggshells and animal bones can serve as worthwhile and long-lasting starting materials to produce HAp NPs. These elements include calcium and phosphorus, two ingredients that are crucial to the formation of HAp. The usage of these natural resources in the synthesis of HAp NPs is discussed below:

6.5.1 Natural Resources of HAp

6.5.1.1 Eggshells

Calcium carbonate ($CaCO_3$) is the main component of eggshells. As HAp NPs are made up of calcium, phosphate, and hydroxide ions, the calcium in eggshells is a crucial precursor for HAp synthesis. Cleaning, drying, and converting eggshells from calcium carbonate to calcium phosphate are required steps in the processing of eggshells before use in HAp NPs production. Acid-base reactions can be used to convert substances. For example, the calcium carbonate in eggshells can be dissolved and changed into calcium chloride by treating them with acid (hydrochloric acid). HAp NPs can be produced through a subsequent reaction with a supply of phosphate.

Eggshells are an easily available and long-lasting source of calcium. They can be used for HAp NPs synthesis and are often considered as waste materials, which reduces the process' impact on the environment.

6.5.1.2 Animal Bones

Animal bones, including those from fish, chicken, and mammals, are rich in calcium and phosphate minerals, which serve as a natural source of calcium and phosphorus for HAp NPs synthesis. Bones must go through a demineralization process to extract the mineral content. Typically, this involves using an acid, like hydrochloric acid, to dissolve the bone minerals and transform them into soluble calcium and phosphate ions. Then, using precipitation or other synthesis techniques, these ions can be utilized to create HAp. Using animal bones as a source of calcium and phosphorus for HAp synthesis helps to improve sustainability because it repurposes waste materials and minimizes the requirement for precursors that must be mined or chemically processed.

6.5.2 BENEFITS OF BIOGENIC SYNTHESIS OF HAp NPs

6.5.2.1 Sustainability

By lowering the demand for non-renewable resources and reducing waste, the utilization of natural sources like eggshells and animal bones supports sustainability. Additionally, it adheres to the fundamentals of the ecological economy.

6.5.2.2 Biocompatibility

Because of their natural origin, HAp NPs produced from natural sources are frequently more biocompatible, making them appropriate for a range of biomedical uses, including bone grafts and dental materials.

6.5.2.3 Eco-friendly

Making use of these natural resources eliminates the need for raw material transportation, chemical processing, and energy-intensive mining, all of which have an adverse effect on the environment.

6.5.2.4 Economic Benefits

Eggshells and animal bones are often inexpensive and easily accessible materials, which makes them profitable for HAp NPs production.

6.5.2.5 Quality Control

As natural sources can introduce impurities, quality control and purification procedures may be necessary to guarantee that the finalized HAp product satisfies the criteria necessary for applications [19].

6.5.3 EXAMPLES OF BIOGENIC SYNTHESIS OF HAp NPs

To promote the development of HAp NPs, biological agents or natural extracts are frequently used in the biogenic manufacturing methods of HAp NPs. These sustainable and eco-friendly techniques have proven effective in a variety of applications.

Here are a few case studies and demonstrations of successful HAp NP biogenic production processes:

6.5.3.1 Synthesis Mediated by Microorganisms

Researchers produced HAp NPs by using bacteria like *Bacillus subtilis*. These microbes generate extracellular polymeric materials that act as templates for the nucleation and development of HAp NPs. The well-defined HAp NPs produced by this biogenic technique have potential uses in medical and materials science [20].

6.5.3.2 Green Tea Extracts

In a study, HAp NPs were made using green tea extracts. Green tea polyphenols stabilize the HAp NPs and give them a biocompatible covering. The use of these HAp NPs in medication delivery systems has shown potential [21].

6.5.3.3 Seashell Extracts

Seashells from marine organisms can be used to extract calcium ions, such extracts were combined with phosphate sources by researchers to create HAp NPs. The procedure is excellent for biomedical applications and is environmental friendly [22].

6.5.3.4 Biopolymer Templates

Silk Fibroin Protein Template: The production of HAp NPs has been carried out using silk fibroin, a protein derived from silkworms. For bone tissue engineering, fibroin acts as a scaffold for HAp crystallization, producing composite materials with high mechanical properties [23].

6.5.3.5 Bio-Based Surfactants

To produce HAp NPs, soybean lecithin, a natural surfactant, was used. Lecithin serves as a stabilizer and promotes the development of nanoscale HAp particles with potential uses in tissue engineering and medication delivery [24].

6.5.3.6 Extracellular Vesicles (EVs)

The usage of EVs, which are tiny, membrane-bound vesicles generated by cells, as templates for HAp production has been investigated by researchers. To produce biocompatible nanomaterials appropriate for medication delivery and regenerative therapy, EVs can control the formation of HAp NPs [25].

6.5.3.7 Coral Skeletons and Marine Organisms

Calcium carbonate, the major component of coral skeletons, has been utilized to make HAp NPs and other calcium phosphate compounds. Coral's naturally porous nature serves as an excellent model for HAp development. These biomimetic materials can be used in bone tissue engineering [26].

6.5.3.8 Growing Factors and Biological Fluids

To promote the growth of HAp NPs, researchers have used saliva as a source of calcium ions. Saliva has proteins and calcium ions that can aid in the nucleation and crystallization of HAp [27].

These case studies and examples highlight the potential of HAp NP biogenic production techniques. They provide sustainable and biocompatible methods that make use of the special qualities of living things, natural extracts, and eco-friendly materials. These techniques can be used in a variety of industries, including biomedicine, medication delivery, tissue engineering, and remediation of the environment.

6.6　SIGNIFICANCE OF SELECTING ECO-FRIENDLY SOLVENTS AND REAGENTS FOR HAP NPS SYNTHESIS

When it comes to HAp NPs synthesis, choosing environmentally friendly reagents and solvents is crucial for minimizing negative effects on the environment, protecting worker health and safety, advancing sustainability, resource efficiency, reducing waste generation, and adhering to regulations. Adhering to the fundamentals of green chemistry, this is a significant move towards establishing a more ecologically conscious and sustainable HAp and HAp-derived product manufacturing pipeline.

6.7　APPLICATIONS OF GREEN SYNTHESIZED HAP NPs

The biocompatibility, low environmental impact, and sustainability of green-synthesized HAp make it a promising material in several applications. The following is a summary of possible uses in the areas of dentistry, environmental remediation, healthcare, and other fields:

6.7.1　Health Care

Eco-friendly HAp NPs are used in a number of healthcare applications because of their mineral composition. They are used in the following applications:

6.7.1.1　Bone Engineering

Due to its structural resemblance to natural bone minerals, green-synthesized HAp NPs can be employed in orthopedic implants and bone transplants. HAp NPs make up a large portion of natural bone. They replicate the mineral makeup of the bone matrix. They promote integration and bone repair. Because green-synthesized HAp preserves HAp's natural biocompatibility, it can be integrated with living bone tissue. By encouraging bone cell adhesion and proliferation, HAp NPs osteoconductive qualities aid in the process of regeneration. Since HAp NPs scaffolds are porous, and they offer an ideal setting for tissue ingrowth, cell adhesion, and proliferation. During green synthesis of HAp NPs, bioactive compounds, growth factors, or pharmaceuticals can be incorporated into the HAp structure for regulated discharge of bioactive substances. Which accelerated the healing process and increased bone regeneration by the controlled release of these bioactive substances from the HAp scaffold. By adjusting the porosity and design of green-synthesized HAp scaffolds, the mechanical and biological performance can be maximized. Optimizing the properties of the scaffold guarantees the best possible cell infiltration and vascularization for effective bone repair. HAp NPs can reconstruct bone; the scaffold will eventually be replaced by newly created bone, preserving the natural structure of the bone. It is possible to alter the surface of HAp scaffolds produced through green synthesis to boost osteo-genic differentiation, increase cell adhesion, or add antibacterial qualities. Surface alterations improve the scaffold's overall ability to regenerate bone. Orthopedic and regenerative medicine could advance significantly with the use of green-synthesized HAp in scaffolds and implants for bone regeneration. This is a sustainable and bio-compatible strategy. It addresses the need for efficient, patient-friendly solutions in bone tissue engineering as well as the environmental issues related to conventional synthesis methods [28].

6.7.1.2 Drug Administration

In cancer therapy and other medical treatments, HAp NPs can act as drug carriers for the regulated and targeted administration of drugs. Green synthesized HAp NP is a potential material for drug delivery applications because of its controlled release features. The efficacy of HAp NP in drug delivery systems is attributed to its distinct features, such as its biocompatibility and capacity to encapsulate and release medici-nal compounds gradually. HAp NPs are biodegradable; the medicine is released from its capsule over time while preventing long-term buildup. Drug encapsulation is made possible by HAp NP's porous structure, and its controlled release characteristics per-mit gradual, sustained drug release. Green synthesis techniques can further enhance the release kinetics, offering a regulated and extended therapeutic benefit. HAp NPs are capable of ion exchange, which permits the regulated release of medications or ions. This characteristic permits the release of therapeutic medicines in reaction to interactions with ions or physiological fluids within the body. For encapsulated phar-maceuticals, HAp offers a protective environment that inhibits deterioration and pre-serves the stability of therapeutic substances during storage and release. This green synthesis strategy may aid in the creation of cutting-edge drug delivery systems with lower environmental impact and more therapeutic efficacy [29].

6.7.1.3 Therapeutic Agents and Imaging

HAp NPs are versatile materials with unique features that find extensive use in the fields of imaging and therapeutic agents. They may encapsulate a range of thera-peutic substances; they are useful drug delivery carriers. The administration of anticancer medications has been investigated using HAp. Because of HAp's con-trolled release characteristics, therapeutic medication levels are maintained, mini-mizing adverse effects, and enhancing therapy effectiveness. HAp NPs also used in the transport of molecules. Proteins and peptides can be delivered with HAp NPs, which helps to facilitate their controlled release and shields them against destruction. This holds particular significance in fields like tissue engineering and regenerative medicine. Genetic material (DNA and RNA) can be delivered via HAp NPs. Gene delivery for uses in genetic engineering or gene therapy can be facilitated by interac-tions between the negatively charged DNA or RNA and the positively charged HAp surface. Therapeutic coatings of HAp NPs are used to enhance bone integration of orthopedic implants. This is especially important while having joint replacement surgery. HAp NPs can function as contrast agents in a variety of imaging methods, such as X-ray imaging. Since HAp has a high density, it improves contrast and can be used to visualize soft tissues like bones. Surface-modified HAp can be used to improve contrast in magnetic resonance imaging (MRI). It is possible to customize functionalized HAp NPs for certain imaging applications. The potential of HAp NPs in photothermal therapy for cancer treatment has been investigated. HAp could produce heat in response to near-infrared light, which can cause localized hyperthermia and the death of cancer cells. It can be applied to ultrasound imaging to improve tissue characterization and visibility [30].

6.7.1.4 Dental Applications

HAp NPs produced by green synthesis have the following dental applications:

Toothpaste and Products for Oral Care:
Remineralization: To aid in the remineralization of tooth enamel, toothpaste and other oral hygiene products can contain green-synthesized HAp NPs. In addition to strengthening teeth, this may lessen tooth sensitivity.

Composites and Fillings in Dentistry: Dental fillings and composites can be made with green-synthesized HAp as a component. With better biocompatibility and the possibility for increased remineralization at the restoration site, the nanoparticles aid in the development of restorative materials.

Implants and Bone Grafts: Bone grafts and implants for oral and maxillofacial surgery can be made using green-synthesized HAp NPs. These materials are appropriate for procedures Dentin like dental implants and periodontal surgery because they encourage bone regeneration and integration with native bone [31].

Medicine Administration for Periodontal Therapy: While treating periodontal disease, green-synthesized HAp NPs may be used as carriers for localized drug delivery. Treatments for periodontal disorders can be more effective when the therapeutic chemicals from HAp NPs are released under regulated conditions and are directed toward the area of concern.

Antimicrobial Coatings: The antibacterial qualities of green-synthesized HAp NPs can be added to coatings for orthodontic products or dental implants. During orthodontic procedures, this aids in infection prevention and the promotion of improved dental health [32].

Dental Imaging: HAp NPs produced in green may be investigated for use as contrast agents in dental imaging methods. In imaging examinations like cone-beam computed tomography (CBCT) or other diagnostic procedures, their ability to improve contrast can be useful [33].

Denture Components: To increase biocompatibility and general denture performance, denture base materials can be combined with green-synthesized HAp NPs. By adding green-synthesized HAp NPs, root canal sealers can become more biocompatible and could encourage remineralization in the root canal area [34].

Preventing Hypersensitivity to Dentin: To treat dentin hypersensitivity, green-synthesized HAp NPs may be utilized as desensitizing agents. Tooth sensitivity may be lessened by their capacity to remineralize dentin. The growing emphasis on sustainability and environmentally friendly dental procedures is consistent with the utilization of green-synthesized HAp NPs in dental applications. These nanoparticles can improve oral health outcomes in a variety of dental operations and treatments, as well as biocompatibility and controlled release characteristics [35].

6.7.2 Environmental Remediation

HAp NPs have special characteristics, a high adsorption capacity, and can interact with pollutants. They have shown potential in a range of environmental remediation applications. HAp NPs are used in the following important remediation applications:

6.7.2.1 Heavy Metal Removal

Heavy metal ions from aqueous solutions can be efficiently absorbed by HAp NPs. The HAp surface chemistry promotes metal ion binding, which helps to eliminate pollutants from water such as lead, cadmium, and mercury [36].

6.7.2.2 Fluoride Elimination

HAp has a significant affinity for fluoride ions. In areas where fluoride levels are high, green-synthesized HAp NPs can be utilized to remove excess fluoride from drinking water, thus addressing the problem of fluorosis [37].

6.7.2.3 Phosphate Removal

By absorbing phosphate ions, HAp NPs can treat eutrophication in water bodies. Overabundance of phosphates can harm aquatic environments by causing algal blooms.

6.7.2.4 Water Waste Management

Dye and medication are examples of organic contaminants that HAp NPs can adsorb from wastewater. They are appropriate for treating industrial effluents due to their enormous surface area and adsorption capability.

6.7.2.5 Remediation of Soil

To immobilize heavy metal pollutants in soil, HAp NPs are applied. By interacting with metal ions, the nanoparticles lessen the ion's mobility and bioavailability in the soil [38].

6.7.2.6 Treatment of Radioactive Waste

The ability of HAp NPs to adsorb radionuclides from contaminated areas, such as uranium and thorium, has been studied. For the remediation of radioactive waste, this use is essential.

6.7.2.7 pH Modification for Water Treatment

During the water treatment process, HAp NPs can serve as pH buffers. The pH of water is stabilized and adjusted by the regulated release of calcium and phosphate ions from HAp, which aids in other remediation procedures.

6.7.2.8 Contaminated Site Nano Remediation

To treat contaminated locations *in situ*, HAp NPs can be employed in nano remediation techniques. To immobilize or sequester contaminants, the nanoparticles can be injected or sprayed directly to the contaminated area.

6.7.2.9 Better Soil Conditions

By offering a slow-release source of calcium and phosphate ions, HAp NPs added to the soil can improve nutrient availability. This helps to improve the quality of the soil for sustainable farming.

6.7.2.10 Water Filtration Systems

To eliminate a range of impurities, such as phosphates, heavy metals, and organic pollutants, HAp NPs can be added to water filtration systems. Decentralized water treatment systems may benefit greatly from this technology. The application of HAp NPs in environmental remediation shows how adaptable and successful they are in solving a variety of problems related to soil and water contamination. Green synthesis techniques are useful instruments in achieving the goal of healthier and cleaner ecosystems because they further improve their sustainability and environmental friendliness [39].

6.7.3 Cosmetics and Skincare

The cosmetics and personal care sector are accepting HAp NPs because of their special qualities and advantages for skin and hair. The following are some important uses for HAp NPs in personal care and cosmetics:

6.7.3.1 Skincare Items

Anti-Aging Formulations: Because HAp NPs improve skin suppleness and lessen the appearance of fine lines and wrinkles, they are employed in anti-aging skincare products. They aid in the skin's general smoothness and firmness.

Moisturizers: To give the skin moisture, HAp NPs can be added to moisturizers. Because of their small particle size, they can penetrate the skin more deeply and provide superior moisturization.

Products for Sunscreen: Because HAp NPs can both scatter and absorb UV light, they can be included in sunscreen compositions. This enhances the shielding action against the damaging effects of sun exposure.

6.7.3.2 Hair Care Items

Hair Conditioners: To strengthen and thicken hair strands, add HAp NPs to hair conditioners. They might help lessen breakage and damage to hair.

Hair Styling Products: HAp NPs in hair styling products can improve the manageability and texture of hair. They might also give hair strands a protective layer.

Nanocosmetry: Active compounds in cosmetics can be delivered by HAp NPs acting as carriers. With the use of nanocosmetics, chemicals can more effectively permeate the skin and have greater effects.

Scar Reduction and Wound Healing: The potential of HAp NPs in wound healing and scar reduction has been investigated. To encourage tissue regeneration and lessen scarring, they can be used in topical preparations [40].

6.7.4 CATALYSIS AND OPTOELECTRONIC DEVICES

HAp NPs have intriguing characteristics that make them useful for optoelectronic devices and catalytic applications. HAp NP's adaptability and potential for a wide range of applications like environmental remediation, biocatalysts, UV photodetector, biosensors, and light emitting device are demonstrated by their incorporation in optoelectronic devices and catalysis. Additional applications for HAp NPs are probably going to be found through continued study and innovation in these domains, which will help to develop optoelectronic materials, environmental technologies, and sensing devices [41].

6.8 CHALLENGES FACED IN GREEN SYNTHESIS OF HAP NPS

Although green-synthesized HAp NPs have many benefits, there are some issues and restrictions that must be resolved before they can be successfully used in a range of applications. The following are some major obstacles and possible solutions:

6.8.1 CONTROL OF PARTICLE SIZE AND MORPHOLOGY

It can be difficult to precisely control the morphology and particle size of green-synthesized HAp NPs. To improve control over particle size and morphology, synthesis factors like reaction temperature, pH, and precursor concentrations can be optimized.

6.8.2 BIOCOMPATIBLE MATERIALS AND CYTOTOXICITY

The synthesis method and precursor materials used in the production of green-synthesized HAp NPs can affect their biocompatibility. Furthermore, synthesis residues or impurities may have an impact on cytotoxicity. Extensive biocompatibility testing is necessary, encompassing both *in-vitro* and *in-vivo* investigations. Using biocompatible precursors and implementing thorough purification procedures will help reduce cytotoxic effects.

6.8.3 AGGREGATION AND STABILITY

The performance of green-synthesized HAp NPs in applications may be impacted by their low stability and propensity to agglomerate. The stability of HAp NPs can be increased by surface modification and functionalization methods. Aggregation can potentially be avoided by using stabilizing agents during synthesis and maintaining appropriate storage conditions.

6.8.4 COMBINING INTO COMPLICATED SYSTEMS

Compatibility and performance issues may arise when integrating green-synthesized HAp NPs into complex structures like 3D-printed scaffolds or drug delivery vehicles. Interdisciplinary teams working together can assist in addressing integration issues. It is imperative to conduct comprehensive testing and optimization within the application context.

6.8.5   INSUFFICIENT STANDARDIZATION

Differences in the characteristics of HAp NPs may arise from the lack of standard operating procedures for green synthesis. The reproducibility and dependability of HAp NPs produced by green synthesis can be enhanced by establishing standardized methods, quality control measures, and certification processes.

6.8.6   COST-RELATED FACTORS

The economic viability of large-scale production may be impacted by some green synthesis techniques being more expensive than conventional techniques, to mitigate costs, investigate affordable green precursors, maximize synthesis parameters for effectiveness, and consider environmentally friendly and sustainable activities.

6.9   CONCLUSION

In conclusion, the eco-friendly synthesis of HAp NPs emerges as a viable and sustainable method with substantial implications for biomedicine. The adoption of environmentally friendly processes not only enhances the overall safety profile and biocompatibility of the synthesized nanoparticles but also addresses environmental concerns. The achievement in producing HAp NPs through environmentally friendly approaches underscores the potential to reduce the environmental footprint associated with conventional synthetic techniques, often characterized by the use of harsh chemicals and energy-intensive procedures. By harnessing the inherent properties of natural materials, researchers have demonstrated the feasibility of producing HAp NPs with controlled size, morphology, and surface characteristics. Moreover, the biomedical applications of these green-synthesized HAp NPs are diverse and promising. Exhibiting noteworthy biocompatibility and bioactivity, these nanoparticles prove highly suitable for diverse medical applications, spanning from drug delivery systems to the formation of bone tissue. The sustainability inherent in the synthesis methods aligns seamlessly with the growing emphasis within the scientific community on adopting eco-friendly procedures. Anticipated advancements in green HAp NP manufacturing are poised to introduce innovative techniques aimed at enhancing the material's properties and broadening its scope in medical applications. Collaboration among researchers, biochemists, and material scientists will be crucial to advance our comprehension of these sustainable synthesis techniques and translate them into practical solutions for medical challenges.

In summary, the production of green HAp NPs introduces novel avenues for biomedical research and applications, offering a more sustainable and environmentally friendly alternative to traditional approaches. Beyond contributing to global sustainability objectives, the commitment to environmentally friendly methods in nanomaterial production facilitates the progress of safer and more effective biomedical technology.

REFERENCES

1. F. Boukhelf et al. *Sustainability (Switzerland)* (2022) **14**, 9353.
2. V. S. Kattimani, S. Kondaka & K. P. Lingamaneni. *Bone and Tissue Regeneration Insights* (2016) **7**, BTRI.S36138.
3. P. N. Chavan, M. M. Bahir, R. U. Mene, M. P. Mahabole & R. S. Khairnar. *Materials Science and Engineering: B* (2010) 168, 224–230.
4. T. Laonapakul. *Engineering and Applied Science Research.* (2015) **42**, 269–275.
5. M. Othmani, H. Bachoua, Y. Ghandour, A. Aissa & M. Debbabi. *Materials Research Bulletin* (2018) **97**, 560–566.
6. N. A. S. Mohd Pu'ad et al. *Materials Today Proceedings.* (2019) **29**, 233–239. Elsevier Ltd.
7. W. S. W. Harun, R. I. M. Asri, A. B. Sulong, S. A. C. Ghani & Z. Ghazalli. *Hydroxyapatite: Advances in Composite Nanomaterials, Biomedical Applications and Its Technological Facets* (2018). InTech. doi: 10.5772/intechopen.71063.
8. S. Mondal, S. V. Dorozhkin & U. Pal. *Wiley Interdisciplinary Reviews: Nanomedicine and Nanobiotechnology* (2018) **10**, 1504.
9. H. Zhou & J. Lee. *Acta Biomaterialia* (2011) **7**, 2769–2781. doi: 10.1016/j.actbio.2011.03.019.
10. S. Balhuc et al. *Crystals* (2021) **11**. doi: 10.3390/cryst11060674.
11. S. Kargozar et al. *Journal of Functional Biomaterials* (2022) **13**. doi: 10.3390/jfb13030100.
12. A. Nayak & B. Bhushan. *Materials Today Proceedings* (2021) **46**, 11029–11034. Elsevier Ltd.
13. S. R. Ghazali, N. H. Rosli, L. S. Hassan, M. Z. Helmi Rozaini & H. Hamzah. *Earth and Environmental Science* (2021) **646**, 012059.
14. J. Zhang, Y. Feng, X. Zhou, Y. Shi & L. Wang. *International Journal of Polymeric Materials and Polymeric Biomaterials* (2021) **70**, 37–53. doi: 10.1080/00914037.2019.1685518.
15. S. Lamkhao et al. *Scientific Reports* (2019) **9**, 4015.
16. S. Liaqat, Z. Ahmed, F. Iqbal & M. U. Umer. *Journal of Food Science* (2023) **88**, doi: 10.1111/1750-3841.16783.
17. M. Ibrahim, M. Labaki, J. M. Giraudon & J. F. Lamonier. *Journal of Hazard Materials* (2019) **383**, 121139.
18. N. A. S. Mohd Pu'ad, P. Koshy, H. Z. Abdullah, M. I. Idris & T. C. Lee. *Heliyon* (2019) **5**. doi: 10.33786/JCPR.2019.v09i03.006.
19. F. Cestari et al. *Nanomaterials* (2021) 11, 1–14.
20. Z. Ashwini, S. Kakade, J. Thorat, S. Patil & A. Bankar *Journal of Current Pharma Research* (2019) **9**, 2878–2885.
21. M. Cazzola et al. *International Journal Molecular Science* (2018) **19**, 2255.
22. K. Dhanaraj & G. Suresh. *Vacuum* (2018) 152, 222–230.
23. H. Kim, L. Che, Y. Ha & W. Ryu. *Materials Science and Engineering* (2014) C40, 324–335.
24. W. Michał, D. Ewa & C. Tomasz. *Colloid and Polymer Science* (2015) **293**, 1561–1568.
25. K. S. Katti, A. H. Ambre, S. Payne & D. R. Katti. *Materials Research Express* (2015) **2**, 045401.
26. S. kumar Balu, S. Andra, J. Jeevanandam & V. Sampath. *Journal of the Mechanical Behavior of Biomedical Materials* (2021) **119**. doi: 10.1016.2021.104523.
27. A. Sobczak-Kupiec, E. Olender, D. Malina & B. Tyliszczak. *Materials Chemistry and Physics* (2018) 206, 158–165.
28. S. Mondal, U. Pal & A. Dey. *Ceramics International* (2016) 42, 18338–18346.
29. S. Mortazavi-Derazkola, M. Salavati-Niasari, H. Khojasteh, O. Amiri & M. Ghoreishi. *Journal of Cleaner Production* (2017) 168, 39–50.

30. S. Mondal et al. *Ceramics International* (2020) 46, 29249–29260.
31. F. Meyer et al. *Nature methods* (2022) **3**, 429–440 https://doi.org/10.1038/s41592-022-01431-4.
32. P. Ming et al. *Frontiers in Bioengineering and Biotechnology* (2022) **10**, 899293.
33. A. E. Melgar et al. *International Journal of Advances in Medical Biotechnology – IJAMB* (2021) **4**, 54–60.
34. B. H. Baras et al. *Materials* (2020), **13**, 1096.
35. S. Steinert et al. *Biomimetics* (2020), 5, 65.
36. S. Pai, S. M Kini, R. Selvaraj & A. Pugazhendhi. *Journal of Water Process Engineering* (2020) **38**, 101574.
37. D. D. Ganta, B. Y. Hirpaye, S. K. Raghavanpillai & S. Y.Menber. *Journal of Chemistry* (2022) **1**, 4917604.
38. J. C. Sin et al. *Journal of Environmental Chemical Engineering* (2021) **9**, 105736.
39. M. ul. Hassan et al. *Journal of Nuclear Materials* (2021) **543**, 152566.
40. M. A. Sliem, R. A. Karas & M. A. Harith. *Journal of Photochemistry and Photobiology B* (2017) 173, 661–671.
41. Z. Xu et al. *ACS Omega* (2019) 4, 21998–22007.

7 Sustainable Green Synthesized Metal Oxide Nanoparticles
Characterization and Applications

*Ishita Kapil, Pinky Yadav, Ritik Kumar,
Sachin Vats, and Ayana Bhaduri*

7.1 INTRODUCTION

The interdisciplinary scientific field of nanotechnology is expanding swiftly with its numerous applications in almost all the domains known to mankind, be it medicine, biology, energy sciences, chemistry, physics, electronics, materials sciences, aviation, food industry, or any other area of scientific investigation [1]. Nanotechnology entails the conversion of bulk materials into nanomaterials that are 1–100 nm in size. Materials with reduced sizes have an increased *'surface-to-volume'* ratio, which gives them new fascinating optical, electrical, chemical, biological, and physical characteristics [2]. Owing nanomaterials' extraordinarily small size, they possess these distinctive material qualities that allow them to influence a wide range of applications including environmental remediation, electronics, energy storage and conversion devices, optical devices, engineering, sensor technology, textiles, medicine and healthcare, and many more [3–5].

Nanotechnology is the result of integrating basic ideas from the physical, chemical, and biological sciences. At the minuscule size of nanometres, several processes take place due to the quantum confinement effects. Nanotechnology connects quantum mechanics with classical mechanics [6].

This mesoscopic technology is used in the medical field to produce nano-assemblies, such as nanomedicine, and nanotools for diagnostic and therapeutic purposes. Nanotechnology-based drugs and diagnostic kits are currently being used to treat diseases that were previously incurable. Additionally, bulk industrial manufacturing and production have been significantly impacted by this technology. Notably, the scientific communities of both the developed and developing worlds concur that nanotechnology will represent the upcoming major advancement in scientific innovation [7–9].

DOI: 10.1201/9781003473749-9

Furthermore, nanotechnology has the potential to be used in all facets of human life [2,8]. It is worth noting that nanoscale structures are prevalent and plentiful in mother nature as well: from milk (a nanoscale colloid), haemoglobin, oxygen-carrying proteins, cells, bacteria, and viruses to photonic crystals made up of melanin in peacock feathers, spider silk (100 nm in diameter, a truly pure natural nanofiber) or special nanostructure on the surface of lotus leaves responsible for ultra-hydrophobicity which results in the self-cleaning *'lotus effect'* [10,11].

Although the wide range of possibilities for nanotechnology seems promising, it is uncertain what risks come with its unrestricted use. In the field of nanotoxicology, more research is needed to find the answers [12]. Current research has demonstrated that inorganic nanoparticles, including but not limited to gold, iron oxide, silver, silica, and titanium oxide, possess the ability to transit from the body into the brain [13]. These airborne metals or metal oxide nanomaterials could readily attach to surfaces containing hazardous chemicals and transfer those contaminants, posing several risks to both human health and the ecosystem [14,15]. Via membranes and cell walls of biological systems, they can infiltrate with extreme ease, and they can stay embedded there for as long as necessary to carry out their intended functions. They can generate a variety of toxicity consequences that are typically unpredictable, including neurotoxicity. They can also cause these effects to be delayed or prolonged [15,16]. Excess nanomaterials can bioaccumulate and set a serious menace to both humans and the ecosystem [17]. Future prosperity is ensured by combining nanotechnology and sustainability.

Recently, a variety of chemical and physical techniques including sol-gel method, hydrothermal/solvothermal method, microemulsion technique, chemical vapour synthesis, chemical reduction with organic and inorganic reducing agents, laser ablation, electro-spraying, ball milling, evaporation-condensation, arc discharge method, and laser pyrolysis are used for producing nanomaterials/nanoparticles [4,18]. However, there are possible challenges to these techniques, like the requirement for harmful solvents, the need for high pressure and temperature, or the generation of potentially dangerous byproducts that pollute the environment. This means that in order to advance nanotechnology, environmentally friendly processes for synthesizing nanomaterials must be developed as a part of environmental remediation [19].

Green nanotechnology is an effective way to lessen the harmful effects of producing and using nanomaterials, thereby reducing the risk associated with nanotechnology [20]. Scientists are exploring green synthesis techniques to create non-toxic, biocompatible, and environmentally friendly metal oxide nanoparticles employing plant systems, such as phytochemicals, and plant extracts. Plants are an excellent source of sustainable and inexhaustible resources for the synthesis of nanoparticles since they convert light energy into chemical energy and can be consumed, harvested, and utilized [21,22]. Green synthesis is considered a gold standard due to its ease of use and plant diversity. Three key prerequisites for green synthesis include choosing a green solvent, a non-toxic reducing agent, and a safe chemical for stabilization [23]. Plant-derived nanoparticles are employed in a variety of industries, such as food science, bioengineering, farming, cosmetics, nanomedicine, and human health protection owing to their lower risk of negative side effects [24].

Recently, metal oxide nanoparticles (MONPs) have drawn a lot of interest due to their very noticeable magnetic, optical, photocatalytic, anti-fungal, and anti-bacterial characteristics [23,25]. Metal oxide nanoparticles produced via green synthesis yield surprisingly good results and can be produced in an environmentally and economically sustainable approach. Among the various MONPs, oxides of titanium are the most extensively researched nanoparticles owing to their stable structure, low cost, low toxicity, biocompatibility, high catalytic activities, and other properties [26]. However, other metal oxide nanoparticles such as oxides of zinc, iron, manganese, nickel, magnesium, tin, copper, indium, and cerium have also widened the horizon of scientific investigation for various applications due to their fascinating properties [27–31]. This work aims to discuss an overview of the environmentally friendly production, characterization, and applications of metal oxide nanoparticles: nickel oxide (NiO), magnesium oxide (MgO), manganese oxide (MnO_2), tin oxide (SnO_2), and zinc oxide (ZnO) nanoparticles.

7.2 EVOLUTION OF GREENER APPROACH TO NANOTECHNOLOGY: PRINCIPLES AND SUSTAINABLE IMPLEMENTATION

The idea of 'green chemistry' in relation to 'sustainable development' has been the subject of extensive study over the last decade. Green chemistry, or sustainable development, aims to balance present needs with future generations' ability to meet them. For industries reliant on chemistry, sustainable growth is essential because it tackles pollution and resource use. Green synthesis of nanoparticles involves using eco-friendly solvents, non-toxic reducing agents, and safe chemicals. Plants, algae, bacteria, and plants are used for this process (Figure 7.1). Green chemistry focuses

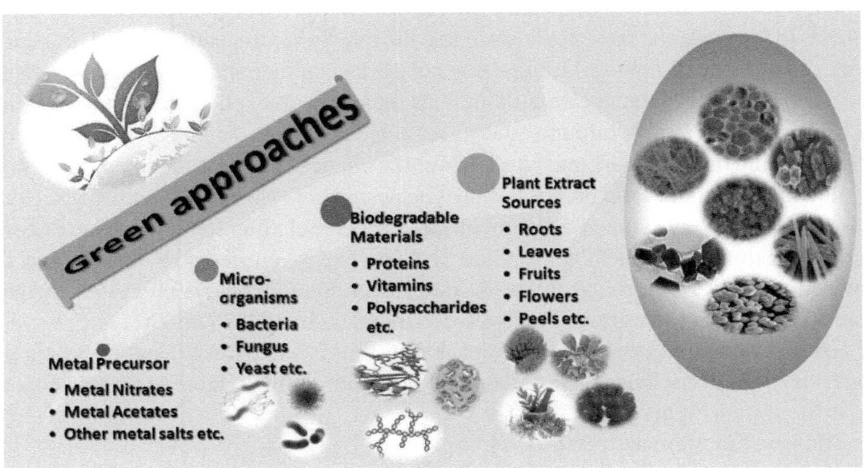

FIGURE 7.1 Schematic of the synthesis of metal oxide nanoparticles by using a green approach.

on reducing hazardous substance usage and generation in design, production, and application of chemical products, with low risk as the prerequisite for execution. The early stage and possible worldwide application of basic green chemistry to the creation of new materials makes it noteworthy. Understanding the relationship between chemical structure and functional groups in nanomaterials might lead to fresh design principles for environmentally friendly materials.

The biosynthesis approach for creating metal oxide nanoparticles (MONPs) uses a low-cost, environmentally friendly reducing agent that requires little reaction. It uses microorganisms, plants, or extracts as stabilizers and reducing agents [32]. Bioactive molecules in microorganisms and active ingredients in plants have antibacterial and antioxidant properties. Biosynthesis is a low energy, cheap, and environmentally friendly process in contrast to chemical methods [33]. However, microbial-based synthesis has shortcomings like long drawn out for the growth of fungi or bacteria, expensive, and biosafety issues [34].

Scientists prefer plant sources for green manufacturing of MONPs because of their characteristics along with bioactive chemicals. Nano-objects with diverse properties are produced by plant parts such as leaves, seeds, bark, roots, and fruits. These plant-based extracts contain bioactive primary compounds like proteins, amino acids, polysaccharides, terpenoids, alkaloids, quinines, tannins, citric acid, flavonoids, phenolic compounds, enzymes, peptides, saponins, etc. having antibacterial and antioxidant properties as well as the ability to reduce metal salts and stabilize nanoparticles [32]. MONPs preparation and properties can be improved via adjusting variables namely pH, temperature, and reagent concentration. Benefits of plant-mediated synthesis include quick, affordable, straightforward operation, a wealth of resources, and good biocompatibility. Plant extracts offer a stable protective layer during plant-mediated synthesis which hinders MONPs accumulation/aggregation and are quick, affordable, and easy to use [35].

The design standards or principles known as the *Twelve Principles of Green Chemistry* provide the basis for sustainable design. Safety precautions like protective gear can reduce risk, but failure in these measures can increase hazard and exposure risks. Minimizing hazards rather than focusing solely on exposure controls can limit risk, even in undesirable circumstances like accidents spills, sabotage, etc. In order to minimize intrinsic dangers and hence lower the chance of accident and damage, safer sustainable chemicals and processes must be designed [36]. Green chemistry seeks to lessen risks at every stage of a chemical's life cycle, including toxicity, physical risks, and environmental risks. This method reduces negative effects on people and the environment by minimizing intrinsic dangers in chemicals and processes through careful design that integrates the Twelve Principles. In 1998, *The Twelve Principles of Green Chemistry*, presented by Paul Anastas and John Warner, served as a guiding framework for scheming new chemical products along with processes throughout their life cycle (Figure 7.2).

The principles are as follows [37]:

- **Prevention of Waste**: A chemical synthesis design that prioritizes prevention over disposal or utilization of waste.
- **Atomic Economy**: Design of synthesis to optimize the ratio of raw components to finished product with little to no waste.

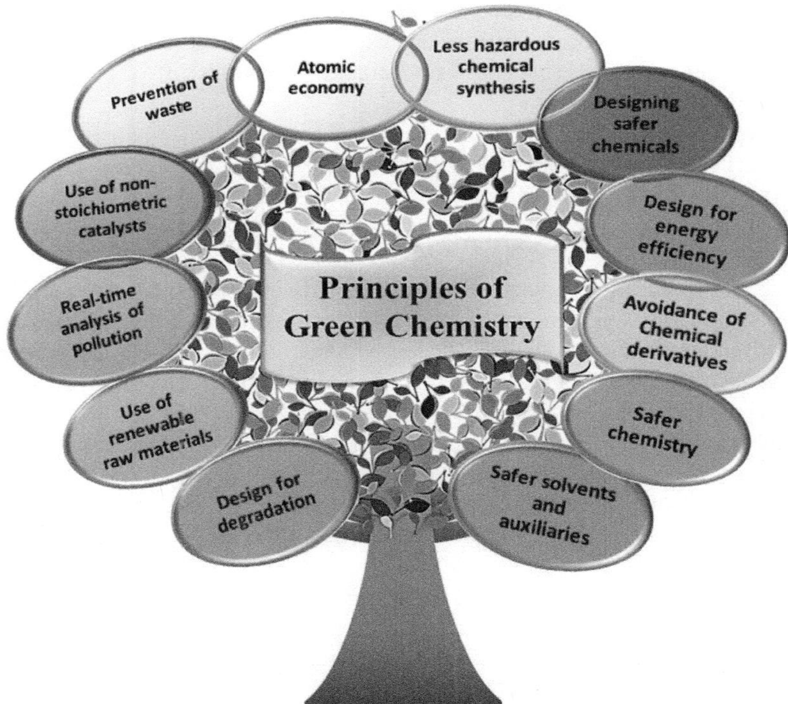

FIGURE 7.2 Schematic of green chemistry principles.

- **Less Hazardous Chemical Synthesis**: Producing and employing materials with negligible or no toxicity.
- **Designing Safer Chemicals**: Non-toxic, yet effective chemicals.
- **Safer Solvents and Auxiliaries**: Reduce or completely avoid using solvents or any other auxiliary chemicals; when used, utilize the safest kind.
- **Design for Energy Efficiency**: Analyse and reduce the effects of energy utilization in chemical synthesis. Whenever feasible, start chemical reactions at concerning pressure and temperature.
- **Utilizing Sustainable Raw Materials**: Waste or agricultural products are sources of renewable raw resources.
- **Avoidance of Chemical Derivatives**: If at all feasible, reduce or completely stop using blocking or protective groups, as well as any short-term alterations, as these procedures call for extra reagents and may result in waste.
- **Use of Non-stoichiometric Catalysts**: Reduces waste through the use of catalytic processes and modest amounts of efficient catalysts that can encourage the reaction repeatedly.
- **Design for Degradation**: Non-persistent, breaking down into harmless chemicals.
- **Real-Time Analysis of Pollution**: Reduction/avoidance of the by-products via manipulating the synthesis approach.

- **Generally Safer Chemistry to Prevent Accidents**: Reducing the likelihood of mishaps like leaks, fires and explosions by creating harmless chemicals and their different states (solid, liquid, or gaseous).

They were condensed into the more practical and enduring acronym PRODUCTIVELY [38].

7.3　GREEN SYNTHESIS OF METAL OXIDE NANOPARTICLES

In the green synthesis technique, various plant extracts (seeds, flower, stem, leaves, and roots) are used to prepare the materials. The extracts are rich in hydrogen, carbon, and nitrogen-containing phytochemicals, including flavonoids, phenolics, alkaloids, and other phytochemicals [39]. Plant extracts combined with metal salts yield nanoparticles with varying sizes, shapes, and surface morphologies [40]. Plant extracts are rich in phytochemicals, which are essential to reduce the metal ions resulting in the formation of metal nanoparticles [39]. The reduction of the metal oxides from metal salts is caused by functional phytochemicals or by the oxygen in the atmosphere. Particle agglomeration results from electrostatic attraction, which also stabilizes the phytochemicals and mediates the formation of metal oxides. The major source of reactive oxygen species (ROS) is superoxide which is inhibited by phenolic compounds containing plant-derived carboxyl and hydroxyl groups. Proteins found in plant extracts are observed to combine with high and low-weight phytochemicals, starch mixtures, and other proteins to form metal nanoparticles [41,42]. Figure 7.3 represents the mechanism of synthesis of MONPs by various greener approaches. Table 7.1 summarizes the various synthesis methods to obtain different MONPs employing several greener methods using plants, microbes, or various phytochemicals. The properties and advantages of some of the metal oxides obtained by greener approaches are discussed in the next section.

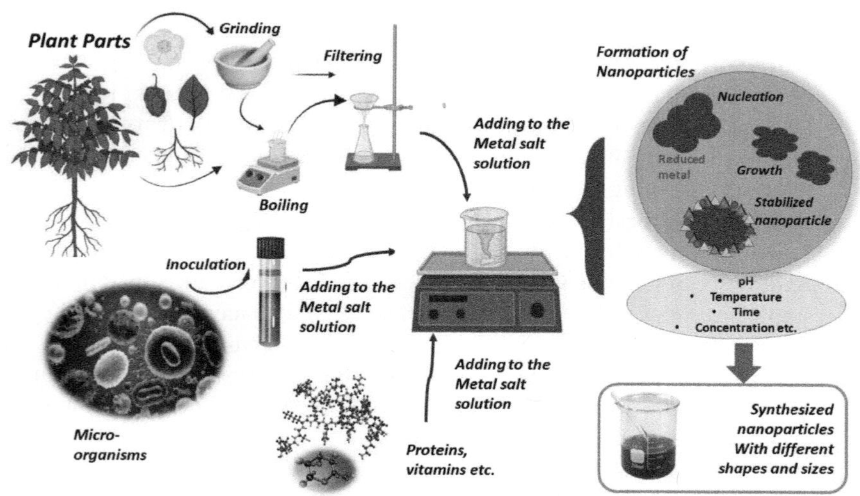

FIGURE 7.3　Mechanism of synthesis MONPs by various green approaches.

TABLE 7.1

Nanomaterials Derived Via Various Green Approaches

Type	Name/Part	MO NPs	Metal Ion Source	Synthesis Method	Reference
Plant source	*Moringa oleifera*/leaf	ZnO	$Zn(NO_3)_2 \cdot 6H_2O$	Green synthesis method	[43]
	Kalopanax pictus/leaf	MnO_2	$KMnO_4$	Reduction method	[44]
	Ixora coccinea/leaf	ZnO	$Zn(CH_3COO)_2 \cdot 2H_2O$	Biological method	[45]
	Artocarpus gomezi-	ZnO	$Zn(NO_3)_3 \cdot 6H_2O$	Solution combustion synthesis	[46]
	Anus/Fruits	SnO_2	$SnCl_4 \cdot xH_2O$	Biogenic synthesis method	[47]
	Saraca indica/Flower	SnO_2	$SnCl_4 \cdot 5H_2O$	Green synthesis method	[45]
	Aspalathus linearis/Leaf	NiO	$NiCl_2 \cdot 6H_2O$	Reduction method	[48]
	Poncirus trifoliate/leaf	NiO	$NiNO_3 \cdot 6H_2O$	Green synthesis method	[49]
	Eucalyptus globulus/leaf	SnO_2	$SnCl_2 \cdot 2H_2O$	Green synthesis method	[50]
	Annona squamosal/Fruits	MgO	$Mg(NO_3)_2$	Chemical method	[51]
	Orange fruit/Peel	MgO	$MgSO_4$	Combustion method	[39]
	Carica papaya/ leaves				
Biodegradable materials	Collagen	ZnO	$[Zn(CH_3COO)_2] \cdot 2H_2O$	Physical and chemical method	[52]
	Glucose	MnO_2	$KMnO_4$	SP-assisted method	[53]
	Sucrose	MgO	$Mg(OH)_2$	Couple synthesis method	[54]
	Glucose	NiO	$NiCl_2 \cdot 6H_2O$	Co-precipitation method	[55]
	Vitamin C	SnO_2	$SnCl_2$	Green synthesis method	[56]
Micro organisms	*Sargassum muticum*	ZnO	$Zn(CH_3COO)_2 \cdot 2H_2O$	Green synthesis method	[57]
	Microbacterium sp. *MRS-1*	NiO	$NiSO_4$	Green synthesis method	[58]
	Erwinia herbicola	SnO_2	$SnCl_2 \cdot 2H_2O$	Green synthesis method	[59]
	P. polymyxa	ZnO	ZnO	Biosynthesis method	[60]
		MgO	MgO		
		MnO_2	MnO_2		

7.3.1 NiO Nanoparticles

Mechanical, chemical, magneto-optical, and *surface-to-volume* ratio differences between nanoscale and microscale materials identified NiO nanoparticles (NiO NPs) as promising candidates for numerous applications [61]. The superconductivity, chemical stability, electro-catalysis, and electron-transfer potential of NiO NPs have drawn the attention of numerous researchers [62]. An intrinsic p-type semiconductor having a broad bandgap of ~3.7–4.0 eV, nanoscale nickel oxide finds extensive application in the biological sciences [63]. Toxic pollutants and dyes have been adsorbed onto these particles [64]. Their anti-inflammatory qualities make them useful in bio-medicines as well [65]. Previous reports on NiO NPs have also demonstrated their cytotoxic effects through the release of Ni^{++} and ROS to oxidative damage [66].

Numerous physical and chemical techniques, including electro-deposition, sol-gel process, galvanostatic anodization, solvothermal route, hydrothermal synthesis, and co-precipitation methods have been employed to generate NiO NPs [67–69]. While using physical methods, a significant amount of energy is required, whereas chemical synthesis typically produces hazardous chemicals that are harmful to the environment and non-biodegradable products [48]. In order to address these issues, scientists have created a more dependable approach that is environmentally sustainable. The desired metal oxide nanoparticles (MONPs), which are safer, more affordable, environmentally benign, and green, can be synthesized using a variety of biological sources, including plants, algae, microorganisms, etc. [70,71]. The employment of plant extracts for green synthesis is growing in popularity due to their rich phytochemical profile, which includes alkaloids, flavonoids, and phenolic compounds [61,72,73]. These phytochemicals are thought to be effective in stabilizing and chelating biogenic nanoparticles. Additionally, plant extract may have the ability to regulate the shape and size of NPs and act as a potent capping and reducing agent [74]. Thus, researchers have focused on the plant-mediated synthesis of metallic nanoparticles. Years have seen the achievement of green NiO NPs synthesis, and interest in NiO NPs synthesis for various biological applications is growing.

NiO NPs have been reported to be synthesized by plants. Nasseri et al. prepared nickel oxide nanoparticles by extracting the flowers of *Tamarix serotine*. The nanoparticles were then characterized by UV spectroscopy, BET surface area measurements, and X-ray diffraction. These nanoparticles were used as a green, recyclable catalyst for quinoline derivative synthesis [75]; Thema et al. synthesized NiO NPs using *Agathosma betulina* (round leaf buchu), which was demonstrated through various methods including HRTEM, XPS, and EDS. The highly crystallized single-phase cubic NiO nanoparticles showed adequate photovoltaics and photodetection responses, with further research to identify the physical and chemical mechanisms of their formation [48]. Sone et al. synthesized NiO NPs using *Callistemon viminalis* (weeping bottlebrush) red flowers aqueous extract, which was confirmed using techniques like SEM, EDS, XRD, FTIR-ATR, Raman, XPS, and UV-Vis-NIR spectrophotometry. The crystalline Bunsenite NiO nanospheres showed quasi-reversible redox processes, potential for pseudocapacitance, and a specific capacitance of 101 F/g. These nano-powders could be used for energy storage in pseudo-capacitors [63]. Ezhilarasi et al. successfully synthesized NiO NPs using

Moringa oleifera (drumstick tree), which exhibits excellent antibacterial properties and cytotoxic activity. The nanoparticles are non-toxic and possess higher antioxidant and antimicrobial potential. They are effective against gram-positive bacterial strains and HT-29 colon cancer cell lines. The active generation of reactive oxygen species (ROS) destroys cell proteins, DNA, and membrane, leading to cell death. This simple and environmentally friendly method allows for the production of antimicrobial coatings for biomedical and environmental applications, demonstrating the potential of NiO nanoparticles in various applications [76], and Iqbal et al. synthesized NiO NPs using *Nephelium lappaceum* (rambutan) leaves extract as a reducing and stabilizing agent. The low toxicity and biocompatibility of the synthesis, which make it appropriate for biomedical applications, were demonstrated through a variety of techniques and biological assays [77]. The leaf extract of *G. wallichianum* has been successfully used as an efficient reducing and stabilizing agent in the development of NiO NPs, given the significance of green nano-chemistry. Gonorrhoea, rheumatism, arthritis, leucorrhoea, and liver and cardiovascular issues can all be treated with *G. wallichianum* [61].

7.3.2 MgO Nanoparticles

Magnesium oxide has distinct electronic, mechanical, thermal, optical, and chemical properties, despite the fact that there are many other types of metal oxides, including ZnO, CuO, TiO_2, and CdO, etc. [78]. With its wide energy bandgap, high melting point, low heat capacity, and high reactivity, this material is very stable and safe to use in a variety of applications. Magnesium oxide nanoparticles (MgO NPs) are the most effective MONPs against bacteria. Breaking down cell membranes is the mechanism that allows antibacterial action to occur.

Chemical precipitation, sol-gel, chemical reduction, and sonochemical are some of the techniques that researchers have used to create MgO nanoparticles. Reagents free of environmentally hazardous compounds are used in green synthesis techniques, such as plant extracts or water solvents.

Previous studies have extensively investigated the application of *Moringa oleifera* leaf extract in the production of MgO NPs. A. Fatiqin et al. successfully synthesized MgO NPs using *Moringa oleifera* leaf extract; green synthesis of MgO nanoparticles using plant extracts is a promising alternative. FTIR spectroscopy, UV-Visible spectroscopy, XRD, and SEM analysis were used to study the optical and structural properties of the synthesized nanoparticles. The spherical shape of the nanoparticles was confirmed, and their antibacterial properties were demonstrated against gram-positive and gram-negative bacteria. MgO NPs produced had strong antibacterial and antioxidant action against the pathogens *S. aureus, S. dysenteriae,* and *E. faecalis* [79]. Researchers Mirhosseini et al. looked into the antibacterial activity of MgO suspension against *Escherichia coli* (*E. coli*) in milk [80]. An electron microscope was used for scanning in order to describe the morphological alterations that the antibacterial treatment caused in *E. coli*. The results show that pressure and the presence of MgO can harm cell membranes, causing the contents of the cell to leak out and, ultimately, the death of the bacterial cells. Using *Bauhinia purpurea* extract as an antibacterial against *Staphylococcus aureus* (*S. aureus*), Das et al. synthesized

MgO nanoparticles. Using an FE-SEM and a fluorescence microscope to analyse colony-forming units, the antibacterial activity of magnesium oxide was investigated. The outcomes demonstrated that the generated magnesium oxide (MgO) exhibited strong antibacterial activity against *S. aureus* at a low dosage (250 µg/mL) [81].

Our group synthesized green MgO nanoparticles using tea extract and used them for photocatalytic removal of brilliant green and crystal violet dyes. The nanoparticles demonstrated high efficacy [(BG:86%) and crystal violet (CV:94%) dyes under visible irradiation in 120 min], making them an attractive candidate for environmental remediation [31].

7.3.3   MnO$_2$ Nanoparticles

The widespread applications of manganese oxide nanoparticles (MnO$_2$ NPs) in various fields, including microelectronics [82], optoelectronics [83], chemical sensing devices, ion-sieves, rechargeable batteries, and catalysis, have drawn attention to them. Furthermore, they are based on earthly excess elements because of their many advantages for a variety of applications; Mn is the twelfth most plentiful element on Earth and the third most common transition element after Ti and Fe [84]. MnO$_2$ NPs, also exhibit strong photocatalytic activity for the degradation of dyes.

The most popular methods for achieving nanoscale synthesis include facile solution-based approaches, sol-gel, chemical precipitation, solvothermal, hydrothermal, solid-state synthesis, and electro/photo chemical reduction techniques [85–87]. For nanoscale synthesis, physical and chemical techniques are feasible, but green fabrication is preferable because it is more environmentally friendly [88]. Plant extracts can be used to prepare nanoscale materials more effectively than other biological processes, such as the microbial route, which requires the time-consuming maintenance of cell cultures [89].

For instance, Majani et al. synthesized MnO$_2$ NPs using *Saraca (indica) asoca* (Ashoka) leaf extract as a reducing agent. The nanoparticles showed high crystallinity and cytotoxicity against MCF7 and MDA-MB-231 breast cancer cell lines, indicating their potential as an anti-cancer agent and highlighting the potential of plant extract-based synthesis as an alternative to traditional therapies [90]. Manganese dioxide NPs (VBLE-MnO$_2$ NPs) have been synthesized by Haibin Lu et al. from the leaf extract of *V. betonicifolia*. These NPs show excellent antibacterial, antifungal, and biofilm inhibition performances against all tested microbial species. These NPs also exhibit strong antioxidant potential, proving a novel environmentally friendly method of creating nanomaterials with improved medicinal qualities. By providing antibacterial coatings for medical equipment, the synthesized VBLE-MnO$_2$ NPs may help lower the frequency of pathogenic bacteriological and mycological infections brought on by biomaterials and medical implants [91].

7.3.4   SnO$_2$ Nanoparticles

SnO$_2$ is one of the materials with great sensitivity for sensor applications, and it is mostly employed for gas reduction. Several methods for producing SnO$_2$ nanoparticles include the sono-chemical, sol-gel, energy ball billing method, chemical precipitation, hydrothermal, solvothermal, and microemulsion methods [14]. With a direct optical energy bandgap between approximately 2.5 and 3.4 eV, SnO_2 is a p-type semiconductor material.

In their natural state, these SnO_2 nanoparticles are frequently amorphous and not crystalline [92]. Numerous morphologies, including hollow spheres, mesoporous structures, nanowires, nanobelts, nanotubes, and nanorods, are present in this SnO_2 nano-plastic [93]. SnO_2 morphologies allow us to determine the possible uses of synthesized nanoparticles, such as sensors that generate a signal that is directly proportional to the concentration of each gas. However, the total volume of gas exposed to an electrochemical sensor during its lifespan, as well as other environmental conditions, has a substantial impact on how long the sensor lasts [94]. For instance, SnO_2 NPs were economically biosynthesized from an aqueous leaf extract of the *Aquilaria malaccensis* plant, also known as *agarwood*, in a tin (IV) chloride pentahydrate solution at room temperature by Ahmad et al. using a green, fast, and non-toxic process. It was discovered that the synthesized NPs agglomerates were evenly dispersed spheres with significant potential for use in optoelectronic devices and catalysis [25]. Elango et al. synthesized SnO_2 NPs using *Persia americana* seed methanolic extract that were found to be environmentally friendly and completely degraded, making them a promising alternative to toxic organic dyes [14].

7.3.5 ZnO Nanoparticles

Since zinc oxide nanoparticles (ZnO NPs) have high specific surface area, biocompatibility, ability to absorb and scatter ultraviolet light, and antibacterial qualities, they are a popular type of nanomaterial having applications in electrochemistry, cosmetics, textile industry, biomedicines, medical devices, etc. The two main approaches for synthesizing ZnO NPs are physical and chemical. The former has the drawbacks of greater energy usage, poor purity, high cost, uneven particle-size distribution, and significant amounts of secondary waste, that cannot be reversed. Methods for green production of ZnO NPs from plant extracts have piqued the interest of researchers in an era of environmentally friendly development due to the benefits of low cost, low complexity, and environmental sustainability. With a focus on factors influencing ZnO NPs morphology and antibacterial properties as well as mechanisms of action presents a green mechanism for ZnO NPs synthesis using plant extracts. The findings suggested that the intrinsic crystallographic morphological characteristics and the conditions under which ZnO NPs were prepared have an impact on morphology [95].

As the idea of environmental preservation is now profoundly ingrained in the expectations of the populace, there is widespread concern over the generation of ZnO NPs using environmentally friendly methods as the number of applications for which they are used rises [96]. For instance, Faisal et al. synthesized ZnO NPs from *Myristica fragrans* fruit extracts and characterized using various techniques such as X-ray diffraction (XRD), FTIR, UV spectroscopy, SEM, TEM, DLS, and TGA. The crystallites were highly pure and confirmed their spherical or elliptical shape. The NPs were found to be highly active against bacterial strains, with *Klebsiella pneumoniae* being the most sensitive. They also displayed inhibitory potential against enzymes like protein kinase, α-amylase, and α-glucosidase. The NPs were also used as photocatalytic agents, demonstrating their eco-friendly synthesis, nontoxicity, and biocompatible nature, making them potential candidates for biomedical and environmental applications [97]. Robles et al. synthesized ZnO NPs using *Hibiscus sabdariffa* flower extract. The sample showed good photocatalytic activity, degrading 97% of MB in 150 min. The ZnO-J8 sample demonstrated a 95% MB degradation in 90 min, resulting in less toxic and waste-free ZnO [98].

7.4 CHARACTERIZATIONS OF GREEN SYNTHESIZED METAL OXIDE NANOPARTICLES

Evaluations of chemical and physical equipment, such as UV-Visible spectroscopy, Fourier transforms infrared spectroscopy (FTIR), transmission electron microscopy (TEM), scanning electron microscopy (SEM), and X-ray diffraction (XRD), can be employed for the characterization of the nanoparticles synthesized by green approaches (schematic in Figure 7.4).

7.4.1 X-Ray Diffraction (XRD)

Figure 7.5a–d displayed the XRD spectra of green synthesized ZnO NPs, MnO$_2$ NPs, NiO NPs, and SnO$_2$ NPs. In Figure 7.5a, XRD patterns of ZnO NPs are shown which were synthesized using tomato, cabbage, onion, and carrot. XRD profiles of

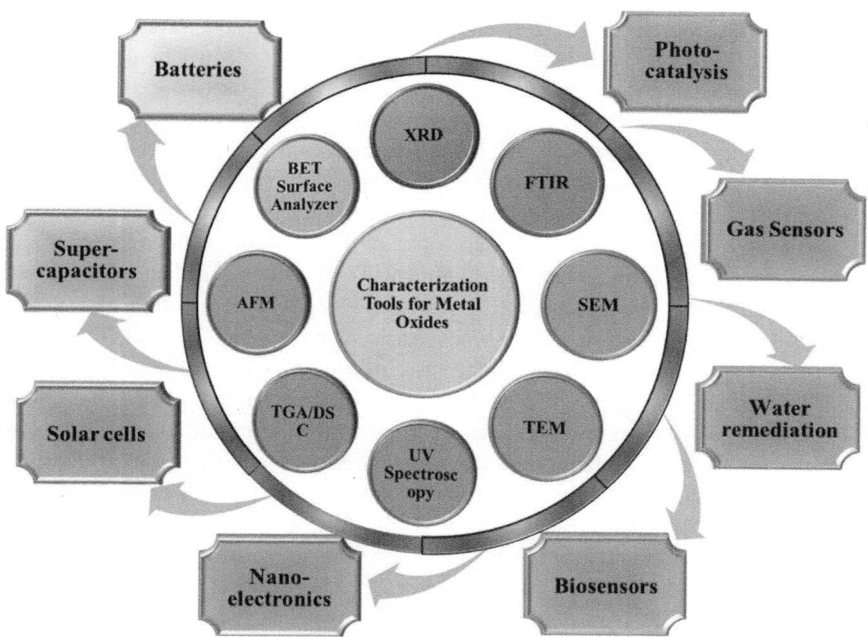

FIGURE 7.4 Schematic of the various characterizing techniques and applications for metal oxide nanoparticles.

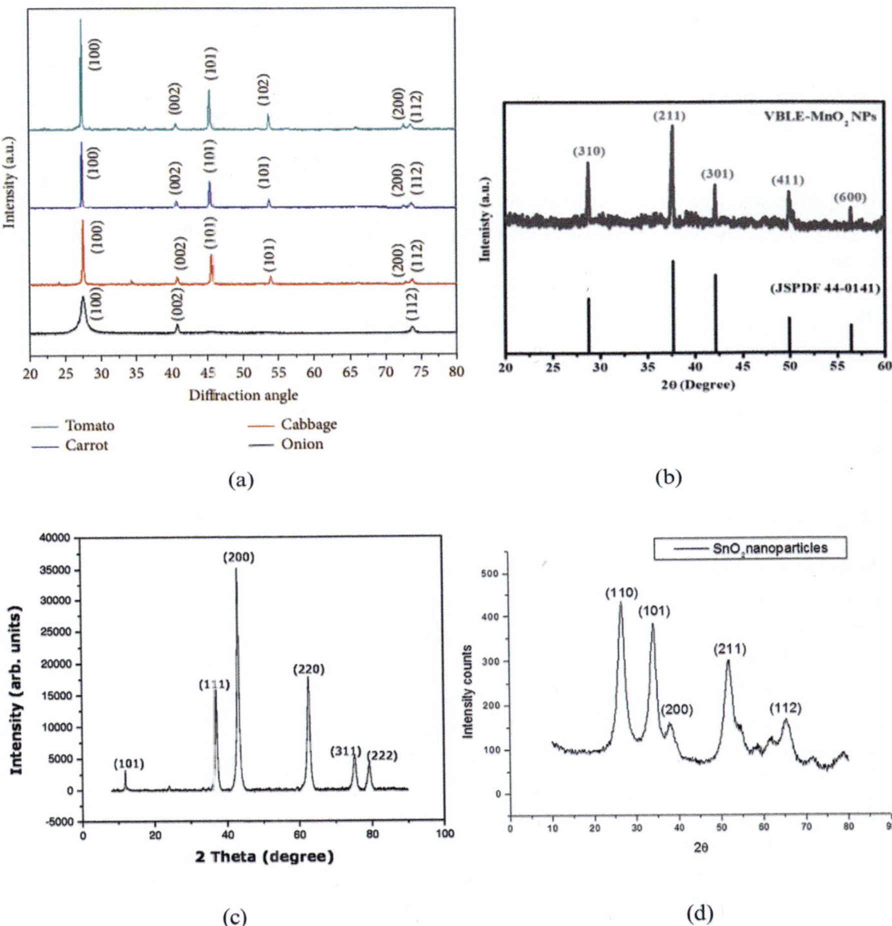

FIGURE 7.5 XRD spectra of (a) ZnO NPs (b) MnO_2 NPs (c) NiO NPs (d) SnO_2 NPs. (Images adapted from Ref. [99] (open access), Ref. [91] (open access), Ref. [100] permission taken from Elsevier, Ref. [14] permission taken from Elsevier.)

obtained products are in accordance with the hexagonal wurtzite structure (*space group P63mc*) of ZnO NPs, with lattice parameters $a = b = 3.249$ Å and $c = 5.206$ Å. At 27.39° and 40.64°, two broad diffraction peaks were seen, corresponding to crystal planes of (100) and (002), respectively. In this case, the onion-extracted nanoparticles only displayed two prominent peaks. Furthermore, in all extracted samples with the exception of onion, peaks at 47.31°, 53.60°, 72.63°, and 73.75° existed with the crystal planes (101), (102), (200), and (112), respectively. The highest intensity peak was found at (100), indicating the preferential growth plane and proving high degree of purity of ZnO NPs. Nanoparticles with average crystallite sizes of 17 nm, 18 nm,

24 nm, and 15 nm, respectively, have been synthesized using onion, cabbage, carrot, and tomato. The results above, suggested that smaller nanoparticles are produced using the tomato extract [99]. MnO_2 NPs were synthesized using *Viola betonicifolia* leaf extract (Figure 7.5b). The XRD profile of VBLE MnO_2 NPs shows the distinct peaks at 28.78°, 37.66°, 42.14°, 49.90°, and 56.44° of 2θ. These peaks are associated with the crystal planes (401), (411), (600), (211), and (310) and showed that the VBLE MnO_2 NPs are highly crystalline, as seen by the peak intensities [91]. Figures 7.5c and d display the XRD spectra of green synthesized NiO NPs by making use of *Aegle marmelos* leaf extract [100] and SnO_2 NPs by making use of *Persia americana* seed methanolic extract [14]. The intense and sharp diffraction peaks of NiO and SnO_2 NPs match perfectly with JCPDS data (4-0835 and 88-0287, respectively) and indicate that both are in crystalline pattern. The crystallite size of these nanoparticles was estimated by Scherrer formula, and the results were 8.15 nm (NiO NPs) and 4 nm (SnO_2 NPs). The extract used in synthesis acts as a fuel to create high surface area and well-dispersible nanoparticles in the nano regime.

7.4.2 UV-VIS SPECTROSCOPY

The most prevalent technique for determining the optical and electrical characteristics of nanoparticles is UV absorption spectroscopy, where the absorption ranges correspond to the metallic nanoparticle diameters. The UV-Visible spectra of NiO NPs synthesized by making use of *Phytolacca dodecandra* L' leaf extract in Figure 7.6a and SnO_2 NPs synthesized by making use of *Persia americana* methanolic extracts [14] are shown in Figure 7.6b. The reaction's absorbance was measured between 200 and 800 nm. When considering the absorbance at the 224 nm peak range, this displays the surface plasmon resonance (SPR). As the time period increases, the combination progressively transitions from a clear solution to a precipitated solution, signifying the formation of SnO_2 nanoparticles. The quantum confinement effect is anticipated for semiconducting materials (SnO_2 NPs), and as particle size decreases, the limit of absorption will shift to a higher energy. The value of the absorption edge is 224 nm. In the case of NiO NPs, the resultant NPs showed the 350–356 nm spectral range distinctive surface plasmon resonance (SPR) band. At 350 nm, the synthesized NiO exhibited its maximum UV-vis absorbance. By extrapolating the linear part of $(\alpha h v)^2$ vs $h v$ to zero, the optical bandgap (E_g) for the absorption peak can be found. For NiO NPs, the found E_g was 3.19 eV [101].

7.4.3 FOURIER TRANSFORM INFRARED SPECTROSCOPY (FTIR)

The characteristic bonds of FTIR spectra, recorded in transmittance mode, are presented in Figure 7.7a–c. Figure 7.7a shows FTIR spectrum results for NiO prepared using *S. persica* extract, significant variations in absorption peaks were observed in the FTIR spectrum for the bio-components of the synthesized nanoparticles. The broad absorption peak, which was ascribed to the polyphenolic constituents

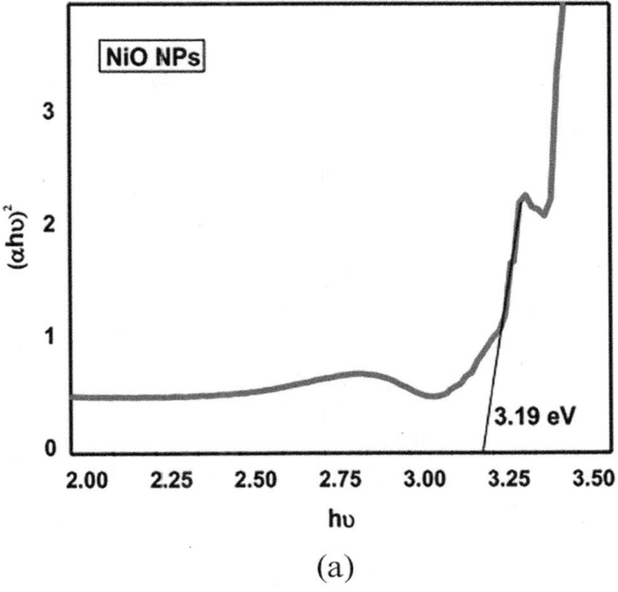

(a)

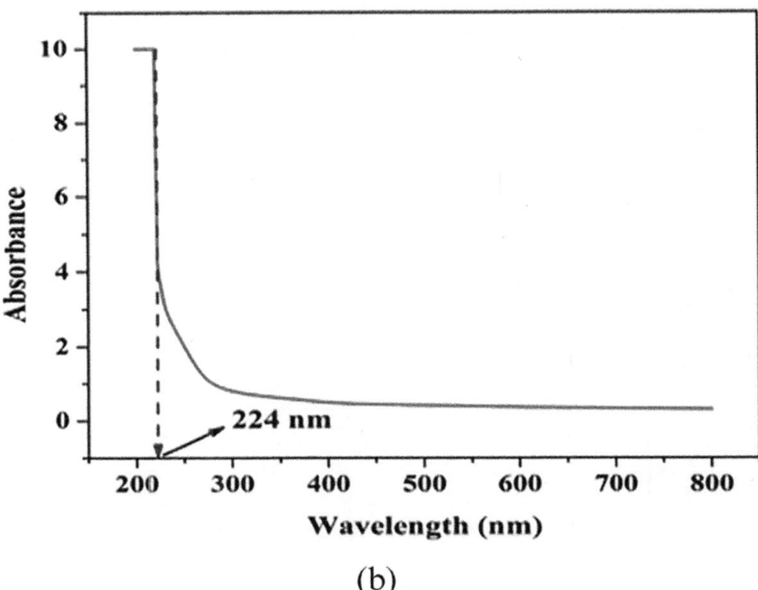

(b)

FIGURE 7.6 UV-Vis spectra of (a) NiO NPs and (b) SnO_2 NPs. (Image adapted from Ref. [101] (open access) and Ref. [14] (permission taken from Elsevier) respectively.)

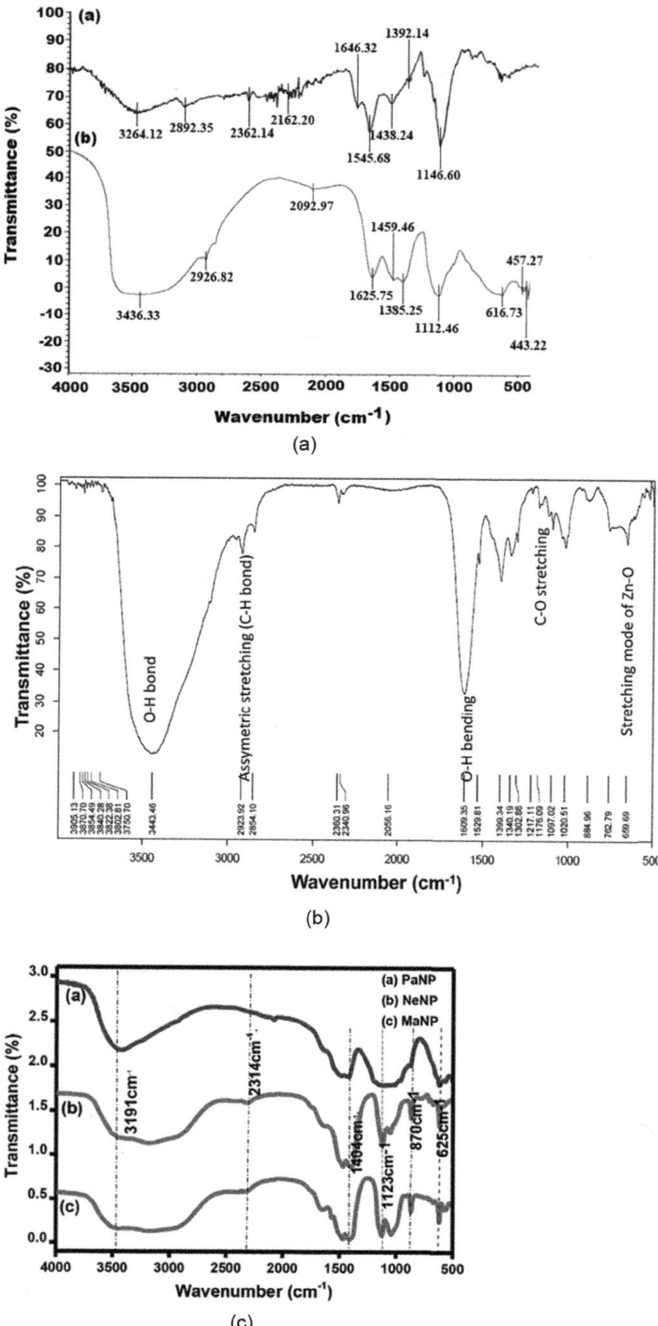

FIGURE 7.7 FTIR spectra of (a) NiO NPs, (b) ZnO NPs, and (c) MgO NPs. (Images adapted from Refs. [102], [103], [39] respectively).

of the *S. persica* extract, matched with O–H stretching of the free hydroxyl group. An absorption band at 3,427.99 cm^{-1} was observed in green ZnO NPs made from pomegranate peel extract. The biogenic SP-NiO NPs included phenols, carboxylic acids, alcohols, ketones, alkenes, alkanes, aromatic constituents, and amines, among other functional moieties. This suggested that these constituents might have served as capping and stabilizing agents as well as contributed to the reduction of Ni(II) [102]. The peak at 659 cm^{-1} represents the stretching mode of ZnO prepared using *Cocos nucifera* leaf extract [103] as shown in Figure 7.7b. It is necessary to perform the FTIR of the extract prepared as FTIR is the high-resolution analytical technique used to determine the bioactive compounds on the functional groups present in them and reveal the structure to get the knowledge that which phytochemical is playing a significant role. Figure 7.7c depicts the FTIR spectra of MgO NPs prepared using papaya (Pa), neem (Ne), and mango (Ma) extracts. Absorption peaks detected at 3,182 cm^{-1}, 3,191 cm^{-1}, and 3,427 cm^{-1} indicated that the O–H bond is being stretched by hydroxyls. Ma-NPs and Ne-NPs exhibit peaks at 2,318 cm^{-1} and 2,314 cm^{-1}, respectively, which are indicative of the aromatic aldehyde's C–H stretching bond. The peak absorbed at 1,615 cm^{-1} in Ma-NP indicates the presence of saturated primary alcohol, while 1,048 cm^{-1} resembles C=O carbonyl groups. The peak absorbed within the range of 1,404 cm^{-1} to 422 cm^{-1} is indicative of αCH$_2$ bending, which is associated with aldehyde, ketones, and aromatic tertiary amine groups. Alcohol, ester, and carboxylic acid are present in Ma-NPs and Ne-NPs, respectively, as indicated by the absorption peaks at 1,137 cm^{-1} and 1,123 cm^{-1}. Between 625 cm^{-1}–611 cm^{-1} and 870 cm^{-1}–861 cm^{-1}, the absorption peak matches Mg–O bond stretching indicating the presence of MgO nanoparticles [39].

7.4.4 High Resolution-Transmission Electron Microscopy (HR-TEM)

HR-TEM images revealed the formation of slightly agglomerated, porous, cubic, and spherical nanoparticles of NiO obtained by using *Aegle marmelos* leaf extract (Figure 7.8a–c). The polycrystalline nature is demonstrated by the selected area electron diffraction (SAED) in Figure 7.8d. The particle size, which fell between 8 and 10 nm, matches the average size of crystallite determined from the XRD pattern quite well. The existence of magnetic interaction and polymeric adherence between the particles is confirmed by their agglomeration [100].

7.4.5 Vibrating Sample Magnetometry (VSM)

Vibrating sample magnetometry (VSM) is a technique that measures the magnetic properties. Figure 7.9 displays VSM studies of NiO NPs prepared by using leaf extract of *Aegle marmelos* [100]. By moving the outside magnetic field in the range of +20 kOe and −20 kOe at 300K, the magnetization behaviour of the prepared NiO NPs was examined. Figure 7.8 displays the variation of magnetization (M) vs magnetic field (H). In low magnetic field areas, the NiO showed small curves, and in higher magnetic field regions, there was a linear portion. Super para-magnetism is confirmed by the observation that the sample did not become saturated, despite increasing magnetic fields of roughly 20 kOe. The magnetic property will shift from

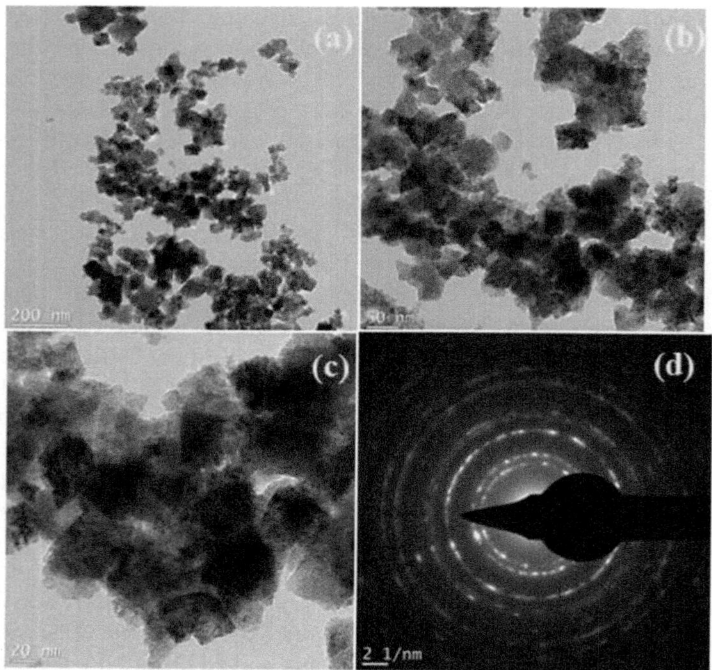

FIGURE 7.8 (a–d) HRTEM of NiO NPs. (Image adapted from Ref. [100] permission taken from Elsevier.)

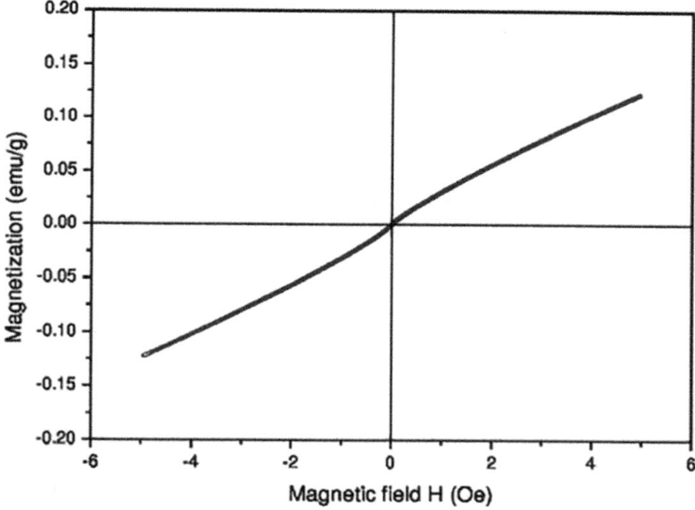

FIGURE 7.9 VSM of NiO NPs. (Image adapted from Ref. [100] permission taken from Elsevier.)

diamagnetism to super para-magnetism if the particle size decreases [104]. Therefore, the decreased particle size (i.e. 8.15 nm) caused the superparamagnetic behaviour that was observed, and the obtained XRD results further support this theory. Even at higher magnetic fields, the presence of superparamagnetic behaviour, which is sensitive to magnetic impurities, indicates the purity of the sample [100].

7.5 TOXICITY OF METAL OXIDE NANOPARTICLES ASPECTS

There are numerous risks to human and environmental health associated with nanotechnology, a significant area of innovative science and economic growth. Surfaces that may bind and transport hazardous chemical pollutants are made possible by the surface characteristics and minuscule size of nanoparticles and nanotubes. These materials may also be hazardous in and of themselves by producing reactive radicals [12]. Applications of metal oxides are numerous in physics, materials science, and chemistry. MONPs toxicity can arise from various mechanisms, including oxidative stress, coordination effects, non-homeostasis effects, genotoxicity, and others. The exposure routes, size, and dissolution of the MONPs are factors that influence them [105]. Three different types of a material's fundamental properties can be impacted by the oxide nanoparticles' particle size. First, there are the cell parameters and lattice symmetry; then, there are the electronic characteristics; and finally, there is the quantum confinement because of the small size. Due to these characteristics, they persist in unique agglomeration states, and the effects of their dispersion in biological fluids can differ.

Metal oxides have the potential to cause toxicity through a number of mechanisms, including exposure, size, and dissolution. Ingestion, dermal exposure, and inhalation are the most frequent ways that nanoparticles are exposed. For instance, although numerous studies revealed size-dependent toxicity, the International Agency for Research on Cancer (IARC) regarded bare zinc oxide (ZnO) as generally safe [96,106]. Nanoparticles of ZnO (20–30 nm) are widely utilized in sunscreen, topical skin care products, and cosmetics. The skin is shielded from damaging ultraviolet (UV) radiation from sunshine by sunscreens or sun blockers. According to the size and mode of exposure, the aforementioned studies concluded that ZnO nanoparticles may be hazardous in large quantities and nano ZnO are extensively utilized in cosmetics like sunscreen. Biogenic ZnO, MnO_2, and MgO nanoparticles with average crystallite sizes of 62.8, 18.8, and 10.9 nm had notably strong inhibitory effects against the *Xoo* strain suggesting lower toxicity of green synthesized MONPs [60]. A detailed study by Gakis et al. on 35 different metal/semimetal oxides NPs and 10 metals NPs showed that MgO, SnO_2, TiO_2 CuO, etc. are harmless to both human and animal cells and microorganisms whereas NiO NPs might be toxic for human cells [107].

7.6 APPLICATIONS

Sustainable, economical, safe, and non-toxic method for producing simple metal oxides is green synthesis [108]. Green synthesis technique-produced nanoparticles have a variety of applications, along with bioactivity, anti-tumour targeting, anti-cancer, anti-inflammatory, and bio-absorption. They are employed in powder metallurgy, photocatalysts, microelectronic devices, transistors, magnetic devices, anticorrosive coatings, wastewater treatment, antibacterial and electrocatalysts in addition to biological applications [120,121]. A few of the applications are listed in Table 7.2 (schematic in Figure 7.4).

TABLE 7.2

Applications of Green Synthesized Metal Oxide Nanoparticles

Sr. No.	Plant Extract	Precursor Salt	Applications	Reference
1.	*Camellia sinensis* leaves	$NiCl_2 \cdot 6H_2O$	Photocatalytic activity for dye degradation.	[109]
2.	*Apium graveolens*	$Ni(CH_3COO)_2$	Removing aliphatic hydrocarbons in crude oil.	[110]
3.	*Ocimum sanctum leaves*	$NiNO_3$ and $NiSO_4$	Used as adsorbent hazardous anionic pollutants, dyes and antimicrobial activity against E-coli and pseudo-monas.	[64]
4.	*Agathosma betulina* (round leaf buchu) leaves	$Ni(NO_3)_2 \cdot 6H_2O$	Electrochemical storage devices.	[48]
5.	*Acacia nilotica* leaves	$Ni(NO_3)_2 \cdot 6H_2O$, in the presence of NaOH	Electrochemical, Antibacterial, haemolytic.	[111]
6.	*Mucuna pruriens* seeds	$(Mg(NO_3)_2 \cdot 6H_2 O)$	Antibacterial, antioxidant and photocatalytic activity.	[112]
7.	*Amaranthus blitum* and *Aloe vera*	$Mg(NO_3)_2$	Water treatment	[113]
8.	*Lactobacillus plantarum*	$Mg(NO_3)_2$	Anticancer study	[114]
9.	*Mushroom* extract	$(Mg (CH_3COO)_2 \cdot 4H_2O)$	Seed germination	[115]
10.	*Orange* peel extract	$Mg(NO_3)_2$	Antimicrobial and antibacterial activity	[51]
11.	*Saraca asoca* leaf	$Mn(NO_3)_2$	Anti- cancer study	[90]
12.	*Phyllanthus amarus* plant	$C_4H_{16}MnO_8$	Fluorescence emitting materials, long wavelength and radiation resistant scintillators.	[83]
13.	*Y. gloriosa* extract	$(CH_3COO)_2Mn \cdot 6H_2O$	Photocatalytic activity for dye degradation.	[116]

(Continued)

TABLE 7.2 (*Continued*)

Applications of Green Synthesized Metal Oxide Nanoparticles

Sr. No.	Plant Extract	Precursor Salt	Applications	Reference
14.	*Persia americana* seed	$SnCl_2$	Degradation of toxic organic dyes	[14]
15.	*Aloe barbadensis miller*	$Sn(NO_3)_2 \cdot 3H_2O$	Antibacterial and antifungal	[117]
16.	*Daphne alpine*	$SnCl_4 \cdot 5H_2O$	Adsorption of Cd^{2+}	[118]
17.	*Persia americana*	$SnCl_2$	Photocatalytic dye degradation	[14]
18.	*M. fragrans* extract	$Zn(NO_3)_2 \cdot 2H_2O$	Antibacterial and antidiabetic activity	[97]
19.	*Hibiscus sabdariffa*	$(Zn(NO_3)_2 \cdot 6H_2O)$	Photocatalytic activity, Cytotoxicity study Antioxidant activity Antifungal activity	[98]
20.	*C. halicacabum*	$Zn(NO_3)_2$	Antibacterial activity	[119]

7.7 MAJOR CHALLENGES AND FUTURE PERSPECTIVE

The field of nanoparticles (NPs) and their potential uses has seen tremendous advancements in recent years. Research on NPs has advanced, utilizing biological sources like plants, bacteria, fungi, and yeast for green synthesis, but still faces challenges in large-scale manufacturing.

Some of the major problems related to green synthesis of MONPs are enumerated below:

- To optimize the size and form of nanoparticles, significant optimization studies on reactants and manufacturing factors like temperature, pH, and rotational speed are required.
- Enhancing NPs' diverse physicochemical characteristics for particular uses should be the focus of research as well.
- Understanding the substances used in synthesis and stabilization, as well as research into extracts attachment to MONPs surfaces,
- It is necessary to thoroughly examine how an individual metabolite found in plant extracts and the cellular elements of microbes contribute to the production of NPs.
- The commercial scale-up of NPs production employing green methods must be closely monitored.
- To boost NPs yield and stability with shorter reaction durations, several reaction parameters must be tuned.

- More research ought to concentrate on the possible environmental risks connected to MONPs.
- Application of environmentally benign, non-cytotoxic nanoparticles with safe exposure threshold values to address current toxicity concerns.

Green synthesis techniques could produce nanoparticles more efficiently and cheaply than traditional methods. However, research is needed on extraction and purification, toxicological analysis, and the use of genetically modified microorganisms for biosynthesis and stabilization. Enhancing metal tolerance could provide a new strategy for the green synthesis of metal oxide nanoparticles.

7.8 CONCLUSION

The chapter discusses the green approaches to metal oxide nanoparticles (MONPs) synthesis. An eco-friendly, non-toxic, and clean method is provided by green synthesis. MONPs can be produced from various plant materials and microorganisms. Sustainable nanoscience development is greatly aided by the low cost, nontoxicity, ease of scaling up, and environmental friendliness of the plant extract-mediated approach for MONPs synthesis, making it suitable for applications in several fields like catalysis, medicine, water treatment, and bioengineering. Significantly, no hazardous chemical reagents are used in the synthesis reaction, and the resultant MONPs are free of contaminants. For this reason, they are particularly valuable in the biomedical field where toxicity is crucial. The size, form, and reaction speed are influenced by experimental parameters. Biosynthesized MONPs have been characterized by employing several techniques including UV-Vis spectroscopy, FTIR, XRD, and HR-TEM, to ascertain their size, shape, and morphology. Nonetheless, it is imperative to consider several factors, including bioavailability, adverse reactions, interactions with cells, bio-dispersion, and degradation. The assemblage of MONPs can cause damage to DNA, membranes, proteins, and mitochondria. Although numerous researches have demonstrated the biological creation of MONPs, additional investigation is necessary to expand their applications and guarantee their successful commercialization as well as the employment of green synthesis method for human society's sustainable development.

AUTHOR CONTRIBUTIONS

Ishita Kapil: Literature survey and writing the original draft, **Pinky Yadav**: Preparation of schematics and revision, **Ritik Kumar** and **Sachin Vats**: Literature survey, **Ayana Bhaduri**: Conceptualization, manuscript preparation, review, and supervision.

REFERENCES

1. C. Buzea, I.I. Pacheco, K. Robbie, Nanomaterials and nanoparticles: sources and toxicity, *Biointerphases*. 2 (2007) MR17–71. https://doi.org/10.1116/1.2815690.
2. N.A. Singh, Nanotechnology innovations, industrial applications and patents, *Environ. Chem. Lett.* 15 (2017) 185–191. https://doi.org/10.1007/s10311-017-0612-8.

3. D. Titus, E. James Jebaseelan Samuel, S.M. Roopan, Chapter 12: Nanoparticle characterization techniques, in: A. K. Shukla, S. Iravani (Eds.), *Green Synth. Charact. Appl. Nanoparticles*, Elsevier, 2019: pp. 303–319. https://doi.org/10.1016/B978-0-08-102579-6.00012-5.

4. G. Pal, P. Rai, A. Pandey, Chapter 1: Green synthesis of nanoparticles: A greener approach for a cleaner future, in: A. K. Shukla, S. Iravani (Eds.), *Green Synth. Charact. Appl. Nanoparticles*, Elsevier, 2019: pp. 1–26. https://doi.org/10.1016/B978-0-08-102579-6.00001-0.

5. K. Maekawa, K. Yamasaki, T. Niizeki, M. Mita, Y. Matsuba, N. Terada, H. Saito, Drop-on-demand laser sintering with silver nanoparticles for electronics packaging, *IEEE Trans. Components, Packag. Manuf. Technol.* 2 (2012) 868–877. https://api.semanticscholar.org/CorpusID:20083080.

6. P.D. Sia, Nanotechnology between classical and quantum scale: Applications of a new interesting analytical model, *Adv. Sci. Lett.* 17 (2012) 82–86.

7. F. Salamanca-Buentello, D.L. Persad, E.B. Court, D.K. Martin, A.S. Daar, P.A. Singer, Nanotechnology and the developing world, *PLoS Med.* 2 (2005) 0383–0386. https://doi.org/10.1371/journal.pmed.0020097.

8. A.M. Waldron, D. Spencer, C.A. Batt, The current state of public understanding of nanotechnology, *J. Nanoparticle Res.* 8 (2006) 569–575. https://doi.org/10.1007/s11051-006-9112-7.

9. M.C. Roco, The long view of nanotechnology development: The National Nanotechnology Initiative at 10 years, *J. Nanoparticle Res.* 13 (2011) 427–445. https://doi.org/10.1007/s11051-010-0192-z.

10. J. Zi, X. Yu, Y. Li, X. Hu, C. Xu, X. Wang, X. Liu, R. Fu, Coloration strategies in peacock feathers, *Proc. Natl. Acad. Sci. USA.* 100 (2003) 12576–12578. https://doi.org/10.1073/pnas.2133313100.

11. M. Singh, D. Vaya, R. Kumar, B. Das, Role of EDTA capped cobalt oxide nanomaterial in photocatalytic degradation of dyes, *J. Serbian Chem. Soc.* 86 (2021) 327–340. https://doi.org/10.2298/JSC200711074S.

12. M.N. Moore, Do nanoparticles present ecotoxicological risks for the health of the aquatic environment? *Environ. Int.* 32 (2006) 967–976. https://doi.org/10.1016/j.envint.2006.06.014.

13. B. Song, Y. Zhang, J. Liu, X. Feng, T. Zhou, L. Shao, Is neurotoxicity of metallic nanoparticles the cascades of oxidative stress? *Nanoscale Res. Lett.* 11 (2016) 291. https://doi.org/10.1186/s11671-016-1508-4.

14. G. Elango, S.M. Kumaran, S.S. Kumar, S. Muthuraja, S.M. Roopan, Green synthesis of SnO2 nanoparticles and its photocatalytic activity of phenolsulfonphthalein dye, *Spectrochim. Acta - Part A Mol. Biomol. Spectrosc.* 145 (2015) 176–180. https://doi.org/10.1016/j.saa.2015.03.033.

15. D.M. Teleanu, C. Chircov, A.M. Grumezescu, R.I. Teleanu, Neurotoxicity of nanomaterials: An up-to-date overview, *Nanomaterials.* 9 (2019) 1–14. https://doi.org/10.3390/nano9010096.

16. S.J.S. Flora, Chapter 8: The applications, neurotoxicity, and related mechanism of gold nanoparticles, in: X. Jiang, H. Gao (Eds.), *Neurotox. Nanomater. Nanomedicine*, Academic Press, 2017: pp. 179–203. https://doi.org/10.1016/B978-0-12-804598-5.00008-8.

17. P. Saxena, V. Sangela, S. Ranjan, V. Dutta, N. Dasgupta, M. Phulwaria, D.S. Rathore, Harish, Aquatic nanotoxicology: impact of carbon nanomaterials on algal flora, *Energy, Ecol. Environ.* 5 (2020) 240–252. https://doi.org/10.1007/s40974-020-00151-9.

18. S. Bayda, M. Adeel, T. Tuccinardi, M. Cordani, F. Rizzolio, The history of nanoscience and nanotechnology: From chemical-physical applications to nanomedicine, *Molecules.* 25 (2020). https://doi.org/10.3390/molecules25010112.

19. S.S. Sanjay, Chapter 2: Safe nano is green nano, in: A. K. Shukla, S. Iravani (Eds.), *Green Synth. Charact. Appl. Nanoparticles*, Elsevier, 2019: pp. 27–36. https://doi.org/10.1016/B978-0-08-102579-6.00002-2.

20. P. Papolu, A. Bhogi, Green synthesis of various metal oxide nanoparticles for the environmental remediation-An overview, *Mater. Today Proc.* 92 (2023) 924–927. https://doi.org/https://doi.org/10.1016/j.matpr.2023.04.544.

21. D. Nath, P. Banerjee, Green nanotechnology - a new hope for medical biology, *Environ. Toxicol. Pharmacol.* 36 (2013) 997–1014. https://doi.org/10.1016/j.etap.2013.09.002.

22. S. Iravani, Green synthesis of metal nanoparticles using plants, *Green Chem.* 13 (2011) 2638–2650. https://doi.org/10.1039/C1GC15386B.

23. M. Ashour, A.T. Mansour, A.M. Abdelwahab, A.E. Alprol, Metal oxide nanoparticles' green synthesis by plants: Prospects in phyto- and bioremediation and photocatalytic degradation of organic pollutants, *Processes.* 11 (2023). https://doi.org/10.3390/pr11123356.

24. M. Das, S. Chatterjee, Chapter 11: Green synthesis of metal/metal oxide nanoparticles toward biomedical applications: Boon or bane, in: A. K. Shukla, S. Iravani (Eds.), *Green Synth. Charact. Appl. Nanoparticles*, Elsevier, 2019: pp. 265–301. https://doi.org/10.1016/B978-0-08-102579-6.00011-3.

25. W. Ahmad, A. Pandey, V. Rajput, V. Kumar, M. Verma, H. Kim, Plant extract mediated cost-effective tin oxide nanoparticles: A review on synthesis, properties, and potential applications, *Curr. Res. Green Sustain. Chem.* 4 (2021) 100211. https://doi.org/10.1016/j.crgsc.2021.100211.

26. S. O'Neill, J.M.C. Robertson, V. Héquet, F. Chazarenc, X. Pang, K. Ralphs, N. Skillen, P.K.J. Robertson, Comparison of titanium dioxide and zinc oxide photocatalysts for the inactivation of Escherichia coli in water using slurry and rotating-disk photocatalytic reactors, *Ind. & Eng. Chem. Res.* 62 (2023) 18952–18959. https://doi.org/10.1021/acs.iecr.3c00508.

27. R.S. Varma, Greener approach to nanomaterials and their sustainable applications, *Curr. Opin. Chem. Eng.* 1 (2012) 123–128. https://doi.org/10.1016/j.coche.2011.12.002.

28. A. Gour, N.K. Jain, Advances in green synthesis of nanoparticles, *Artif. Cells, Nanomed, Biotechnol.* 47 (2019) 844–851. https://doi.org/10.1080/21691401.2019.1577878.

29. S. Rehman, S.M. Asiri, F.A. Khan, B.R. Jermy, V. Ravinayagam, Z. Alsalem, R. Al Jindan, A. Qurashi, Anticandidal and in vitro anti-proliferative activity of sonochemically synthesized indium tin oxide nanoparticles, *Sci. Rep.* 10 (2020) 3228. https://doi.org/10.1038/s41598-020-60295-w.

30. I. Hussain, N.B. Singh, A. Singh, H. Singh, S.C. Singh, Green synthesis of nanoparticles and its potential application, *Biotechnol. Lett.* 38 (2016) 545–560. https://doi.org/10.1007/s10529-015-2026-7.

31. P. Yadav, M. Batra, N. Yadav, A. Bhaduri, Photocatalytic approach of camellia sinensis extract-mediated green-synthesized magnesium oxide nanoparticles, in: Z. H. Khan, M. Jackson, N. A. Salah (Eds.), *Recent Adv. Nanotechnol.*, Springer Nature Singapore, Singapore, 2023: pp. 555–564.

32. K.B. Narayanan, N. Sakthivel, Green synthesis of biogenic metal nanoparticles by terrestrial and aquatic phototrophic and heterotrophic eukaryotes and biocompatible agents, *Adv. Colloid Interface Sci.* 169 (2011) 59–79. https://doi.org/10.1016/j.cis.2011.08.004.

33. P. Sevilla, M. Hernández, E. Corda, G.R. García-Ramos, C. Domingo, Caracterización molecular de nanoportadores de fármacos mediante espectroscopía intensificada por plasmones localizados: Fluorescencia (SEF) y Raman (SERS), *Opt. Pura y Apl.* 46 (2013) 111–119. https://doi.org/10.7149/OPA.46.2.111.

34. V. Kumar, S.K. Yadav, Plant-mediated synthesis of silver and gold nanoparticles and their applications, *J. Chem. Technol. Biotechnol.* 84 (2009) 151–157. https://doi.org/10.1002/jctb.2023.

35. X. Wang, L. Yuan, H. Deng, Z. Zhang, Structural characterization and stability study of green synthesized starch stabilized silver nanoparticles loaded with isoorientin, *Food Chem.* 338 (2021) 127807. https://doi.org/10.1016/j.foodchem.2020.127807.

36. J.C. Warner, A.S. Cannon, K.M. Dye, Green chemistry, *Environ. Impact Assess. Rev.* 24 (2004) 775–799. https://doi.org/10.1016/j.eiar.2004.06.006.

37. P. Anastas, N. Eghbali, Green chemistry: Principles and practice, *Chem. Soc. Rev.* 39 (2010) 301–312. https://doi.org/10.1039/b918763b.

38. S.L.Y. Tang, R.L. Smith, M. Poliakoff, Principles of green chemistry: Productively, *Green Chem.* 7 (2005) 761–762. https://doi.org/10.1039/b513020b.

39. R.B. Rotti, D. V. Sunitha, R. Manjunath, A. Roy, S.B. Mayegowda, A.P. Gnanaprakash, S. Alghamdi, M. Almehmadi, O. Abdulaziz, M. Allahyani, A. Aljuaid, A.A. Alsaiari, S.S. Ashgar, A.O. Babalghith, A.E. Abd El-Lateef, E.B. Khidir, Green synthesis of MgO nanoparticles and its antibacterial properties, *Front. Chem.* 11 (2023) 1–13. https://doi.org/10.3389/fchem.2023.1143614.

40. M. Shah, D. Fawcett, S. Sharma, S.K. Tripathy, G.E.J. Poinern, Green synthesis of metallic nanoparticles via biological entities, *Materials.* (2015). https://doi.org/10.3390/ma8115377.

41. A.K. Mittal, Y. Chisti, U.C. Banerjee, Synthesis of metallic nanoparticles using plant extracts, *Biotechnol. Adv.* 31 (2013) 346–356. https://doi.org/10.1016/j.biotechadv.2013.01.003.

42. S. Shabina, S. Gaurav, A.M. Irfan, M. Sarmad, Green nanotechnology: A review on nanomedicinal potential and green synthesis of silver nanoparticles (ag-np's), *Res. J. Biotechnol.* 15 (2020) 177–187.

43. N. Matinise, X.G. Fuku, K. Kaviyarasu, N. Mayedwa, M. Maaza, ZnO nanoparticles via Moringa oleifera green synthesis: Physical properties & mechanism of formation, *Appl. Surf. Sci.* 406 (2017) 339–347. https://doi.org/10.1016/j.apsusc.2017.01.219.

44. S.A. Moon, B.K. Salunke, B. Alkotaini, E. Sathiyamoorthi, B.S. Kim, Biological synthesis of manganese dioxide nanoparticles by Kalopanax pictus plant extract, *IET Nanobiotechnol.* 9 (2015) 220–225. https://doi.org/10.1049/iet-nbt.2014.0051.

45. S.H. Gebre, M.G. Sendeku, New frontiers in the biosynthesis of metal oxide nanoparticles and their environmental applications: an overview, *SN Appl. Sci.* 1 (2019) 1–28. https://doi.org/10.1007/s42452-019-0931-4.

46. D. Suresh, R.M. Shobharani, P.C. Nethravathi, M.A. Pavan Kumar, H. Nagabhushana, S.C. Sharma, Artocarpus gomezianus aided green synthesis of ZnO nanoparticles: Luminescence, photocatalytic and antioxidant properties, *Spectrochim. Acta - Part A Mol. Biomol. Spectrosc.* 141 (2015) 128–134. https://doi.org/10.1016/j.saa.2015.01.048.

47. V.K. Vidhu, D. Philip, Biogenic synthesis of SnO2 nanoparticles: Evaluation of antibacterial and antioxidant activities, *Spectrochim. Acta - Part A Mol. Biomol. Spectrosc.* 134 (2015) 372–379. https://doi.org/10.1016/j.saa.2014.06.131.

48. F.T. Thema, E. Manikandan, A. Gurib-Fakim, M. Maaza, Single phase Bunsenite NiO nanoparticles green synthesis by Agathosma betulina natural extract, *J. Alloys Compd.* 657 (2016) 655–661. https://doi.org/10.1016/j.jallcom.2015.09.227.

49. S. Saleem, B. Ahmed, M.S. Khan, M. Al-Shaeri, J. Musarrat, Inhibition of growth and biofilm formation of clinical bacterial isolates by NiO nanoparticles synthesized from Eucalyptus globulus plants, *Microb. Pathog.* 111 (2017) 375–387. https://doi.org/10.1016/j.micpath.2017.09.019.

50. S.M. Roopan, S.H.S. Kumar, G. Madhumitha, K. Suthindhiran, Biogenic-production of SnO2 nanoparticles and its cytotoxic effect against hepatocellular carcinoma cell line (HepG2), *Appl. Biochem. Biotechnol.* 175 (2015) 1567–1575. https://doi.org/10.1007/s12010-014-1381-5.

51. S.K. Moorthy, C.H. Ashok, K.V. Rao, C. Viswanathan, Synthesis and characterization of Mgo nanoparticles by neem leaves through green method, *Mater. Today Proc.* 2 (2015) 4360–4368. https://doi.org/10.1016/j.matpr.2015.10.027.

52. S. Vijayakumar, B. Vaseeharan, Antibiofilm, anti-cancer and ecotoxicity properties of collagen based ZnO nanoparticles, *Adv. Powder Technol.* 29 (2018) 2331–2345. https://doi.org/10.1016/j.apt.2018.06.013.

53. H.M. Kim, N. Saito, D.W. Kim, Solution plasma-assisted green synthesis of MnO_2 adsorbent and removal of cationic pollutant, *J. Chem.* 2019 (2019). https://doi.org/10.1155/2019/7494292.

54. Q. Zhou, J.W. Yang, Y.Z. Wang, Y.H. Wu, D.Z. Wang, Preparation of nano-MgO/Carbon composites from sucrose-assisted synthesis for highly efficient dehydrochlorination process, *Mater. Lett.* 62 (2008) 1887–1889. https://doi.org/10.1016/j.matlet.2007.10.031.

55. S.K. Mohamed, A.M. Elhgrasi, O.I. Ali, Facile synthesis of mesoporous nano Ni/NiO and its synergistic role as super adsorbent and photocatalyst under sunlight irradiation, *Environ. Sci. Pollut. Res.* (2022) 64792–64806. https://doi.org/10.1007/s11356-022-19970-w.

56. J. Yang, K. qing Yang, L. Qiu, Biosynthesis of vitamin C stabilized tin oxide nanoparticles and their effect on body weight loss in neonatal rats, *Environ. Toxicol. Pharmacol.* 54 (2017) 48–52. https://doi.org/10.1016/j.etap.2017.06.013.

57. S. Azizi, M.B. Ahmad, F. Namvar, R. Mohamad, Green biosynthesis and characterization of zinc oxide nanoparticles using brown marine macroalga Sargassum muticum aqueous extract, *Mater. Lett.* 116 (2014) 275–277. https://doi.org/10.1016/j.matlet.2013.11.038.

58. S. Sathyavathi, A. Manjula, J. Rajendhran, P. Gunasekaran, Extracellular synthesis and characterization of nickel oxide nanoparticles from Microbacterium sp. MRS-1 towards bioremediation of nickel electroplating industrial effluent, *Bioresour. Technol.* 165 (2014) 270–273. https://doi.org/10.1016/j.biortech.2014.03.031.

59. N. Srivastava, M. Mukhopadhyay, Biosynthesis of SnO2 nanoparticles using bacterium erwinia herbicola and their photocatalytic activity for degradation of dyes, *Ind. Eng. Chem. Res.* 53 (2014) 13971–13979. https://doi.org/10.1021/ie5020052.

60. S.O. Ogunyemi, M. Zhang, Y. Abdallah, T. Ahmed, W. Qiu, M.A. Ali, C. Yan, Y. Yang, J. Chen, B. Li, The bio-synthesis of three metal oxide nanoparticles (ZnO, MnO2, and MgO) and their antibacterial activity against the bacterial leaf blight pathogen, *Front. Microbiol.* 11 (2020) 1–14. https://doi.org/10.3389/fmicb.2020.588326.

61. B.A. Abbasi, J. Iqbal, T. Mahmood, R. Ahmad, S. Kanwal, S. Afridi, Plant-mediated synthesis of nickel oxide nanoparticles (NiO) via Geranium wallichianum: Characterization and different biological applications, *Mater. Res. Express.* 6 (2019). https://doi.org/10.1088/2053-1591/ab23e1.

62. B. Sasi, K.G. Gopchandran, P.K. Manoj, P. Koshy, P. Prabhakara Rao, V.K. Vaidyan, Preparation of transparent and semiconducting NiO films, *Vacuum.* 68 (2002) 149–154. https://doi.org/10.1016/S0042-207X(02)00299-3.

63. B.T. Sone, X.G. Fuku, M. Maaza, Physical & electrochemical properties of green synthesized bunsenite NiO nanoparticles via Callistemon Viminalis' extracts, *Int. J. Electrochem. Sci.* 11 (2016) 8204–8220. https://doi.org/10.20964/2016.10.17.

64. C.J. Pandian, R. Palanivel, S. Dhananasekaran, Green synthesis of nickel nanoparticles using Ocimum sanctum and their application in dye and pollutant adsorption, *Chinese J. Chem. Eng.* 23 (2015) 1307–1315. https://doi.org/10.1016/j.cjche.2015.05.012.

65. S. Sudhasree, A. Shakila Banu, P. Brindha, G.A. Kurian, Synthesis of nickel nanoparticles by chemical and green route and their comparison in respect to biological effect and toxicity, *Toxicol. Environ. Chem.* 96 (2014) 743–754. https://doi.org/10.1080/02772 248.2014.923148.

66. N. Gong, K. Shao, W. Feng, Z. Lin, C. Liang, Y. Sun, Biotoxicity of nickel oxide nanoparticles and bio-remediation by microalgae Chlorella vulgaris, *Chemosphere.* 83 (2011) 510–516. https://doi.org/10.1016/j.chemosphere.2010.12.059.

67. Y. Zhang, Thermal oxidation fabrication of NiO film for optoelectronic devices, *Appl. Surf. Sci.* 344 (2015) 33–37. https://doi.org/10.1016/j.apsusc.2015.03.099.

68. M. Kundu, L. Liu, Binder-free electrodes consisting of porous NiO nanofibers directly electrospun on nickel foam for high-rate supercapacitors, *Mater. Lett.* 144 (2015) 114–118. https://doi.org/10.1016/j.matlet.2015.01.032.

69. R.A. Soomro, Z.H. Ibupoto, Sirajuddin, M.I. Abro, M. Willander, Electrochemical sensing of glucose based on novel hedgehog-like NiO nanostructures, *Sens. Actuators, B Chem.* 209 (2015) 966–974. https://doi.org/10.1016/j.snb.2014.12.050.

70. F.T. Thema, P. Beukes, A. Gurib-Fakim, M. Maaza, Green synthesis of Monteponite CdO nanoparticles by Agathosma betulina natural extract, *J. Alloys Compd.* 646 (2015) 1043–1048. https://doi.org/10.1016/j.jallcom.2015.05.279.

71. N. Thovhogi, A. Diallo, A. Gurib-Fakim, M. Maaza, Nanoparticles green synthesis by Hibiscus Sabdariffa flower extract: Main physical properties, *J. Alloys Compd.* 647 (2015) 392–396. https://doi.org/10.1016/j.jallcom.2015.06.076.

72. J. Iqbal, B.A. Abbasi, T. Mahmood, S. Kanwal, B. Ali, S.A. Shah, A.T. Khalil, Plant-derived anticancer agents: A green anticancer approach, *Asian Pac. J. Trop. Biomed.* 7 (2017) 1129–1150. https://doi.org/10.1016/j.apjtb.2017.10.016.

73. J. Iqbal, B.A. Abbasi, R. Batool, T. Mahmood, B. Ali, A.T. Khalil, S. Kanwal, S.A. Shah, R. Ahmad, Potential phytocompounds for developing breast cancer therapeutics: Nature's healing touch, *Eur. J. Pharmacol.* 827 (2018) 125–148. https://doi.org/10.1016/j.ejphar.2018.03.007.

74. A. Kar, Synthesis of nano-spherical nickel by templating hibiscus flower petals, *Am. J. Nanosci. Nanotechnol.* 2 (2014) 17. https://doi.org/10.11648/j.nano.20140202.11.

75. M.A. Nasseri, F. Ahrari, B. Zakerinasab, A green biosynthesis of NiO nanoparticles using aqueous extract of Tamarix serotina and their characterization and application, *Appl. Organomet. Chem.* 30 (2016) 978–984. https://doi.org/10.1002/aoc.3530.

76. A.A. Ezhilarasi, J.J. Vijaya, K. Kaviyarasu, M. Maaza, A. Ayeshamariam, L.J. Kennedy, Green synthesis of NiO nanoparticles using Moringa oleifera extract and their biomedical applications: Cytotoxicity effect of nanoparticles against HT-29 cancer cells, *J. Photochem. Photobiol. B Biol.* 164 (2016) 352–360. https://doi.org/10.1016/j.jphotobiol.2016.10.003.

77. J. Iqbal, B.A. Abbasi, T. Mahmood, S. Hameed, A. Munir, S. Kanwal, Green synthesis and characterizations of Nickel oxide nanoparticles using leaf extract of Rhamnus virgata and their potential biological applications, *Appl. Organomet. Chem.* 33 (2019) 1–16. https://doi.org/10.1002/aoc.4950.

78. N.Y. Abdou, M.M. Farag, W.M. Abd-Allah, Thermoluminescent properties of nano-magnesium phosphate ceramic for radiation dosimetry, *Eur. Phys. J. Plus.* 135 (2020) 1–12. https://doi.org/10.1140/epjp/s13360-020-00310-1.

79. A. Fatiqin, H. Amrulloh, W. Simanjuntak, Green synthesis of MgO nanoparticles using moringa oleifera leaf aqueous extract for antibacterial activity, *Bull. Chem. Soc. Ethiop.* 35 (2021) 161–170. https://doi.org/10.4314/BCSE.V35I1.14.

80. M. Mirhosseini, M. Afzali, Investigation into the antibacterial behavior of suspensions of magnesium oxide nanoparticles in combination with nisin and heat against Escherichia coli and Staphylococcus aureus in milk, *Food Control.* 68 (2016) 208–215. https://doi.org/10.1016/j.foodcont.2016.03.048.

81. S.L. Bhadargade, M. Kaushik, G. Joshi, A study of factors influencing the problem-solving skills of engineering students, *J. Eng. Educ. Transform.* 33 (2020) 8–19. https://doi.org/10.16920/jeet/2020/v33i4/143655.

82. T.D. Dang, M.A. Cheney, S. Qian, S.W. Joo, B.K. Min, A novel rapid one-step synthesis of manganese oxide nanoparticles at room temperature using poly(dimethylsiloxane), *Ind. Eng. Chem. Res.* 52 (2013) 2750–2753. https://doi.org/10.1021/ie302971g.

83. K.S. Prasad, A. Patra, Green synthesis of MnO2 nanorods using Phyllanthus amarus plant extract and their fluorescence studies, *Green Process. Synth.* 6 (2017) 549–554. https://doi.org/10.1515/gps-2016-0166.

84. H. Veeramani, D. Aruguete, N. Monsegue, M. Murayama, U. Dippon, A. Kappler, M.F. Hochella, Low-temperature green synthesis of multivalent manganese oxide nanowires, *ACS Sustain. Chem. Eng.* 1 (2013) 1070–1074. https://doi.org/10.1021/sc400129n.

85. A.M. Abdelgawad, M.E. El-Naggar, W.H. Eisa, O.J. Rojas, Clean and high-throughput production of silver nanoparticles mediated by soy protein via solid state synthesis, *J. Clean. Prod.* 144 (2017) 501–510. https://doi.org/10.1016/j.jclepro.2016.12.122.

86. G. Sangeetha, S. Rajeshwari, R. Venckatesh, Green synthesis of zinc oxide nanoparticles by aloe barbadensis miller leaf extract: Structure and optical properties, *Mater. Res. Bull.* 46 (2011) 2560–2566. https://doi.org/10.1016/j.materresbull.2011.07.046.

87. V. Hoseinpour, M. Souri, N. Ghaemi, A. Shakeri, Optimization of green synthesis of ZnO nanoparticles by Dittrichia graveolens (L.) aqueous extract, *Heal. Biotechnol. Biopharma.* 1 (2017) 39–49. DOI: 10.22034/HBB.2017.10.

88. D. Gnanasangeetha, D. Saralathambavani, One pot synthesis of zinc oxide nanoparticles via chemical and green method, *Res. J. Mater. Sci.* 1 (2013) 1–8.

89. C. Vidya, C. Manjunatha, M.N. Chandraprabha, M. Rajshekar, M.A.L. Antony Raj, Hazard free green synthesis of ZnO nano-photo-catalyst using Artocarpus Heterophyllus leaf extract for the degradation of Congo red dye in water treatment applications, *J. Environ. Chem. Eng.* 5 (2017) 3172–3180. https://doi.org/10.1016/j.jece.2017.05.058.

90. S.S. Majani, S. Sathyan, M.V. Manoj, N. Vinod, S. Pradeep, C. Shivamallu, V. K.N, S.P. Kollur, Eco-friendly synthesis of MnO2 nanoparticles using Saraca asoca leaf extract and evaluation of in vitro anticancer activity, *Curr. Res. Green Sustain. Chem.* 6 (2023) 100367. https://doi.org/10.1016/j.crgsc.2023.100367.

91. H. Lu, X. Zhang, S.A. Khan, W. Li, L. Wan, Biogenic synthesis of MnO2 nanoparticles with leaf extract of viola betonicifolia for enhanced antioxidant, antimicrobial, cytotoxic, and biocompatible applications, *Front. Microbiol.* 12 (2021). https://doi.org/10.3389/fmicb.2021.761084.

92. Y. Wang, L. Tan, L. Wang, Hydrothermal synthesis of SnO2 nanostructures with different morphologies and their optical properties, *J. Nanomater.* 2011 (2011). https://doi.org/10.1155/2011/529874.

93. Z.F. Sun, Y. Chang, N. Xia, Recent development of nanomaterials-based cytosensors for the detection of circulating tumor cells, *Biosensors.* 11 (2021) 281. https://doi.org/10.3390/bios11080281.

94. P.R. Chung, C.T. Tzeng, M.T. Ke, C.Y. Lee, Formaldehyde gas sensors: A review, *Sensors (Switzerland).* 13 (2013) 4468–4484. https://doi.org/10.3390/s130404468.

95. T.H. Chen, C.C. Lin, P.J. Meng, Zinc oxide nanoparticles alter hatching and larval locomotor activity in zebrafish (Danio rerio), *J. Hazard. Mater.* 277 (2014) 134–140. https://doi.org/10.1016/j.jhazmat.2013.12.030.

96. S. Lopes, F. Ribeiro, J. Wojnarowicz, W. Lojkowski, K. Jurkschat, A. Crossley, A.M.V.M. Soares, S. Loureiro, Zinc oxide nanoparticles toxicity to Daphnia magna: Size-dependent effects and dissolution, *Environ. Toxicol. Chem.* 33 (2014) 190–198. https://doi.org/10.1002/etc.2413.

97. S. Faisal, H. Jan, S.A. Shah, S. Shah, A. Khan, M.T. Akbar, M. Rizwan, F. Jan, Wajidullah, N. Akhtar, A. Khattak, S. Syed, Green synthesis of zinc oxide (ZnO) nanoparticles using aqueous fruit extracts of myristica fragrans: Their characterizations and biological and environmental applications, *ACS Omega.* 6 (2021) 9709–9722. https://doi.org/10.1021/acsomega.1c00310.

98. C.A. Soto-Robles, P.A. Luque, C.M. Gómez-Gutiérrez, O. Nava, A.R. Vilchis-Nestor, E. Lugo-Medina, R. Ranjithkumar, A. Castro-Beltrán, Study on the effect of the concentration of Hibiscus sabdariffa extract on the green synthesis of ZnO nanoparticles, *Results Phys.* 15 (2019) 102807. https://doi.org/10.1016/j.rinp.2019.102807.

99. A. Degefa, B. Bekele, L.T. Jule, B. Fikadu, S. Ramaswamy, L.P. Dwarampudi, N. Nagaprasad, K. Ramaswamy, Green synthesis, characterization of zinc oxide nanoparticles, and examination of properties for dye-sensitive solar cells using various vegetable extracts, *J. Nanomater.* 2021 (2021). https://doi.org/10.1155/2021/3941923.

100. A. Angel Ezhilarasi, J. Judith Vijaya, K. Kaviyarasu, L. John Kennedy, R.J. Ramalingam, H.A. Al-Lohedan, Green synthesis of NiO nanoparticles using Aegle marmelos leaf extract for the evaluation of in-vitro cytotoxicity, antibacterial and photocatalytic properties, *J. Photochem. Photobiol. B Biol.* 180 (2018) 39–50. https://doi.org/10.1016/j.jphotobiol.2018.01.023.

101. S.G. Firisa, G.G. Muleta, A.A. Yimer, Synthesis of nickel oxide nanoparticles and copper-doped nickel oxide nanocomposites using *Phytolacca dodecandra* L'Herit leaf extract and evaluation of its antioxidant and photocatalytic activities, *ACS Omega.* 7 (2022) 44720–44732. https://doi.org/10.1021/acsomega.2c04042.

102. H. Balto, M. Amina, R.S. Bhat, H.M. Al-Yousef, S.H. Auda, A. Elansary, Green synthesis of nickel nanoparticles using Salvadora persica and their application in antimicrobial activity against oral microbes, *Microbiol. Res. (Pavia).* 14 (2023) 1879–1893. https://doi.org/10.3390/microbiolres14040128.

103. F. Rahman, M.A. Majed Patwary, M.A. Bakar Siddique, M.S. Bashar, M.A. Haque, B. Akter, R. Rashid, M.A. Haque, A.K.M. Royhan Uddin, Green synthesis of zinc oxide nanoparticles using Cocos nucifera leaf extract: characterization, antimicrobial, antioxidant and photocatalytic activity, *R. Soc. Open Sci.* 9 (2022). https://doi.org/10.1098/rsos.220858.

104. P. Li, L. Chen, R. Qihe, G. Li, Magnetic crossover of NiO nanocrystals at room temperature, *Appl. Phys. Lett.* 89 (2006) 15–18. https://doi.org/10.1063/1.2357562.

105. K. Girigoswami, Toxicity of metal oxide nanoparticles, *Adv. Exp. Med. Biol.* 1048 (2018) 99–122. https://doi.org/10.1007/978-3-319-72041-8_7.

106. D. Sahu, G.M. Kannan, R. Vijayaraghavan, Size-dependent effect of zinc oxide on toxicity and inflammatory potential of human monocytes, *J. Toxicol. Environ. Heal. - Part A Curr. Issues.* 77 (2014) 177–191. https://doi.org/10.1080/15287394.2013.853224.

107. G.P. Gakis, I.G. Aviziotis, C.A. Charitidis, Metal and metal oxide nanoparticle toxicity: moving towards a more holistic structure-activity approach, *Environ. Sci. Nano.* 10 (2023) 761–780. https://doi.org/10.1039/d2en00897a.

108. N. Patil, R. Bhaskar, V. Vyavhare, R. Dhadge, V. Khaire, Y. Patil, Overview on methods of synthesis of nanoparticles, *Int. J. Curr. Pharm. Res.* 13 (2021) 11–16. https://doi.org/10.22159/ijcpr.2021v13i2.41556.

109. I. Bibi, S. Kamal, A. Ahmed, M. Iqbal, S. Nouren, K. Jilani, N. Nazar, M. Amir, A. Abbas, S. Ata, F. Majid, Nickel nanoparticle synthesis using Camellia Sinensis as reducing and capping agent: Growth mechanism and photo-catalytic activity evaluation, *Int. J. Biol. Macromol.* 103 (2017) 783–790. https://doi.org/10.1016/j.ijbiomac.2017.05.023.

110. A.M. Sadaa, Z.T. Al Abdullah, Green synthesis of nickel nanoparticles and their application of removal of aliphatic hydrocarbons from crude oil, *Iraqi J. Sci.* 62 (2021) 4333–4341. https://doi.org/10.24996/ijs.2021.62.11(SI).14.

111. S. Hussain, M. Ali Muazzam, M. Ahmed, M. Ahmad, Z. Mustafa, S. Murtaza, J. Ali, M. Ibrar, M. Shahid, M. Imran, Green synthesis of nickel oxide nanoparticles using Acacia nilotica leaf extracts and investigation of their electrochemical and biological properties, *J. Taibah Univ. Sci.* 17 (2023). https://doi.org/10.1080/16583655.2023.2170162.

112. S. Rahmani-Nezhad, S. Dianat, M. Saeedi, A. Hadjiakhoondi, Characterization and catalytic activity of plant-mediated MgO nanoparticles using Mucuna Pruriens L. seed extract and their biological evaluation, *J. Nanoanalysis.* 4 (2017) 290–298. https://doi.org/10.22034/jna.2017.540020.

113. D. Hooren, Comparative study of green synthesized magnesium oxide nanoparticles from amaranthus blitum and aloe vera, Int. J. Curr. Res. 10 (2018) 69395–69399. https://www.journalcra.com.

114. V. Mohanasrinivasan, C. Subathra Devi, A. Mehra, S. Prakash, A. Agarwal, E. Selvarajan, S. Jemimah Naine, Biosynthesis of MgO nanoparticles using Lactobacillus Sp. and its activity against human leukemia cell lines HL-60, *Bionanoscience.* 8 (2018) 249–253. https://doi.org/10.1007/s12668-017-0480-5.

115. K. Jhansi, N. Jayarambabu, K.P. Reddy, N.M. Reddy, R.P. Suvarna, K.V. Rao, V.R. Kumar, V. Rajendar, Biosynthesis of MgO nanoparticles using mushroom extract: effect on peanut (Arachis hypogaea L.) seed germination, *3 Biotech.* 7 (2017). https://doi.org/10.1007/s13205-017-0894-3.

116. V. Hoseinpour, M. Souri, N. Ghaemi, Green synthesis, characterisation, and photocatalytic activity of manganese dioxide nanoparticles, *Micro Nano Lett.* 13 (2018) 1560–1563. https://doi.org/10.1049/mnl.2018.5008.

117. F.M. Eissa, Green synthesis, antibacterial, and antifungal activities of 1,3,4–oxadiazines, *J. Heterocycl. Chem.* 55 (2018) 1479–1483. https://doi.org/10.1002/jhet.3194.

118. S. Haq, W. Rehman, M. Waseem, M. Shahid, M.U. Rehman, K.H. Shah, M. Nawaz, Adsorption of Cd2+ ions on plant-mediated SnO2 nanoparticles, *Mater. Res. Express.* 3 (2016) 1–9. https://doi.org/10.1088/2053-1591/3/10/105019.

119. K. Nithya, S. Kalyanasundharam, Effect of chemically synthesis compared to biosynthesized ZnO nanoparticles using aqueous extract of C. halicacabum and their antibacterial activity, *OpenNano.* 4 (2019). https://doi.org/10.1016/j.onano.2018.10.001.

120. I. Ijaz, E. Gilani, A. Nazir, A. Bukhari, Detail review on chemical, physical and green synthesis, classification, characterizations and applications of nanoparticles, *Green Chem. Lett. Rev.* 13 (2020) 59–81. https://doi.org/10.1080/17518253.2020.1802517.

121. S.S. Salem, A. Fouda, Green synthesis of metallic nanoparticles and their prospective biotechnological applications: an overview, *Biol. Trace Elem. Res.* 199 (2021) 344–370. https://doi.org/10.1007/s12011-020-02138-3.

8 Meshfree Radial Basis Function Pseudo-Spectral Computation of Nano-Fuels in Triple-Layer Magnetic Interfacial Flow

Rajesh Kumar Chandrawat, Gurpreet Singh Bhatia, Deepak Kumar, O. Anwar Bég, and S. Kuharat

8.1 INTRODUCTION

Nanotechnology has emerged as a significant contributor to various fields of thermal sciences and engineering, achieving significant successes in a range of technologies including energy, propulsion, batteries, fuel cells, etc. A highly promising application of nanotechnology involves the production of nanoparticles possessing high thermal conductivity, which are then blended with base fluids to create nanofluids. The combination of nanoparticles with the base fluid results in remarkable improvements in the thermal properties of the fluid, which can lead to elevating heat transfer and energy conversion. Nanofluids are beneficial in energy conversion systems, enhancing heat transfer efficiency in systems such as heat exchangers and engine cooling. They can be used strategically in thermal energy storage systems, and their precision synthesis and deployment have led to high heat transfer rates and reduced equipment size, resulting in significant energy savings. Therefore, scientists and researchers are continuing to explore the potential of nanofluids in nano-energy and developing new and innovative applications. Zhang and Liu [1] provided a thorough analysis of the fundamental characteristics of nano-liquid metal, asserting that the newly proposed substance represents a novel avenue for the development of highly potent energy materials. Park and Song [2] developed a bio-photovoltaic cell that employed hydrodynamic flow to improve the mass transfer efficiency of electrochemical devices in which a conductive nanofluid was developed with carbon nanotubes (CNTs) utilized as a filler. Mahian et al. [3] evaluated the efficiency of a solar still with a thermal transfer device using nanofluids, obtaining an experimental as well as a theoretical efficiency for freshwater yield and conducting both energy and exergy analyses. Cheedarala et al. [4] focused on extracting energy from time-dependent fluctuating water flow with a non-wettable polymer coating on the interior surface of flexible tubing. Salari et al. [5] investigated three-dimensional solar energy conversion system transport with integrated nanofluids and phase change materials. They focused on the use of nanofluids with a blend of nano-sized magnesium oxide, multiwall carbon nanotubes, and a hybrid composite dispersed in pure water. The study employed a mixture model instead of the more popular single-phase Tiwari-Das nanofluid model under laminar, incompressible conditions. They showed that the nature of the base fluid, the concentration of the nanoparticles, and the thickness of the phase change layer all exert a substantial influence on thermal performance. Eisapour et al. [6] investigated the useful work potential and energy yield of corrugated tubes thermal solar panels employing suspended material in a liquid coolant. They determined the optimum performance of the system by analyzing the exergy destruction and energy efficiency of the nano-slurry coolant fluid. Kaneez et al. [7] deployed a finite element method (FEM) to compute the dusty viscoelastic hybrid nanofluid flow from a stretching sheet with viscous dissipation using both copper oxide and aluminum oxide nanoparticles. Tonekaboni et al. [8] investigated nanofluid-based solar CCHP (combined cooling, heating, and power systems) using a combination of porous materials and copper oxide/alumina nanoparticles, noting that both collector energy efficiency and exergy efficiency are improved. Subhedar et al. [9] demonstrated in their research that the use of nano-coolants in machining processes has several benefits, including an increase in the tool life of cutting tools due to the improved thermal conductivity and lubricity of the coolant. Additionally, the use of nano-coolants has been found to reduce the surface roughness of machined parts, resulting in improved surface finish, mitigation of surface wear, and reduced production costs. These findings suggest that nano-coolants have the potential to significantly enhance the efficiency and quality of machining processes in various industries. Zhang et al. [10] developed stable nano-emulsions of PCM (phase change material)-water, which were prepared with n-hexadecane by changing the key amalgamation factors, specifically the emulsifying agent mixtures and processing factors. Krishna et al. [11] concentrated on the recent developments in heat transfer fluids, focusing on the stability of nanomaterial suspensions. They drew comparisons between different heat transfer fluids that are presently employed in parabolic trough collectors with those that are enhanced with nanoparticles and highlighted that the latter attain properties superior to traditional heat transfer fluids. Gui [12] provided an overview of recent advances in research on systems that harness and convert energy at the micro and nano-scale for practical applications. These systems are typically designed to operate in highly efficient and compact environments, such as microelectronic devices, sensors, and energy storage systems. They can be deployed effectively to enhance and compliment a wide range of energy sources, including thermal, mechanical, and electromagnetic energy, and can convert them into different forms of energy, such as electrical power or light.

Dusty fluids, also known as *particle-laden fluids or fluid-particle suspensions*, are fluids that contain solid particles dispersed in a viscous medium. These complex liquids can also contribute certain unique benefits to nano-energy applications. Similar to nanofluids, the presence of solid particles in dusty fluids can enhance heat transfer in energy systems such as heat exchangers, engine cooling systems, and refrigeration systems. The particles can act as nucleation sites for boiling or condensation, leading to improved heat transfer efficiency. The addition of solid particles in energy conversion systems such as solar thermal power plants and thermoelectric generators can lead to a sustained increase in energy conversion efficiency. The particles can absorb solar radiation or improve electrical conductivity, leading to a marked elevation in energy conversion efficiency. Exploration of suspended fluids in energy is an ongoing endeavor, with scientists and researchers pushing the boundaries to uncover new and innovative applications. A key driver for this is the sustainability of the planet, which is experiencing dramatic issues including global warming, climate change, proliferation of natural disasters all of which are being influenced by the current dominance of unclean energy generation (coal, gas, oil, etc.). The motivation for cleaner, sustainable renewable systems deploying optimized combinations of nanofluids and dusty fluids is therefore very high. In parallel with experimental studies, many excellent mathematical and numerical studies of dusty fluids in a range of energy and industrial applications have been communicated. The seminal investigation into the laminar flow of dusty media (two-phase fluid-particle suspensions) was conducted by Saffman at Caltech in the early 1960s [13]. He examined the stability of the flow of dusty fluid that had uniformly dispersed particles. Gnaneswara Reddy and Ferdows [14] investigated the magnetized radiative micropolar dusty boundary layer flow external to a parabolic geometry, observing that with an increment in Erngen vortex viscosity number, both fluid and particle velocity are enhanced whereas the micro-rotation (angular velocity) is suppressed. Gnaneswara Reddy et al. [15] deployed a non-Fourier heat flux model to investigate the boundary layer flow of carbon nanotube (CNT) hybrid dusty nanofluid from a stretching surface adjacent to a Darcy-Forchheimer porous medium. They noted that greater wall Nusselt numbers are computed for the hybrid dusty nanofluid relative to the non-dusty nanofluid. Ali et al. [16] scrutinized the hydromagnetic thermal convection in dusty Casson fluid flow in a horizontal duct. Chandrawat et al. [17] utilized a stable numerical interface tracking algorithm to simulate the unsteady flow of stratified flow in a horizontal duct containing an upper Saffman dusty fluid layer and a lower Newtonian fluid layer. Mahanthesh et al. [18] simulated the two-phase Marangoni boundary layer radiating dusty nanofluid under a temperature gradient with exponential space-dependent heat generation. Muthuraj et al. [19] used a perturbation technique to derive solutions for thermo-capillary (Marangoni) convection in thermo-solutal peristaltic pumping of dusty fluid in a deformable channel. Kalpana et al. [20] used a finite difference to investigate the MHD two-phase Saffman fluid flow in a porous medium.

The above research has emphasized the need for elucidating as accurately as possible, the behavior of immiscible fluids with varying densities and viscosities in energy systems, as such studies can help to better understand the dynamics of multi-phase flows and improve the design and efficiency of such power generation designs. Furthermore, various industrial processes, such as oil extraction and electromagnetic purification, chemical production, and materials processing prominently feature interfacial transport phenomena. Immiscible fluids are two or more fluids that do not mix together and form distinct boundaries or interfaces between them. Examples of immiscible fluids include oil and water, mercury and water, and air and water. The connective layer between immiscible fluids is exhibited by the surface tension of the fluids, which causes the fluids to resist mixing and maintain their distinct boundaries. Immiscible fluids are often studied in energy, materials, and chemical engineering to improve technical designs. For example, immiscible (interfacial) dynamics can assist in understanding the fate of oil spills on water or the dynamics of emulsions, paints, gels, and slurries. Abd Elmaboud et al. [21] scrutinized the electromagnetic flow of two interfacial fluids with different densities in an inclined duct, considering also thermal conductivity and viscosity effects. They considered the scenario in which the duct features two distinct regions: one contained a permeable matrix saturated with a Newtonian fluid, while the other contained pure fluid. Chen and Jian [22] focused on the entropy generation in a conjugate heat transfer system involving two immiscible fluids. Salari et al. [23] presented the computational analysis of fluid motion driven by temperature differences for two immiscible fluids namely air and Al_2O_3-water nanofluid within a horizontal duct, the heat input is provided from the duct walls. Dong et al. [24] studied the wavy interface instability of two immiscible fluids in a parallel plate channel using the lattice Boltzmann method (LBM). Chandrawat and Joshi [25] motivated by biomechanical transport, developed a fuzzy solution for a dual immiscible fluids channel (Newtonian and micropolar liquids). Wang et al. [26] examined the displacement of immiscible fluids at microscale regimes, where slip lengths are comparable to the characteristic sizes of the system, noting that Cox's law is violated.

In the study of fluid dynamics, the temporal evolution of a flow is a critical aspect that cannot be ignored, as nearly all real-world fluid flows exhibit *time-dependent* behavior. Therefore, there is a compelling need to conduct time-dependent flow analysis in order to accurately capture and understand the dynamic nature of fluid flows and their implications in various fields such as aerospace heat pipes, environmental hydrodynamics (fate transport of chemicals), industrial slurry processing, biowaste hazard transport, nuclear duct thermal management and biotechnological manufacturing (dermatological creams, gels, disinfectants for COVID-19, etc.). Unsteady flow characteristics (pressure, vorticity, shear stress, etc.) vary with time and space and indeed apply to both compressible and incompressible regimes. This is in contrast to steady flow, where the fluid properties do not change with time at any point in the flow field. Unsteady flow is commonly observed in fluid systems where there are time-dependent boundary conditions, such as flow around a moving object, unsteady heat transfer, or in pulsating flow in the human circulatory system. Unsteady flow is often analyzed using computational fluid dynamics (CFD) or experimental techniques to understand the complex flow patterns and their effects on the system being studied. The mathematical simulation of time-dependent flows in a hydraulic system consisting of a pressure tank was examined by Firkowski et al. [27]. The investigation involved practical trials for examining the impact of unsteady friction on the simulation results. Asjad et al. [28] discussed the time-dependent flow of a fractional Casson fluid in the presence of microbial convection. They modeled the field equations with fractional coefficients of derivatives and extracted Laplace transform solutions for the non-dimensional problem.

Another key area of development in the 21st century is nanofluids. These comprise nano-sized particles suspended in base liquids such as water, ethylene glycol, oils, etc. This achieves significant elevation in thermal performance for a range of applications from biomedicine to energy storage. Most studies of nanofluids have utilized single-phase approaches. However, multi-phase models are more accurate since they feature a framework for simulating nanoparticle diffusion, and these include Buongiorno's two-component model, mixture models, and Safmann fluid-particle two-phase models. Mburu et al. [29] utilized the Buongiorno two-component nanoscale model to examine the impact of radiative heat flux and chemical reaction on the irreversible dissipative unsteady nanofluid flow in an inclined channel. Dogonchi et al. [30] computed the magnetized hybrid nanoliquid in a wavy enclosure containing a porous medium and three circular cylinders, also conducting a second law of thermodynamic optimization. Nikodijevic et al. [31] examined the unsteady magneto-convective flow through a duct containing a porous medium. Some further studies of nanofluid unsteady transport in ducts are provided in references [32–37].

Multi-phase and nanofluid duct transport extensively feature nonlinear coupled differential equation systems with complex boundary conditions. These boundary value problems are very challenging to solve analytically, if not intractable. Engineers have developed therefore an extensive range of numerical methods which provide a practical means of extracting approximate but accurate solutions for such problems. Many methods have evolved in the past 50 years, notably finite element, finite difference, and spectral methods. An alternative approach is the group of methods known as Radial Basis Function (RBF) techniques. These methods are also very robust for accommodating coupled, nonlinear, multi-degree partial differential equations (PDEs). The key idea behind RBF methods is to represent the solution to the PDE as a *weighted sum of radial basis functions* centered at certain points in the domain. RBF methods have several advantages over other numerical methods such as finite difference or finite element methods. For example, RBF methods do not require a mesh to be constructed, which can be time-consuming and difficult for complex geometries. RBF methods therefore fall into the category known as "meshless methods" which also include SPH (smoothed particle hydrodynamics). RBF methods are also able to handle irregular and scattered data, which makes them useful for problems where the data is not evenly distributed. American engineer, E.J. Kansa [38] based at Lawrence Livermore National Laboratory, California introduced radial basis function (RBF) methods to solve PDEs in the early 1990s. Since then, RBF methods have become an increasingly popular and powerful tool for solving a wide range of PDEs. More insights on RBFs can be seen in Refs. [39–41]. The radial basis function pseudo-spectral method (RBFPs) [42] combines the accuracy of spectral methods with the flexibility of RBF methods, making it an optimized numerical tool for solving PDEs. In this method, the solution of a PDE is approximated as a sum of RBFs multiplied by time-dependent coefficients. The coefficients are determined by solving an ordinary differential equation obtained by applying a spectral method to the PDE. This method can be seen as a variant of the pseudo-spectral method.

An inspection of the scientific literature has revealed that thus far the *magneto-convective transport in a triple-layered horizontal duct containing three immiscible fluids (nanofluid, Safmann dusty fluid and Newtonian fluid), and featuring viscous dissipation and three distinct pressure gradients*, has, thus far not been simulated numerically. This is the objective of the present work which is motivated by emerging applications in nano-fuel cells, nanomagnetic batteries, and smart energy applications. The nanofluid consists of nanoparticle $(CuO, Al_2O_3, Ag,$ and $TiO_2)$ suspended in an ethylene glycol-water mixture based on the delayed exothermic behavior. This study also aims to explore, both theoretically and computationally, the combined impacts of ion slip, viscous heating, and Hall current on the unsteady flow of these three immiscible fluids. The RBFPS method is utilized to compute all key flow characteristics, and the obtained results are compared to those obtained using a strong stability preserving differential quadrature (DQ) numerical scheme. A lucid discourse on the physics of the flow phenomena is also included with some future pathways for extending the simulations. Applications of the current study include nuclear duct interfacial MHD flows, nano-fuel, magnetic two-phase heat exchangers, electromagnetic petroleum electromagnetic purification and thermomagnetic duct materials processing.

8.2 MATHEMATICAL MODEL FOR TRIPLE IMMISCIBLE FLUID LAYERED MHD DUCT FLOW

The physical regime to be studied is depicted in Figure 8.1.

The flow of three different fluids, namely, nanofluid, Saffman dusty fluid, and Newtonian fluid, is analyzed in this study. The flow is electrically conducting, unsteady, fully developed, laminar, unidirectional, immiscible, and incompressible. The duct comprises two horizontal solid parallel plates (no wall suction/injection at the plates) and is visualized in Figure 8.1. Magnetic induction effects are ignored since magnetic Reynolds number is sufficiently small for the magnetic field lines to remain undistorted. Maxwell displacement current and Joule (Ohmic) magnetic heating are also ignored. Based on these assumptions, the governing equations are formulated for each stratum of the three fluids.

8.2.1 Nanofluid (Lower Layer)

The field equations governing the lower nanofluid layer take the form [43]:
Continuity equation:

$$(\rho_{nf})_t - \nabla \cdot \left(\rho_{nf}\overrightarrow{u_{nf}}\right) = 0. \tag{8.1}$$

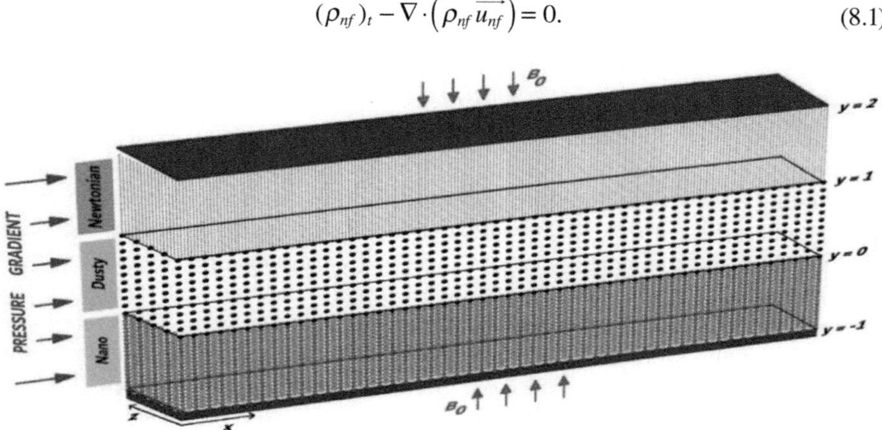

FIGURE 8.1 Electromagnetic duct transport of immisicible nanofluid, dusty fluid and Newtonian fluid.

Fluid and Particulate phase momenta equations:

$$\rho_{nf}\left(\frac{\partial \overrightarrow{u_{nf}}}{\partial t}+\left(\overrightarrow{u_{nf}}\cdot\nabla\right)\overrightarrow{u_{nf}}\right)=-\nabla p+\mu_{nf}\nabla^2\overrightarrow{u_{nf}}+\overrightarrow{f_b},\tag{8.2a}$$

$$\rho_{nf}c_{nf}\cdot\left(\frac{\partial \overrightarrow{T_{nf}}}{\partial t}+\left(\overrightarrow{u_{nf}}\cdot\nabla\right)\overrightarrow{T_{nf}}\right)=\kappa_{nf}\nabla^2\overrightarrow{T_{nf}}+\mu_{nf}\left(\nabla\overrightarrow{u_{nf}}\right)^2.\tag{8.2b}$$

8.2.2 DUSTY SAFFMAN FLUID (INTERMEDIATE LAYER)

The field equations governing the dusty (Saffman) intermediate fluid are as follows ([44,46]):

Continuity equation.

$$\varphi_t-\nabla\cdot\left(\left(1-\varphi\right)\overrightarrow{u_d}\right)=0,\tag{8.3}$$

$$\varphi_t-\nabla\cdot\left(\varphi\overrightarrow{u_{dp}}\right)=0.\tag{8.4}$$

Fluid and Particulate phase momenta equations:

$$\rho_d\left(1-\varphi\right)\cdot\left(\frac{\partial \overrightarrow{u_d}}{\partial t}+\left(\overrightarrow{u_d}\cdot\nabla\right)\overrightarrow{u_d}\right)=\left(1-\varphi\right)\left(-\nabla p+\mu_d\nabla^2\overrightarrow{u_d}\right)-M\rho_{dp}\varphi\left(\overrightarrow{u_d}-\overrightarrow{u_{dp}}\right),\tag{8.5}$$

$$\rho_{dp}\varphi\left(\frac{\partial \overrightarrow{u_{dp}}}{\partial t}+\left(\overrightarrow{u_{dp}}\cdot\nabla\right)\overrightarrow{u_{dp}}\right)=M\rho_{dp}\varphi\left(\overrightarrow{u_d}-\overrightarrow{u_{dp}}\right).\tag{8.6}$$

$$\rho_d c_p\left(1-\varphi\right)\cdot\left(\frac{\partial \overrightarrow{T_d}}{\partial t}+\left(\overrightarrow{u_d}\cdot\nabla\right)\overrightarrow{T_d}\right)=\left(1-\varphi\right)\kappa_d\nabla^2\overrightarrow{T_d}$$

$$+\left(1-\varphi\right)\mu_d\left(\nabla\overrightarrow{u_d}\right)^2-\rho_{dp}c_{dp}\frac{\varphi\left(\overrightarrow{T_d}-\overrightarrow{T_{dp}}\right)}{\gamma_T}+M\rho_{dp}\varphi\left(\overrightarrow{u_d}-\overrightarrow{u_{dp}}\right)^2,\tag{8.7}$$

$$\rho_{dp}c_{dp}\varphi\left(\frac{\partial \overrightarrow{T_{dp}}}{\partial t}+\left(\overrightarrow{u_{dp}}\cdot\nabla\right)\overrightarrow{T_{dp}}\right)=-\rho_{dp}c_{dp}\frac{\varphi\left(\overrightarrow{T_{dp}}-\overrightarrow{T_d}\right)}{\gamma_T}.\tag{8.8}$$

8.2.3 PURELY NEWTONIAN FLUID (UPPER LAYER)

The field equations governing the *non-dusty Newtonian uppermost stratum in the duct* are [47,48]:

Continuity equation:

$$\left(\rho_n\right)_t-\nabla\cdot\left(\rho_n\overrightarrow{u_n}\right)=0,\tag{8.9}$$

$$\rho_n\left(\frac{\partial \vec{u}_n}{\partial t} + \left(\vec{u}_n \cdot \nabla\right)\vec{u}_n\right) = -\nabla p + \mu_n \nabla^2 \vec{u}_n + \vec{f}_p \tag{8.10}$$

$$\rho_n c_n \cdot \left(\frac{\partial \vec{T}_n}{\partial t} + \left(\vec{u}_d \cdot \nabla\right)\vec{T}_n\right) = \kappa_n \nabla^2 \vec{T}_n + \mu_n \left(\nabla \vec{u}_n\right)^2. \tag{8.11}$$

Here $\rho_{nf} = (1-\Psi)\rho_m + \Psi\rho_{np}$ is the density of the nanofluid, ρ_m: density of mixture of ethylene glycol (Eg-Water), ρ_{np}: density of the nanoparticles, Ψ: the volume fraction function for the nanoparticles, $\mu_{nf} = \dfrac{\mu_m}{(1-\Psi)^{2.5}}$: viscosity of the nanofluid, μ_m: viscosity of the ethylene glycol mixture, $\vec{u}_{nf}$: velocity of the nanofluid, $-\nabla p$: pressure term, $\vec{f}_b$: body force, $\vec{T}_{nf}$: temperature of the nanofluid, $c_{nf} = (1-\Psi)c_m + \Psi c_{np}$: specific heat of nanofluid with c_m: specific heat of mixture, c_{np}: specific heat of nanoparticles, $\kappa_{nf} = \dfrac{k_{np} + 2k_m - 2\left(k_m - k_{np}\right)\Psi}{k_{np} + 2k_m + \left(k_m - k_{np}\right)\Psi}$: thermal conductivity of nanofluid with k_m: thermal conductivity of mixture, k_{np}: thermal conductivity of nanoparticles, φ: volume fraction function for dusty particles, $\vec{u}_d$: velocity of the dusty fluid, $\vec{u}_{dp}$: velocity of the dust particle, ρ_d: density of the dusty fluid, μ_d: viscosity of the dusty particle, M: volume transfer coefficient (see Equation 8.13), ρ_{dp}: density of dust particle, $\vec{T}_d$: temperature of the dusty-fluid, c_p: specific heat of dusty fluid, c_{dp}: specific heat of dust particle, κ_d: thermal conductivity of dusty fluid, γ_T: temperature relaxation parameter, ρ_n: density of the Newtonian fluid, $\vec{u}_n$: velocity of Newtonian fluid, μ_n: viscosity of the Newtonian fluid, $\vec{T}_n$: temperature of the Newtonian fluid, c_n: specific heat of Newtonian fluid, κ_n: thermal conductivity of Newtonian fluid.

8.3 FORMULATION OF TRIPLE-LAYER DUCT TRANSPORT PROBLEM

8.3.1 UNI-DIRECTIONAL NON-DIMENSIONAL MODELS

As visualized in Figure 8.1, the lower zone of the duct, i.e., **zone 1** $(-1 \leq y \leq 0)$, is occupied by nanofluid, which has a velocity of $\left(u_{nf}(y,t),0,0\right)$, temperature $\left(T_{nf}(y,t),0,0\right)$, a density ρ_{nf}, and a viscosity μ_{nf}. The middle space or **zone 2** $(0 \leq y \leq 1)$ is occupied by Saffman dusty fluid with a velocity of $\left(u_d(y,t),0,0\right)$, temperature $\left(T_d(y,t),0,0\right)$, a density ρ_d, and a viscosity μ_d. The dust particles in the Saffman dusty fluid have a velocity $\left(u_{dp}(y,t),0,0\right)$, temperature $\left(T_{dp}(y,t),0,0\right)$, a density ρ_{dp}, and a viscosity μ_{dp}. The upper zone or **zone 3** $(1 \leq y \leq 2)$ is occupied by Newtonian fluid with a velocity $\left(u_n(y,t),0,0\right)$, temperature $\left(T_n(y,t),0,0\right)$, a density ρ_n, and a viscosity μ_n. The transport properties are the same in all the zones, and a consistent pressure is applied in the horizontal (X) i. e. axial direction.

When electrically conductive fluids are subjected to a magnetic field, they experience magnetohydrodynamic effects due to the interaction between the magnetic field and the moving charged particles in the fluid. These effects can result in the generation of electrical currents, pressure gradients, and changes in fluid flow patterns. In the presence of a transverse magnetic field, the fluid experiences resistance to movement due to the Lorentz force, which generates drag on the fluid velocities $(\vec{u_l} = \vec{u_{nf}}, \vec{u_d}, \vec{u_n})$. The walls of the channel (duct) through which the fluids flow are typically designed to act as electrical shields, thereby minimizing any unwanted effects on the fluid behavior [44]:

$$J \times B_0 = \frac{\sigma B_0^2 (1 + Bi \cdot Be)\vec{u_l}}{(1 + Bi \cdot Be)^2 + Be^2}$$ (8.12)

Here J is current density, B_0 is the external magnetic field, σ is the electric conductivity, Bi is the ion slip parameter, Be is the Hall current parameter. In *the dusty region of the fluid*, the dust particles are non-deformable and uniformly distributed, with identical size, mass, and density. The dust particles have a particle velocity u_{dp}, density ρ_{dp}, and average mass m_p and are characterized by a particle volume fraction function, $\varnothing$. The particle phase is assumed to be sufficiently dilute such that interparticle interactions can be ignored, and the dust particle size is negligible. Consequently, the net effect of the dust particles on the fluid can be described as an additional force per unit volume, which is a function of the particle volume fraction and is written as [44,46]:

$$D_f = M\rho_{dp}\varphi\left(\vec{u_d} - \vec{u_{dp}}\right)$$ (8.13)

Here $M = \dfrac{6\pi r\mu_d}{m_p}$ is the volume transfer coefficient with radius r. Within the dusty fluid, dust particles absorb thermal energy (heat) from the surrounding fluid via conduction heat transfer across their surfaces. The heat transfer rate depends on the temperature relaxation parameter γ_T, which exhibits an inverse relationship with the Stoke drag coefficient $K^* = 6\pi r\mu_d U_0$ as elaborated in Refs. [44,46]:

$$\gamma_T = \frac{3}{2}\frac{\rho_p \cdot c_p \cdot \mu_d}{K^* \cdot N \cdot k_d}$$ (8.14)

It is assumed that the flow rate and shear are stable at the interface between the three fluids.

Introducing the non-dimensional parameters $\bar{x} = \dfrac{x}{k}, \bar{y} = \dfrac{y}{k}, \overline{u_{nf}} = \dfrac{u_{nf}}{U_0}, \overline{u_d} = \dfrac{u_d}{U_0},$

$\overline{u_{dp}} = \dfrac{u_{dp}}{U_0}, \overline{u_n} = \dfrac{u_n}{U_0}, \bar{p} = \dfrac{p}{\rho_1 U_0^2}, \bar{t} = \dfrac{tU_0}{k}, R = \dfrac{K^* Nh^2}{\mu_d}$, and, $R_1^* = \dfrac{\rho_m}{\rho_{np}}$ is the ratio

of the density of the fluid mixture and nano particle, $R_2^* = \dfrac{k_{np}}{k_m}$ is the ratio of the

thermal conductivity of the nano particle and fluid mixture, $R_3^* = \dfrac{(c_p)_{np}}{(c_p)_m}$ is the ratio of the specific heat of the nano particle and fluid mixture, $\text{Re} = \dfrac{\rho_m U_0 h^2}{\mu_m}$ is the Reynolds number, $\text{Ha}^2 = \dfrac{\sigma B_0^2 h^2}{\mu_m}$ is the Hartmann number, $\text{Ec} = \dfrac{u_0^2}{\left(c_p\right)_m \left(T_2 - T_1\right)}$ is the Eckert number which is the ratio of the kinetic energy of the flow to the enthalpy difference (viscous dissipation), $R = \dfrac{k^* \rho_p h^2}{\mu_d m_p}$ is particle concentration parameter, $\text{Pr} = \dfrac{\mu_m \left(c_p\right)_m}{k_m}$ is the Prandtl number,

$$R_4 = \frac{\rho_d}{\rho_m}, \ R_6 = \frac{\rho_d}{\rho_{dp}}, \ R_{10} = \frac{\rho_n}{\rho_d}, \ R_5 = \frac{\mu_d}{\mu_m}, \ R_{11} = \frac{\mu_n}{\mu_d}, \ R_8 = \frac{k_d}{k_m}, \ R_{13} = \frac{k_n}{k_d}, \ R_7 = \frac{\left(c_p\right)_d}{\left(c_p\right)_m},$$

$$R_9 = \frac{\left(c_p\right)_d}{\left(c_p\right)_{dp}}, \ R_{12} = \frac{\left(c_p\right)_m}{\left(c_p\right)_d}$$ are the ratio of viscosity, density, thermal conductivity and

specific heat of the three liquids. $-\nabla p = -\partial p / \partial x = Ge(t)$ is the *externally applied time-dependent axial pressure gradient*. As in the lower zone of the channel, various nanoparticles are incorporated in the ethylene glycol-water mixture. The densities of water, and ethylene glycol are considered as 997.1, and 1,114 kg/m³, respectively, similarly the thermal conductivities are 0.613 and 0.252 W/m K, and the specific heat capacities are 4,179 and 2,415 J/kg K [49].

The mixture of ethylene glycol-water base fluid for the lower layer nanofluid stratum obeys the following relation [49]:

$$\zeta_{\text{mix}} = 0.573\zeta_{EG} + \left(1 - 0.573\right)\zeta_{\text{Water}}, \tag{8.15}$$

where ζ is any fluid attribute, such as density, specific heat, etc. Table 8.1 provides the values of different ratios R_1^*, R_2^* and R_3^* for various nanoparticles. The velocities and temperature profiles are computed based upon these ratios for different nanoparticle cases; subsequently, one nanoparticle material is selected for all subsequent computations.

For the numerical analysis three different cases for $Ge(t)$ are considered:

Case-I: $Ge(t) = 10$ (constant pressure gradient).

Case-II: $Ge(t) = 10 * \sin\left(\alpha t\right)$ (with an oscillating parameter α).

Case-III: $Ge(t) = 10 * e^{-\lambda t}$ (with decaying parameter λ).

TABLE 8.1

Property Ratios, R_1^*, R_2^* and R_3^* for Various Nanoparticles [49]

Nanoparticle	Density $\left(\text{kg}/\text{m}^3\right)$	Specific Heat Capacity $\left(\text{J}/\text{kg K}\right)$	Thermal Conductivity $\left(\text{W}/\text{m K}\right)$	R_1^*	R_2^*	R_3^*
CuO	6,510	540	18	0.16345	44.31892	0.17044
Al₂O₃	3,970	765	40	0.26803	98.48650	0.24146
TiO₂	4,250	686.2	8.9528	0.25037	22.04325	0.21658
Ag	10,500	235	429	0.10134	1056.26780	0.07417

After removing the bars and introducing the aforementioned non-dimensional parameters along with appropriate initial, interfacial, and boundary conditions based on the three fluid flow scheme, the equations can be expressed as:

Zone-I (Nano-fluid: $-1 \leq y \leq 0$)

$$\frac{\partial u_{nf}}{\partial t} = \frac{Ge(t)}{(1-\Psi)\left(1+\frac{\Psi}{1-\Psi}\cdot\frac{1}{R_1^*}\right)} + \frac{1}{Re}\frac{\partial^2 u_{nf}}{\partial y^2}\frac{1}{(1-\varphi)^{3.5}\left(1+\frac{\Psi}{1-\Psi}\cdot\frac{1}{R_1^*}\right)}$$

$$-\frac{Ha^2}{Re}\left(\frac{1+BiBe}{(1+BiBe)^2+Re^2}\frac{1}{\frac{\Psi}{R_1^*}+(1-\Psi)}\right)u_{nf}, \qquad (8.16)$$

$$\frac{\partial T_{nf}}{\partial t}$$

$$= \frac{1}{Re\,Pr}\left(\frac{R_2^*(1+2\Psi)+2(1-\Psi)}{(R_2^*(1-\Psi)+(2+\Psi))(1-\Psi)^2\left\{1+\left(\frac{\Psi}{1-\Psi}\right)\frac{1}{R_1^*}\right\}\left(1+\left(\frac{\Psi}{1-\Psi}\right)R_3^*\right)}\right)\frac{\partial^2 T_{nf}}{\partial y^2}+$$

$$\frac{Ec}{Re}\left(\frac{1}{(1-\Psi)^{4.5}}\right)\left(\frac{1}{1+\left(\frac{\Psi}{1-\Psi}\right)\frac{1}{R_1^*}}\right)\left(\frac{1}{(1+\left(\frac{\Psi}{1-\Psi}\right)R_3^*}\right)\left(\frac{\partial u_{nf}}{\partial y}\right)^2$$

$$+\left(\frac{Ha^2 Ec(1+BiBe)}{(Re(1+BiBe)^2+Be^2)\left(1+\left(\frac{\Psi}{1-\Psi}\right)\frac{1}{R_1^*}\right)}\right)\left(\frac{1}{1+\left(\frac{\Psi}{1-\Psi}\right)R_3^*}\right)u_{nf}^2. \qquad (8.17)$$

Zone-II (Dusty fluid: $0 \leq y \leq 1$)

$$\frac{\partial u_d}{\partial t} = \frac{Ge(t)}{R_4}+\frac{R_5}{Re\,R_4}\frac{\partial^2 u_d}{\partial y^2}-\left(\frac{1+BiBe}{(1+BiBe)^2+Be^2}\right)\frac{Ha^2}{Re\,R_4}u_d-\frac{RR_5}{Re\,R_4}(u_d-u_{dp}),$$

$$(8.18)$$

$$\frac{\partial u_{dp}}{\partial t} = \frac{RR_5 R_6}{Re\,R_4}(u_d-u_{dp}), \qquad (8.19)$$

$$\frac{\partial T_d}{\partial t} = \frac{R_8}{\Pr \operatorname{Re} R_4 R_7} \frac{\partial^2 T_d}{\partial y^2} + \frac{Ec R_5}{\operatorname{Re} R_4 R_7} \left(\frac{\partial u_d}{\partial y} \right)^2 + \frac{2 R R_8}{3 \Pr \operatorname{Re} R_4 R_7} \left(T_{dp} - T_d \right)$$

$$+ \frac{Ec\, Ha^2}{\operatorname{Re} R_4 R_7} \left(\frac{1 + BiBe}{\left(1 + BiBe\right)^2 + Be^2} \right) u_d^2 + \frac{R Ec R_5}{\operatorname{Re} R_4 R_7} \left(u_d - u_{dp} \right)^2, \quad (8.20)$$

$$\frac{\partial T_{dp}}{\partial t} = \frac{2 R R_8 R_9 R_6}{3 \Pr \operatorname{Re} R_4 R_7} \left(T_d - T_{dp} \right). \quad (8.21)$$

Zone-III (Newtonian fluid: $1 \le y \le 2$)

$$\frac{\partial u_n}{\partial x} = \frac{Ge(t)}{R_4 R_{10}} + \frac{R_5 R_{11}}{\operatorname{Re} R_{10} R_4} \frac{\partial^2 u_n}{\partial y^2} - \frac{Ha^2}{\operatorname{Re}} \left(\frac{1 + BiBe}{\left(1 + BiBe\right)^2 + Be^2} \right) \left(\frac{1}{R_4 R_{10}} \right) u_n, \quad (8.22)$$

$$\frac{\partial T_n}{\partial t} = \frac{R_{13} R_8}{\Pr \operatorname{Re} R_4 R_7 R_{10} R_{12}} \frac{\partial^2 T_n}{\partial y^2} + \frac{Ec R_5 R_{11}}{\operatorname{Re} R_7 R_{12} R_4 R_{10}} \left(\frac{\partial u_n}{\partial y} \right)^2$$

$$+ \frac{Ha^2\, Ec}{\operatorname{Re} R_4 R_7 R_{10} R_{12}} \left(\frac{1 + BiBe}{\left(1 + BiBe\right)^2 + Be^2} \right) u_{nf}^2. \quad (8.23)$$

Initial Conditions:

$$\left. \begin{aligned}
u_{nf}\left(y, 0\right) &= 0; \\[4pt]
u_d\left(y, 0\right) &= 0; \\[4pt]
u_{dp}\left(y, 0\right) &= 0; \\[4pt]
u_n\left(y, 0\right) &= 0; \\[4pt]
T_{nf}\left(y, 0\right) &= 0 \text{ or } 100^\circ C; \\[4pt]
T_d\left(y, 0\right) &= 0; \\[4pt]
T_{dp}\left(y, 0\right) &= 0; \\[4pt]
T_n\left(y, 0\right) &= 0.
\end{aligned} \right\} \qquad (8.24)$$

Boundary Conditions:

$$
\left.\begin{aligned}
& u_{nf}(-1,t) = 0; \\
& u_{nf}(2,t) = 0; \\
& T_{nf}(-1,t) = Tw_1 = 0; \\
& T_n(2,t) = Tw_2 = 1; \\
& u_{nf}(0,t) = u_d(0,t); \\
& u_d(1,t) = u_d(1,t); \\
& T_{nf}(0,t) = T_d(0,t); \\
& T_d(1,t) = T_n(1,t); \\
& \left(\frac{\partial u_{nf}}{\partial y}\right)_{y=0} = \frac{\mu_d}{\mu_{nf}}\left(\frac{\partial u_d}{\partial y}\right)_{y=0}; \\
& \left(\frac{\partial u_{nf}}{\partial y}\right)_{y=0} = R_3(1-\Psi)^{2.5}\left(\frac{\partial u_d}{\partial y}\right)_{y=0} \\
& \left(\frac{\partial u_d}{\partial y}\right)_{y=1} = R_{11}\left(\frac{\partial u_n}{\partial y}\right)_{y=1}; \\
& u_{dp}(0,t) = u_{nf}(0,t) = u_d(0,t); \\
& u_{dp}(1,t) = u_d(1,t) = u_n(1,t); \\
& T_{dp}(0,t) = T_d(0,t); \\
& T_{dp}(1,t) = T_d(1,t); \\
& \left(\frac{\partial T_{af}}{\partial y}\right)_{y=0} = \frac{R_8\left(R_2^*(1-\Psi)+(2-\Psi)\right)}{R_2^*(1+2\Psi)+(2-2\Psi)}\left(\frac{\partial T_d}{\partial y}\right)_{y=0}; \\
& \left(\frac{\partial T_d}{\partial y}\right)_{y=1} = R_{13}\left(\frac{\partial T_n}{\partial y}\right)_{y=1}.
\end{aligned}\right\} \qquad (8.25)
$$

8.3.2 DIMENSIONLESS DUCT WALL CHARACTERISTICS

8.3.2.1 Skin Friction Coefficients

As the fluid moves along the duct (channel) walls, it generates a frictional force that acts upon the plate boundaries, hindering the flow and generating shear stress. This wall shear stress is also known as *surface skin friction*, as originally propounded in Prandtl's laminar boundary layer theory.

At the *lower wall*, the nanofluid shearing can be analyzed with the following expressions for dimensional skin friction coefficients:

$$C_{f1} = \frac{\mu_{nf}}{\rho_{nf} U_0^2} \left(\frac{\partial U_{nf}}{\partial y} \right)_{y=-1}, \tag{8.26}$$

$$C_{f1} = \frac{\mu_m}{\rho_m U_0^2} \cdot \left(\frac{1}{(1-\Psi)^{3.5} \left(1 + \frac{\Psi}{(1-\Psi) R_1^*} \right)} \right) \left(\frac{\partial U_{nf}}{\partial y} \right)_{y=-1}. \tag{8.27}$$

For specific Al_2O_3-Eg-water nanofluid, the particle density $\rho_{np} = 3{,}970$ kg/m^3 and Eg-water mixture density $\rho_m = 1{,}064.084$ kg/m^3 [49]. The dimensionless skin friction emerges therefore as:

$$C_{f1} = \frac{1}{Re} \cdot \left(\frac{1}{(1-\Psi)^{3.5} \left(1 + \frac{\Psi}{(1-\Psi) R_1^*} \right)} \right) \left(\frac{\partial U_{nf}}{\partial y} \right)_{y=-1}. \tag{8.28}$$

Here, Re is the Reynolds number.

At the *upper wall*, the Newtonian fluid shearing results in the following skin friction coefficients:

$$C_{f2} = \frac{\mu_n}{\rho_n U_0^2} \left(\frac{\partial U_n}{\partial y} \right)_{y=2}. \tag{8.29}$$

$$C_{f2} = \frac{1}{Re} \frac{R_3 R_5}{R_2 R_6} \left(\frac{\partial U_n}{\partial y} \right)_{y=2}. \tag{8.30}$$

8.3.2.2 Nusselt Number

The Nusselt number represents the ratio of convective to conductive heat transfer at the boundary (duct wall). It also computes the gradient of temperature at the boundary or rate of heat transfer from the boundary to the adjacent fluid. The expression of the Nusselt *number* at both plates is calculated as, respectively:

At the lower wall:

$$N_{u_1} = \left(-\frac{\partial T_{nf}}{\partial y} \right)_{y=-1} \tag{8.31}$$

At the upper wall:

$$N_{u_2} = \left(-\frac{\partial T_n}{\partial y} \right)_{y=2} \tag{8.32}$$

8.4 RADIAL BASIS FUNCTION PSEUDO-SPECTRAL (RBFPS) NUMERICAL SOLUTION

To study the velocity and temperature characteristics of each of the 3 immiscible liquids, the whole domain $[-1, 2]$ is divided into three zones: zone-I $[-1, 0]$ representing nanofluid, zone-II $[0, 1]$ representing dusty-fluid and zone-III $[1,2]$ representing Newtonian fluid. Each zone is discretized by N^* nodes denoted by $y_j = 1,2,3,\ldots N^*$. The solution of the interpolation problem for a set of N^* scattered nodes y_j is obtained by expressing the interpolant as a linear combination of radial basis functions as:

$$w(y,t) = \sum_{i=1}^{N^*} \lambda_i(t)\psi_i^* \left(\|y - y_i\| \right). \tag{8.33}$$

The interpolation coefficients λ_i are used in the equation along with the radial basis function ψ_i^*, which has a center at y_i and is defined using the distance norm $\|\cdot\|$. The selection of the RBFs can significantly affect the accuracy and efficiency of the solution. To address this, the compactly supported *Wendland* RBF is utilized in this study. It is a variant of the Wendland RBF that has finite support and is expressed in the following form:

$$\psi^*(r) = (1 - \varepsilon r)_-^{6+} \left(35\varepsilon r^2 + 18\varepsilon r + 3 \right). \tag{8.34}$$

Here the variable $r = \|y\|$ represents the radial distance between the center of the RBF and the point at which the function is being evaluated, $\|\cdot\|$ represents the distance norm between the two points, ε denotes the shape parameter that determines the width of the basis function and controls the smoothness of the interpolation and $(x)_-^+ = \max(0,x)$ is the positive part function. Collocating on all the nodes we have:

$$w(y_j,t) = \sum_{i=1}^{N^*} \lambda_i(t)\psi_i^* \left(\|y_j - y_i\| \right), \quad 1 \le j \le N^*. \tag{8.35}$$

$$
\begin{bmatrix} w(y_1,t) \\ w(y_2,t) \\ \vdots \\ w(y_{N^*},t) \end{bmatrix} = \begin{bmatrix} \psi_1^*\left(\|y_1-y_1\|\right) & \psi_2^*\left(\|y_1-y_2\|\right) & \cdots & \psi_{N^*}^*\left(\|y_1-y_{N^*}\|\right) \\ \psi_1^*\left(\|y_2-y_1\|\right) & \psi_2^*\left(\|y_2-y_2\|\right) & \cdots & \psi_{N^*}^*\left(\|y_2-y_{N^*}\|\right) \\ \vdots & \vdots & \ddots & \vdots \\ \psi_1^*\left(\|y_{N^*}-y_1\|\right) & \psi_2^*\left(\|y_{N^*}-y_2\|\right) & \cdots & \psi_{N^*}^*\left(\|y_{N^*}-y_{N^*}\|\right) \end{bmatrix}
$$

$$(8.36)$$

In reduced matrix-vector notation, the above Equation can be written as:

$$\mathbf{W} = \Psi\Lambda \tag{8.37}$$

The matrix Ψ is an $N^* \times N^*$ interpolation matrix which is constructed with entries given by $\Psi_{ji} = \psi_i^*\left(\|y_j-y_i\|\right)$, while $\Lambda = \left[\lambda_1,\lambda_2,\lambda_3\ldots\lambda_{N^*}\right]'$ is a vector of N^* coefficients of the RBF approximation. In order to compute the derivative of $\mathbf{W}$ with respect to y, one can simply by differentiating Equation (8.35). This involves differentiating the RBF function and then re-evaluating it at each node as per the following:

$$\frac{d}{dy_j}w(y_j,t) = \sum_{i=1}^{N^*}\lambda_i(t)\frac{d}{dy_j}\psi_i^*\left(\|y_j-y_i\|\right),\ 1\le j\le N^* \tag{8.38}$$

$$
\begin{bmatrix} \dfrac{\partial}{\partial y}w(y_1,t) \\[2mm] \dfrac{\partial}{\partial y}w(y_2,t) \\ \vdots \\ \dfrac{\partial}{\partial y}w(y_{N^*},t) \end{bmatrix}
$$

$$
= \begin{bmatrix} \dfrac{d}{dy}\psi_1^*\left(\|y_1-y_1\|\right) & \dfrac{d}{dy}\psi_2^*\left(\|y_1-y_2\|\right) & \cdots & \dfrac{d}{dy}\psi_{N^*}^*\left(\|y_1-y_{N^*}\|\right) \\[2mm] \dfrac{d}{dy}\psi_1^*\left(\|y_2-y_1\|\right) & \dfrac{d}{dy}\psi_2^*\left(\|y_2-y_2\|\right) & \cdots & \dfrac{d}{dy}\psi_{N^*}^*\left(\|y_2-y_{N^*}\|\right) \\ \vdots & \vdots & \ddots & \vdots \\ \dfrac{d}{dy}\psi_1^*\left(\|y_{N^*}-y_1\|\right) & \dfrac{d}{dy}\psi_2^*\left(\|y_{N^*}-y_2\|\right) & \cdots & \dfrac{d}{dy}\psi_{N^*}^*\left(\|y_{N^*}-y_{N^*}\|\right) \end{bmatrix} \begin{bmatrix} \lambda_1 \\ \lambda_2 \\ \vdots \\ \lambda_N \end{bmatrix}
$$

$$(8.39)$$

The matrix form can be written as:

$$w_y = \Psi_y\Lambda \tag{8.40}$$

The matrix Ψ_y is obtained by differentiating the interpolation matrix and entries as $\frac{d}{dy}\psi_i^*\left(\|y-y_i\|\right)\Big|_{y=y_j}$ with $1\le j,i\le N^*$. As the interpolation matrix is invertible, from Equation (8.37) it follows that:

$$w_y = \Psi_y \Psi^{-1} w = K_y w \qquad (8.41)$$

$$\Psi_y \Psi^{-1} = K_y$$

$$
= \begin{bmatrix}
\begin{bmatrix}
\dfrac{d}{dy}\psi_1^*\left(\|y_1 - y_1\|\right) & \dfrac{d}{dy}\psi_2^*\left(\|y_1 - y_2\|\right) & \cdots & \dfrac{d}{dy}\psi_{N^*}^*\left(\|y_1 - y_{N^*}\|\right) \\[2mm]
\dfrac{d}{dy}\psi_1^*\left(\|y_2 - y_1\|\right) & \dfrac{d}{dy}\psi_2^*\left(\|y_2 - y_2\|\right) & \cdots & \dfrac{d}{dy}\psi_{N^*}^*\left(\|y_2 - y_{N^*}\|\right) \\[2mm]
\vdots & \vdots & \ddots & \vdots \\[2mm]
\dfrac{d}{dy}\psi_1^*\left(\|y_{N^*} - y_1\|\right) & \dfrac{d}{dy}\psi_2^*\left(\|y_{N^*} - y_2\|\right) & \cdots & \dfrac{d}{dy}\psi_{N^*}^*\left(\|y_{N^*} - y_{N^*}\|\right)
\end{bmatrix} \\[4mm]
\times \begin{bmatrix}
\psi_1^*\left(\|y_1 - y_1\|\right) & \psi_2^*\left(\|y_1 - y_2\|\right) & \cdots & \psi_{N^*}^*\left(\|y_1 - y_{N^*}\|\right) \\[2mm]
\psi_1^*\left(\|y_2 - y_1\|\right) & \psi_2^*\left(\|y_2 - y_2\|\right) & \cdots & \psi_{N^*}^*\left(\|y_2 - y_{N^*}\|\right) \\[2mm]
\vdots & \vdots & \ddots & \vdots \\[2mm]
\psi_1^*\left(\|y_{N^*} - y_1\|\right) & \psi_2^*\left(\|y_{N^*} - y_2\|\right) & \cdots & \psi_{N^*}^*\left(\|y_{N^*} - y_{N^*}\|\right)
\end{bmatrix}
\end{bmatrix}
$$

$$(8.42)$$

This is termed the *first order differentiation matrix*. Similarly, *the second order differentiation matrix* can be obtained as:

$$w_{yy} = \Psi_{yy} \Psi^{-1} w = K_{yy} w. \qquad (8.43)$$

$$\Psi_{yy} \Psi^{-1} = K_{yy}$$

$$
= \begin{bmatrix}
\begin{bmatrix}
\dfrac{d^2}{dx^2}\psi_1^*\left(\|y_1 - y_1\|\right) & \dfrac{d^2}{dy^2}\psi_2^*\left(\|y_1 - y_2\|\right) & \cdots & \dfrac{d^2}{dy^2}\psi_{N^*}^*\left(\|y_1 - y_{N^*}\|\right) \\[2mm]
\dfrac{d^2}{dy^2}\psi_1^*\left(\|y_2 - y_1\|\right) & \dfrac{d^2}{dy^2}\psi_2^*\left(\|y_2 - y_2\|\right) & \cdots & \dfrac{d^2}{dy^2}\psi_{N^*}^*\left(\|y_2 - y_{N^*}\|\right) \\[2mm]
\vdots & \vdots & \ddots & \vdots \\[2mm]
\dfrac{d^2}{dy^2}\psi_1^*\left(\|y_{N^*} - y_1\|\right) & \dfrac{d^2}{dy^2}\psi_2^*\left(\|y_{N^*} - y_2\|\right) & \cdots & \dfrac{d^2}{dy^2}\psi_{N^*}^*\left(\|y_{N^*} - y_{N^*}\|\right)
\end{bmatrix} \\[4mm]
\times \begin{bmatrix}
\psi_1^*\left(\|y_1 - y_1\|\right) & \psi_2^*\left(\|y_1 - y_2\|\right) & \cdots & \psi_{N^*}^*\left(\|y_1 - y_{N^*}\|\right) \\[2mm]
\psi_1^*\left(\|y_2 - y_1\|\right) & \psi_2^*\left(\|y_2 - y_2\|\right) & \cdots & \psi_{N^*}^*\left(\|y_2 - y_{N^*}\|\right) \\[2mm]
\vdots & \vdots & \ddots & \vdots \\[2mm]
\psi_1^*\left(\|y_{N^*} - y_1\|\right) & \psi_2^*\left(\|y_{N^*} - y_2\|\right) & \cdots & \psi_{N^*}^*\left(\|y_{N^*} - y_{N^*}\|\right)
\end{bmatrix}
\end{bmatrix}
$$

$$(8.44)$$

This is termed the *second-order differentiation matrix*. After substituting all the approximations, the given *partial* differential equations are effectively rendered into a system of *ordinary* differential equations. The reduced system of ordinary differential equations can be written as follows:

Zone-I (Nano-fluid: $-1 \leq y \leq 0$)

$$u_{nf_t} = \frac{Ge(t)}{(1-\Psi)\left(1 + \dfrac{\Psi}{1-\Psi} \cdot \dfrac{1}{R_1^*}\right)} + \frac{1}{Re}\frac{1}{(1-\Psi)^{3.5}\left(1 + \dfrac{\Psi}{1-\Psi} \cdot \dfrac{1}{R_1^*}\right)}K_{yy}u_{nf}$$

$$- \frac{Ha^2}{Re}\left(\frac{1+BiBe}{(1+BiBe)^2 + Re^2\dfrac{\Psi}{R_1^*} + (1-\Psi)}\right)u_{nf}. \tag{8.45}$$

$$T_{nf_t} = \frac{1}{Re\,Pr\left(\dfrac{R_2^*(1+2\Psi) + 2(1-\Psi)}{\left(R_2^*(1-\Psi)+(2+\Psi)\right)(1-\Psi)^2\left(1+\left(\dfrac{\Psi}{1-\Psi}\right)\dfrac{1}{R_1^*}\right)\left(1+\left(\dfrac{\Psi}{1-\Psi}\right)R_3^*\right)}\right)}K_{yy}T_{nf}$$

$$+ \frac{Ec}{Re\left(\dfrac{1}{(1-\Psi)^{4.5}}\right)\left(\dfrac{1}{1+\left(\dfrac{\Psi}{1-\Psi}\right)\dfrac{1}{R_1^*}}\right)\left(\dfrac{1}{1+\left(\dfrac{\Psi}{1-\Psi}\right)R_3^*}\right)}(K_y u_{nf})^2$$

$$\left(\frac{Ha^2 Ec(1+BiBe)}{\left(Re(1+BiBe)^2 + Be^2\right)1+\left(\dfrac{\Psi}{1-\Psi}\right)\dfrac{1}{R_1^*}}\right)\left(\frac{1}{1+\left(\dfrac{\Psi}{1-\Psi}\right)R_3^*}\right)u_{nf}^2.$$

Zone-II (Dusty fluid: $0 \leq y \leq 1$)

$$u_{d_t} = \frac{Ge(t)}{R_4} + \frac{R_5}{Re\,R_4}K_{yy}u_d - \left(\frac{1+BiBe}{(1+BiBe)^2 + Be^2}\right)\frac{Ha^2}{Re\,R_4}u_d - \frac{RR_5}{Re\,R_4}\left(u_d - u_{dp}\right), \tag{8.46}$$

$$u_{dp_t} = \frac{RR_5 R_6}{Re\,R_6}\left(u_d - u_{dp}\right), \tag{8.47}$$

$$T_{d_t} = \frac{R_8}{\Pr\,\mathrm{Re}\;R_4 R_7} K_{yy} T_d + \frac{Ec R_5}{\mathrm{Re}\,R_4 R_7}\left(K_y u_d\right)^2 + \frac{2 R R_8}{3\Pr\,\mathrm{Re}\,R_4 R_7}\left(T_{dp} - T_d\right)$$

$$+ \frac{Ec\,\mathrm{Ha}^2}{\mathrm{Re}\,R_4 R_7}\left(\frac{1 + BiBe}{\left(1 + BiBe\right)^2 + Be^2}\right) u_d^2 \cdots + \frac{R Ec R_5}{\mathrm{Re}\,R_4 R_7}\left(u_d - u_{dp}\right)^2, \qquad (8.48)$$

$$T_{dp_t} = \frac{2 R R_8 R_9 R_6}{3\Pr\,\mathrm{Re}\,R_4 R_7}\left(T_{dp} - T_d\right). \qquad (8.49)$$

Zone-III (Newtonian fluid: $1 \le y \le 2$)

$$u_{n_t} = \frac{Ge(t)}{R_4 R_{10}} + \frac{R_5 R_{11}}{R_4 R_{10} R_4} K_{yy} u_n - \frac{\mathrm{Ha}^2}{\mathrm{Re}}\left(\frac{1 + BiBe}{\left(1 + BiBe\right)^2 + Be^2}\right)\left(\frac{1}{R_4 R_{10}}\right) u_n, \quad (8.50)$$

$$T_{n_t} = \frac{R_{13} R_8}{\Pr\,\mathrm{Re}\,R_4 R_7 R_{10} R_{12}} K_{yy} T_n + \frac{Ec R_5 R_{11}}{\mathrm{Re}\,R_7 R_{12} R_4 R_{10}}\left(K_y u_n\right)^2$$

$$+ \frac{\mathrm{Ha}^2\,Ec}{\mathrm{Re}\,R_4 R_7 R_{10} R_{12}}\left(\frac{1 + BiBe}{\left(1 + BiBe\right)^2 + Be^2}\right) u_n^2. \qquad (8.51)$$

Any available ODE solver in MATLAB® or other time-stepping techniques can be utilized for the solution of the above reduced system of ODEs. The choice of time-stepping technique and ODE solver for solving the time-dependent problem depends on the specific problem at hand and the desired level of accuracy and computational efficiency for the solution. In this study, the reduced system of linear equations is solved by SSP-RK43 scheme. Details are only listed for the first two time steps here and steps 3 and 4 are omitted for brevity.

At the first step:
Zone-I (Nano-fluid: $-1 \le y \le 0$)

$$u_{nf_1} = u_{nf_0} + \frac{\Delta t}{2}\left(\frac{Ge(t)}{\left(1 - \Psi\right)\left(1 + \dfrac{\Psi}{1 - \Psi}\cdot\dfrac{1}{R_1^*}\right)} + \frac{1}{\mathrm{Re}}\frac{1}{\left(1 - \Psi\right)^{3.5}\left(1 + \dfrac{\Psi}{1 - \Psi}\cdot\dfrac{1}{R_1^*}\right)} K_{yy} u_{nf_0} \right.$$

$$\left. - \frac{\mathrm{Ha}^2}{\mathrm{Re}}\left(\frac{1 + BiBe}{\left(1 + BiBe\right)^2 + \mathrm{Re}^2}\frac{1}{\dfrac{\Psi}{R_1^*} + \left(1 - \Psi\right)}\right) u_{nf_0} \right), \qquad (8.52)$$

$$T_{nf_1} = T_{nf_0} + \frac{\Delta t}{2}\left(\frac{1}{\mathrm{Re}} \frac{1}{\mathrm{Pr}} \right.$$

$$\left(\frac{R_2^*\left(1+2\Psi\right)+2\left(1-\Psi\right)}{\left(R_2^*\left(1-\Psi\right)+\left(2+\Psi\right)\right)\left(1-\Psi\right)^2\left(1+\left(\frac{\Psi}{1-\Psi}\right)\frac{1}{R_1^*}\right)\left(1+\left(\frac{\Psi}{1-\Psi}\right)R_3^*\right)} \right) K_{yy} T_{nf_0} +$$

$$\frac{\mathrm{Ec}}{\mathrm{Re}}\left(\frac{1}{\left(1-\Psi\right)^{4.5}} \right)\left(\frac{1}{1+\left(\frac{\Psi}{1-\Psi}\right)\frac{1}{R_1^*}} \right)\left(\frac{1}{1+\left(\frac{\Psi}{1-\Psi}\right)R_3^*} \right)\left(K_y u_{nf_0}\right)^2$$

$$\left. +\left(\frac{\mathrm{Ha}^2\,\mathrm{Ec}\left(1+BiBe\right)}{\mathrm{Re}\left(1+BiBe\right)^2+Be^2\right)\left(1+\left(\frac{\Psi}{1-\Psi}\right)\frac{1}{R_1^*}\right)} \right)\left(\frac{1}{1+\left(\frac{\Psi}{1-\Psi}\right)R_3^*} \right)u_{nf_0}^2 \right) \tag{8.53}$$

Zone-II (Dusty fluid: $0 \leq y \leq 1$)

$$u_{d_1} = u_{d_0} +$$

$$\frac{\Delta t}{2}\left(\frac{Ge(t)}{R_4} + \frac{R_5}{\mathrm{Re}\,R_4} K_{yy} u_{d_0} - \left(\frac{1+BiBe}{\left(1+BiBe\right)^2+Be^2} \right)\frac{\mathrm{Ha}^2}{\mathrm{Re}\,R_4} u_{d_0} - \frac{RR_5}{\mathrm{Re}\,R_4}\left(u_{d_0}-u_{dp_0}\right) \right),$$

$$\tag{8.54}$$

$$u_{dp_1} = u_{dp_0} + \frac{\Delta t}{2}\left(\frac{RR_5 R_6}{\mathrm{Re}\,R_4}\left(u_{d_0}-u_{dp_0}\right) \right), \tag{8.55}$$

$$T_{d_1} = T_{d_0} + \frac{\Delta t}{2}\left(\begin{array}{l} \dfrac{R_8}{\mathrm{Pr}\,\mathrm{Re}\,R_4 R_7}K_{yy}T_{d_0} + \dfrac{EcR_5}{\mathrm{Re}\,R_4 R_7}\left(K_y u_{d_0}\right)^2 + \dfrac{2RR_8}{3\mathrm{Pr}\,\mathrm{Re}\,R_4 R_7}\left(T_{dp_0}-T_{d_0}\right)+ \\[3mm] \dfrac{Ec\,\mathrm{Ha}^2}{\mathrm{Re}\,R_4 R_7}\left(\dfrac{1+BiBe}{\left(1+BiBe\right)^2+Be^2} \right)u_{d_0}^2 + \dfrac{REcR_5}{\mathrm{Re}\,R_4 R_7}\left(u_{d_0}-u_{dp_0}\right)^2, \end{array} \right)$$

$$\tag{8.56}$$

$$T_{dp_1} = T_{dp_0} + \frac{\Delta t}{2}\left(\frac{2RR_8 R_9 R_6}{3\mathrm{Pr}\,\mathrm{Re}\,R_4 R_7}\left(T_{dp_0}-T_{d_0}\right) \right). \tag{8.57}$$

Zone-III (Newtonian fluid: $1 \leq y \leq 2$)

$$u_{n_1} = u_{n_0} + \frac{\Delta t}{2}\left(\frac{Ge(t)}{R_4 R_{10}} + \frac{R_5 R_{11}}{\mathrm{Re}\,R_{10} R_4}K_{yy}u_{n_0} - \frac{\mathrm{Ha}^2}{\mathrm{Re}}\left(\frac{1+BiBe}{\left(1+BiBe\right)^2+Be^2} \right)\left(\frac{1}{R_4 R_{10}} \right)u_{n_0} \right),$$

$$\tag{8.58}$$

$$T_{n_1} = T_{n_0} + \frac{\Delta t}{2}\left(\frac{R_{13}R_8}{\Pr \operatorname{Re} R_4 R_7 R_{10} R_{12}} K_{yy} T_{n_0} + \frac{EcR_5 R_{11}}{\operatorname{Re} R_7 R_{12} R_4 R_{10}}\left(K_y u_{n_0}\right)^2 + \frac{\operatorname{Ha}^2 \operatorname{Ec}}{\operatorname{Re} R_4 R_7 R_{10} R_{12}}\left(\frac{1 + BiBe}{(1 + BiBe)^2 + Be^2} \right) u_{n_0}^2 \right) \tag{8.59}$$

After the first step, the boundary conditions are applied accordingly.

At the 2nd step:

Zone-I (Nano-fluid: $-1 \leq y \leq 0$)

$$u_{nf_2} = u_{nf_1} + \frac{\Delta t}{2}\left(\frac{Ge(t)}{(1-\Psi)\left(1 + \dfrac{\Psi}{1-\Psi} \cdot \dfrac{1}{R_1^*}\right)} + \frac{1}{\operatorname{Re}} \frac{1}{(1-\Psi)^{3.5}\left(1 + \dfrac{\Psi}{1-\Psi} \cdot \dfrac{1}{R_1^*}\right)} K_{yy} u_{nf_1} - \frac{\operatorname{Ha}^2}{\operatorname{Re}}\left(\frac{1+BiBe}{(1+BiBe)^2 + \operatorname{Re}^2} \frac{1}{\dfrac{\Psi}{R_1^*} + (1-\Psi)} \right) u_{nf_1} \right), \tag{8.60}$$

$$T_{nf_2} = T_{nf_1} + \frac{\Delta t}{2}\left(\frac{1}{\operatorname{Re}\Pr} \frac{R_2^*(1+2\Psi) + 2(1-\Psi)}{\left(R_2^*(1-\Psi) + (2+\Psi)\right)(1-\Psi)^2\left(1 + \left(\dfrac{\Psi}{1-\Psi}\right)\dfrac{1}{R_1^*}\right)\left(1 + \left(\dfrac{\Psi}{1-\Psi}\right)R_3^*\right)} K_{yy} T_{nf_1} + \right.$$

$$\frac{\operatorname{Ec}}{\operatorname{Re}}\left(\frac{1}{(1-\Psi)^{4.5}}\right)\left(\frac{1}{1 + \left(\dfrac{\Psi}{1-\Psi}\right)\dfrac{1}{R_1^*}}\right)\left(\frac{1}{1 + \left(\dfrac{\Psi}{1-\Psi}\right)R_3^*}\right)\left(K_y u_{nf_1}\right)^2 + \frac{\operatorname{Ha}^2 \operatorname{Ec}(1+BiBe)}{(\operatorname{Re}(1+BiBe)^2 + Be^2)\left(1 + \left(\dfrac{\Psi}{1-\Psi}\right)\dfrac{1}{R_1^*}\right)}\left(\frac{1}{1 + \left(\dfrac{\Psi}{1-\Psi}\right)R_3^*}\right), \tag{8.61}$$

Zone-II (Dusty fluid: $0 \leq y \leq 1$)

$$u_{d_2} = u_{d_1} + \frac{\Delta t}{2}\left(\frac{Ge(t)}{R_4} + \frac{R_5}{\operatorname{Re} R_4} K_{yy} u_{d_1} - \left(\frac{1+BiBe}{(1+BiBe)^2 + Be^2} \right)\frac{\operatorname{Ha}^2}{\operatorname{Re} R_4} u_{d_1} - \frac{RR_5}{\operatorname{Re} R_4}\left(u_{d_1} - u_{dp_1}\right) \right), \tag{8.62}$$

$$u_{dp2} = u_{dp1} + \frac{\Delta t}{2}\left(\frac{RR_5 R_6}{\mathrm{Re}\, R_6}\left(u_{d_1} - u_{dp1}\right) \right),$$

(8.63)

$$T_{d_2} = T_{d_1} + \frac{\Delta t}{2}\left(\frac{R_8}{\mathrm{Pr}\,\mathrm{Re}\, R_4 R_7} K_{yy} T_{d_1} + \frac{EcR_5}{\mathrm{Re}\, R_4 R_7}\left(K_y u_{d_1}\right)^2 + \frac{2RR_8}{3\mathrm{Pr}\,\mathrm{Re}\, R_4 R_7}\left(T_{dp1} - T_{d_1}\right) + \frac{Ec\,\mathrm{Ha}^2}{\mathrm{Re}\, R_4 R_7}\left(\frac{1+BiBe}{(1+BiBe)^2 + Be^2} \right)u_{d_1}^2 + \frac{REcR_5}{\mathrm{Re}\, R_4 R_7}\left(u_{d_1} - u_{dp1}\right) \right)$$

(8.64)

$$T_{dp2} = T_{dp1} + \frac{\Delta t}{2}\left(\frac{2RR_8 R_9 R_6}{3\mathrm{Pr}\,\mathrm{Re}\, R_4 R_7}\left(T_{dp1} - T_{d_1}\right) \right).$$

(8.65)

Zone-III (Newtonian fluid: $1 \leq y \leq 2$)

$$u_{n_2} = u_{n_1} + \frac{\Delta t}{2}\left(\frac{Ge(t)}{R_4 R_{10}} + \frac{R_5 R_{11}}{\mathrm{Re}\, R_{10} R_4} K_{yy} u_{n_1} - \frac{\mathrm{Ha}^2}{\mathrm{Re}}\left(\frac{1+BiBe}{(1+BiBe)^2 + Be^2} \right)\left(\frac{1}{R_4 R_{10}} \right)u_{n_1} \right),$$

(8.66)

$$T_{n_2} = T_{n_1} + \frac{\Delta t}{2}\left(\frac{R_8 R_{13}}{\mathrm{Pr}\,\mathrm{Re}\, R_4 R_7 R_{10} R_{12}} K_{yy} T_{n_1} + \frac{EcR_5 R_{11}}{\mathrm{Re}\, R_7 R_{12} R_4 R_{10}}\left(K_y u_{n_1}\right)^2 + \frac{\mathrm{Ha}^2\,Ec}{\mathrm{Re}\, R_4 R_7 R_{10} R_{12}}\left(\frac{1+BiBe}{(1+BiBe)^2 + Be^2} \right)u_{n_1}^2 \right)$$

(8.67)

After the 2nd step, the associated boundary conditions are also applied accordingly. After the final step, boundary conditions are again implemented. Hence, using the RK43 scheme in combination with the RBFPS method, within the MATLAB symbolic software, the velocity and temperature profiles of each of the three immiscible fluids in the duct can be obtained numerically.

8.5   NUMERICAL RESULTS, VALIDATION AND DISCUSSION

Computational solutions obtained by the RBFPS approach are visualized graphically in Figures 8.12 to 8.15. Computations are performed for the three immiscible fluids in the horizontal channel with different external pressure gradient cases. All appropriate data utilized is given in the figure captions. In the duct, the Saffman dusty fluid is sandwiched between lower nanofluid and upper Newtonian fluid. Figures 8.2–8.9 **illustrate velocity profiles.** Figures 8.10–8.15 **depict temperature profiles.** The formulated coupled partial differential equations are further solved through RBFPS method for different compositions of nanoparticles. The linear translation momentum profile for suitable suspended nanoparticle obtained by RBFPS method are compared with a numerically stable differential quadrature method (DQM) which deploys modified cubic B-spline functions with appropriate knots. This latter algorithm has been described and validated extensively in previous studies by the authors. The reader is referred to Chandrawat et al. [17]. Inspection of Figure 8.2 confirms that the RBFPS method consistently demonstrated a remarkable level of accuracy in predicting the behavior of fluid systems under varying pressure gradient conditions. In fact, the results obtained using this method have been shown to closely match those of the highly accurate differential quadrature (DQM) method for all three pressure gradient scenarios (Figure 8.2). Such close agreement between the two methods serves as

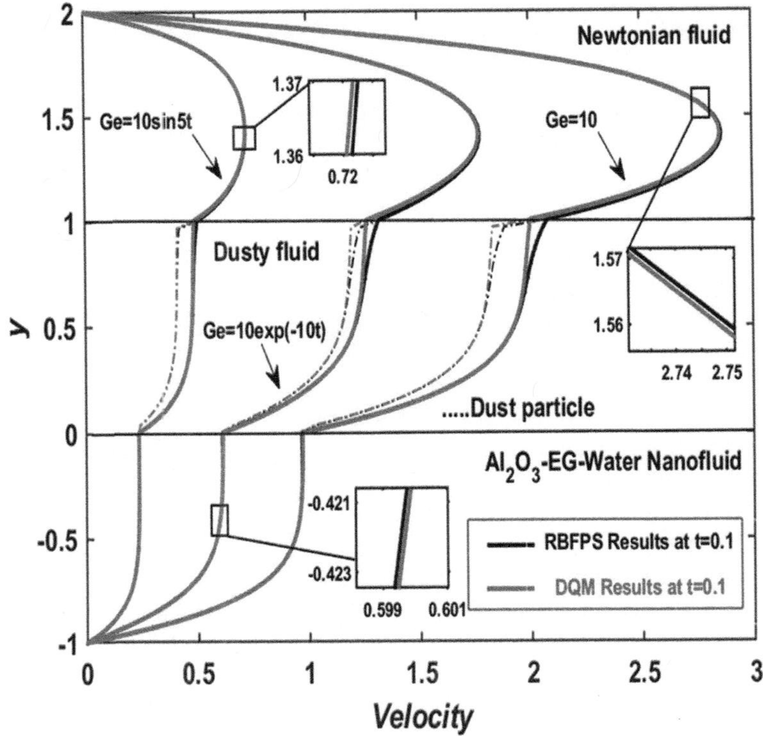

FIGURE 8.2   Comparision of RBFPS code with DQM (above left).

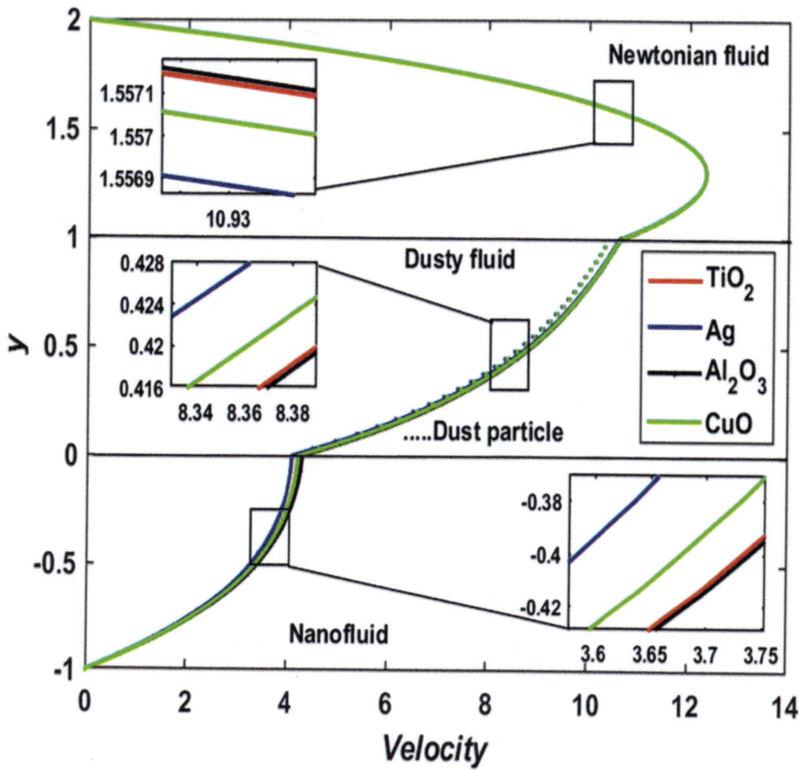

FIGURE 8.3 Velocity profiles of fluids computed using RBFPS with different nanoparticles, when $R_1^* = 0.268031158$, $R_2^* = 98.4865086$, $R_3^* = 0.241459895$ (above right).

a testament to the effectiveness of the RBFPS approach in simulating complex fluid dynamics and confirms the accuracy of the present numerical approach.

Figure 8.3 has been plotted to compare the velocity profiles of the three fluids namely nanofluid (CuO, Al$_2$O$_3$, Ag, TiO$_2$- ethylene glycol/water) in zone 1, Saffman dusty in zone 2 and Newtonian fluids in zone 3 are considered in horizontal duct using the RBFPS scheme for constant pressure gradient. The aim of this graph is to identity the efficient nanofluid mixture. It is observed that the magnitude of velocity profile in case of Al$_2$O$_3$ is slightly greater as compared to CuO, Ag and TiO$_2$. The viscosity modification (reduction) is more prominent for alumina as compared with other nanoparticles. This reduces the shearing resistance to the nanofluid and accentuates velocity i. e. it induces duct flow acceleration. A similar trend is observed in the velocity distribution for the Saffman dusty fluid along with dust particle and Newtonian fluid present in zone 2 and zone 3, respectively. At both the nanofluid/ Saffman dusty interface and the Saffman dusty/Newtonian interface, a sharp change in velocity is computed. The maximum overall velocity is computed clearly for the Newtonian (upper layer) case. The presence of dust particles in zone 2 and nanoparticles in zone 1 therefore clearly produces a deceleration overall relative to the

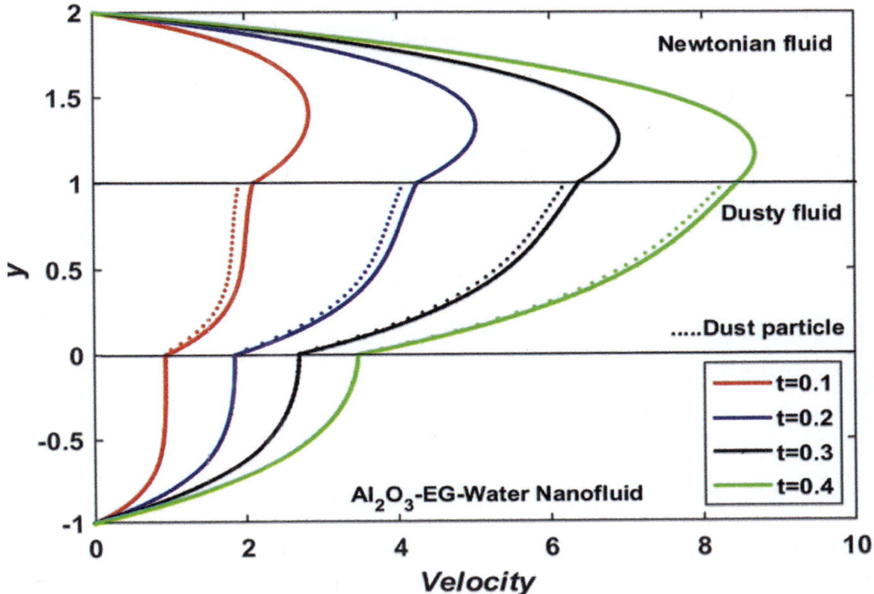

FIGURE 8.4 Velocity profiles of fluids computed using RBFPS with varying time, when $Ge = 10$, $\Psi = 0.02, Re = 2, Ha^2 = 0.2, Bi = 2, Be = 0.2, R_1^* = 0.268031158, R_2^* = 98.4865086, R_3^* = 0.241459895, R_4 = 0.5, R_5 = 0.6, R_6 = 100, R_7 = 0.4, R_8 = 0.3, R_9 = 0.55, R_{10} = 0.65, R_{11} = 0.7, R_{12} = 0.75, R_{13} = 0.8, Ec = 0.5, Pr = 2$ and $R = 2$ (above left).

Newtonian case. However, magnitudes of velocity are significantly greater in zone 2 than zone 1, indicating that nanofluids achieve the lowest duct flow acceleration. The key objective of nanofluids is enhanced heat transfer, not flow acceleration. While the dusty fluid produces lower velocities at any location in the duct relative to Newtonian fluid (zone 3), the results are more realistic for actual fuel cells since real working fluids are not purely Newtonian. The Navier-Stokes model is not adequate to correctly represent practical fuel cell hydrodynamics. It produces a significant over-estimate in duct velocity magnitudes, as indicated clearly in the computations. Furthermore, it is interesting to note that the skewed parabolic distribution characterizing duct flows is only captured for the Newtonian case. The dusty case (zone 2) and nanofluid case (zone 1) exhibit monotonic growth from the lower duct wall ($y=0$) to the upper duct wall ($y=1$) and the peak velocity is always computed at the latter; it is not located within the central section of the layer, as apparent in the Newtonian case (zone 3). Figure 8.4 visualizes the effect of time and constant pressure gradient ($Ge = 10$) on the velocity profile of all fluids in the three zones of the horizontal duct. It is observed that velocity profiles are parabolic in nature and increase with time for all zones of the duct, eventually reaching a stable state. The magnitude of the velocity profiles of nanofluid in zone 1 is also less than that of dusty fluid in zone 2, which is less than that of Newtonian fluid in zone 3. This is attributable to the fluid density

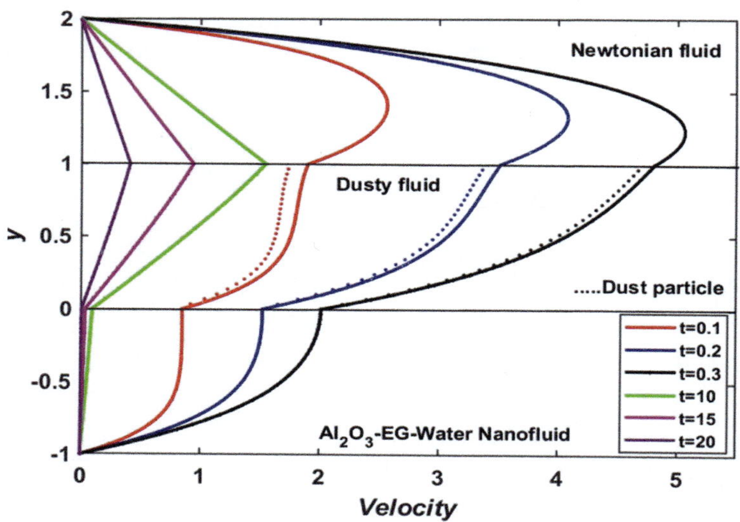

FIGURE 8.5 Velocity profiles of fluids computed using RBFPS with varying time, when $Ge = 10\exp(-2t)$, for $\Psi = 0.02, \mathrm{Re} = 2, \mathrm{Ha}^2 = 0.2, \mathrm{Bi} = 2, \mathrm{Be} = 0.2, R_1^* = 0.268031159, R_2^* = 98.4865086, R_3^* = 0.241459895, R_4 = 0.5, R_5 = 0.6, R_6 = 100, R_7 = 0.4, R_8 = 0.3, R_9 = 0.55, R_{10} = 0.65, R_{11} = 0.7, R_{12} = 0.75, R_{13} = 0.8, \mathrm{Ec} = 0.5, \mathrm{Pr} = 2$ and $R = 2$ (above right)

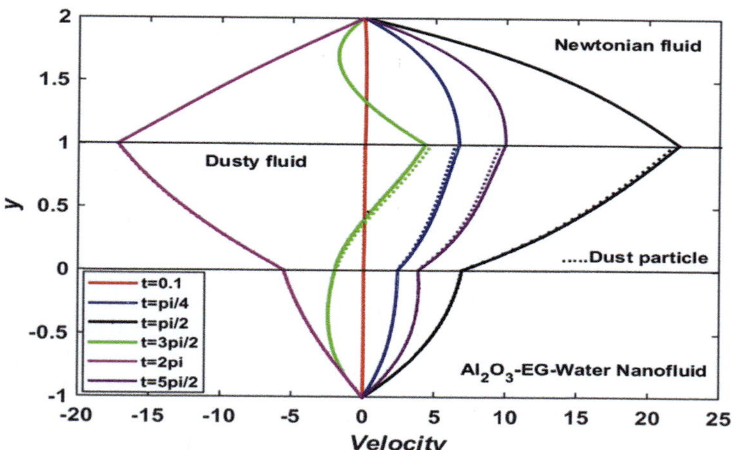

FIGURE 8.6 Velocity profiles of fluids computed using RBFPS with varying time, when $Ge = 10\sin t$, for $\Psi = 0.02, \mathrm{Re} = 2, \mathrm{Ha}^2 = 0.2, \mathrm{Bi} = 2, \mathrm{Be} = 0.2, R_1^* = 0.268031159, R_2^* = 98.4865086, R_3^* = 0.241459895, R_4 = 0.5, R_5 = 0.6, R_6 = 50, R_7 = 0.4, R_8 = 0.3, R_9 = 0.55, R_{10} = 0.65, R_{11} = 0.7, R_{12} = 0.75, R_{13} = 0.8, \mathrm{Ec} = 0.5, \mathrm{Pr} = 2$ and $R = 2$ (above left).

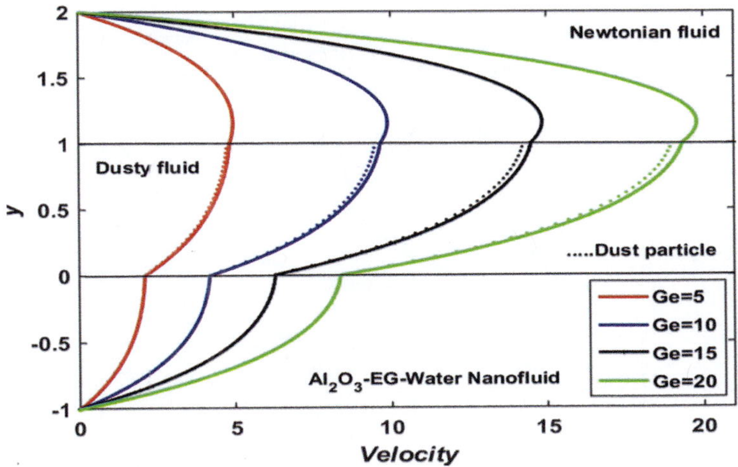

FIGURE 8.7 Velocity profiles of fluids computed using RBFPS with a range of constant pressure gradients, when $t = 0.5, \Psi = 0.02, \mathrm{Re} = 2, \mathrm{Ha}^2 = 0.2, \mathrm{Bi} = 2, \mathrm{Be} = 0.2, R_1^* = 0.268031159, R_2^* = 98.4865086, R_3^* = 0.241459895, R_4 = 0.5, R_5 = 0.6, R_6 = 700, R_7 = 0.4, R_8 = 0.3, R_9 = 0.55, R_{10} = 0.65, R_{11} = 0.7, R_{12} = 0.75, R_{13} = 0.8, \mathrm{Ec} = 0.5, \mathrm{Pr} = 2$ and $R = 0.25$ (above right).

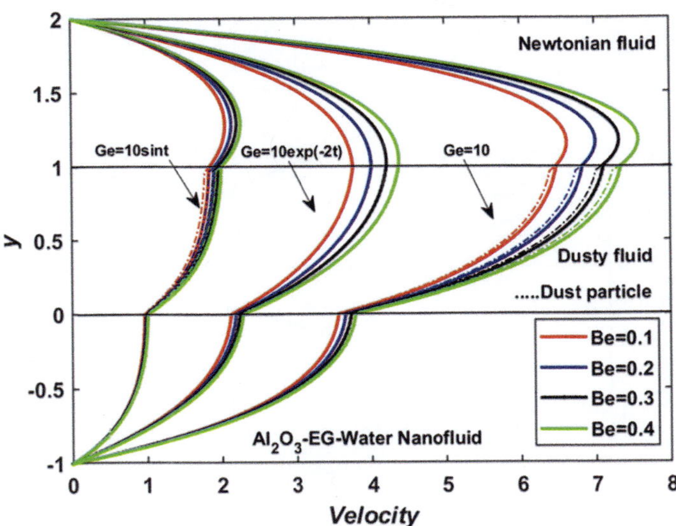

FIGURE 8.8 Velocity profiles of fluids computed using RBFPS with varying Be, when $t = 0.5, \Psi = 0.02, \mathrm{Re} = 2, \mathrm{Ha}^2 = 2, \mathrm{Bi} = 2, R_1^* = 0.268031159, R_2^* = 98.4865086, R_3^* = 0.2414 59895, R_4 = 0.5, R_5 = 0.6, R_6 = 700, R_7 = 0.4, R_8 = 0.3, R_9 = 0.55, R_{10} = 0.65, R_{11} = 0.7, R_{12} = 0.75, R_{13} = 0.8, \mathrm{Ec} = 0.5, \mathrm{Pr} = 2$ and $R = 0.25$ (above left).

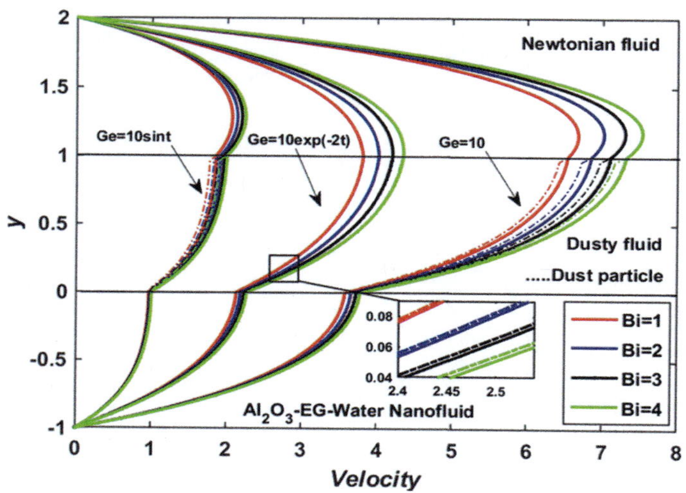

FIGURE 8.9   Velocity profiles of fluids computed using RBFPS with varying Bi, when $t = 0.5, \mathrm{Re} = 2, \Psi = 0.02, \mathrm{Ha}^2 = 2, Be = 0.2, R_1^* = 0.268031159, R_2^* = 98.4865086, R_3^* = 0.241459895, R_4 = 0.5, R_5 = 0.6, R_6 = 500, R_7 = 0.4, R_8 = 0.3, R_9 = 0.55, R_{10} = 0.65, R_{11} = 0.7, R_{12} = 0.75, R_{13} = 0.8, \mathrm{Ec} = 0.5, \mathrm{Pr} = 2$ and $R = 0.25$ (above right).

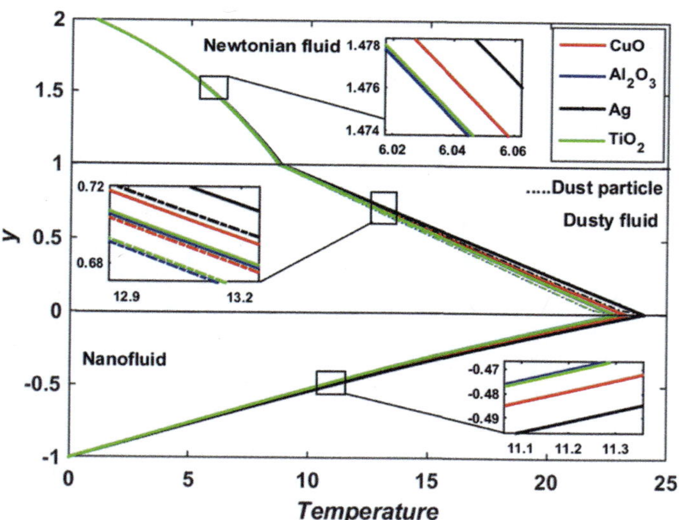

FIGURE 8.10   Temperature profiles of fluids computed using RBFPS with varying Bi, when $\Psi = 0.01, \mathrm{Re} = 2, \mathrm{Ha}^2 = 0.2, Be = 0.2, R_1^* = 0.163453717, R_2^* = 44.3189289, R_3^* = 0.170442279, R_4 = 0.5, R_5 = 0.6, R_6 = 100, R_7 = 0.5, R_8 = 0.5, R_9 = 0.5, R_{10} = 0.5, R_{11} = 0.5, R_{12} = 0.5, R_{13} = 0.8, \mathrm{Ec} = 0.5, \mathrm{Pr} = 2$ and $R = 2$ (above left).

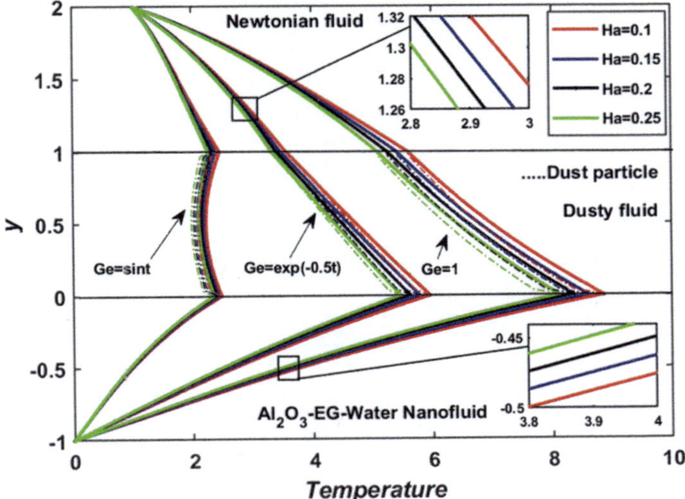

FIGURE 8.11 Temperature profiles of fluids computed using RBFPS with varying Ha^2, when $t = 0.5$, $\Psi = 0.02$, $Re = 2$, $Bi = 2$, $Be = 0.2$, $R_1^* = 0.268031159$, $R_2^* = 98.4865086$, $R_3^* = 0.241459895$, $R_4 = 0.5$, $R_5 = 0.5$, $R_6 = 200$, $R_7 = 0.5$, $R_8 = 0.5$, $R_9 = 0.5$, $R_{10} = 0.5$, $R_{11} = 0.5$, $R_{12} = 0.5$, $R_{13} = 0.5$, $Ec = 2$, $Pr = 2$ and $R = 0.25$ (above right).

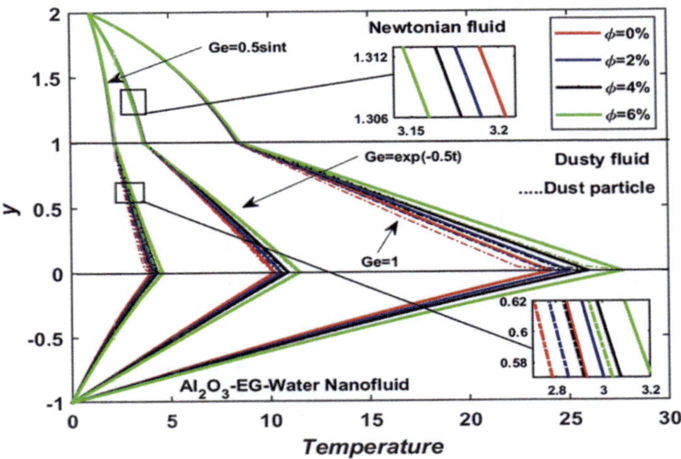

FIGURE 8.12 Temperature profiles of fluids computed using RBFPS with varying Ψ, when $t = 0.5$, $\Psi = 0.02$, $Re = 2$, $Bi = 2$, $Be = 0.2$, $R_1^* = 0.268031159$, $R_2^* = 98.4865086$, $R_3^* = 0.241459895$, $R_4 = 0.5$, $R_5 = 0.5$, $R_6 = 200$, $R_7 = 0.5$, $R_8 = 0.5$, $R_9 = 0.5$, $R_{10} = 0.5$, $R_{11} = 0.5$, $R_{12} = 0.5$, $R_{13} = 0.5$, $Ec = 2$, $Pr = 2$ and $R = 0.25$ (above left).

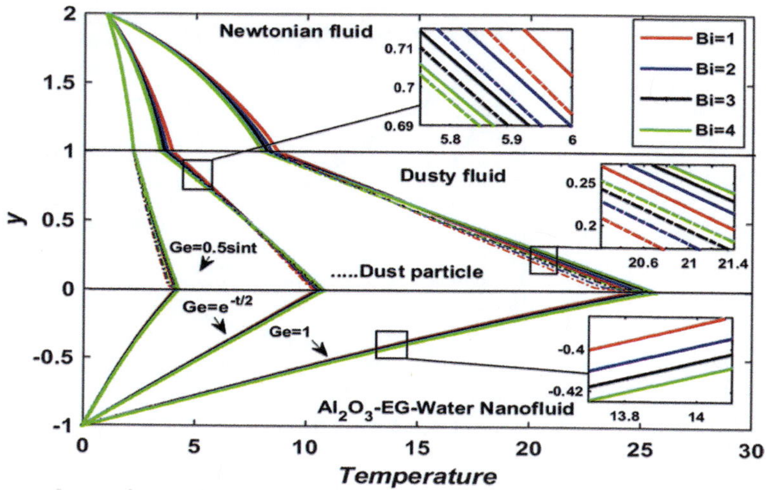

FIGURE 8.13 Temperature profiles of fluids computed using RBFPS with varying Bi, when $t = 2, \mathrm{Ha}^2 = 0.2, \Psi = 0.02, \mathrm{Re} = 2.5,\ Be = 0.2, R_1^* = 0.268031159, R_2^* = 98.4865086, R_3^* = 0.2414$ $59895, R_4 = 0.5, R_5 = 0.5, R_6 = 100, R_7 = 0.5, R_8 = 0.5, R_9 = 0.5, R_{10} = 0.5, R_{11} = 0.5, R_{12} = 0.5,$ $R_{13} = 0.5, \mathrm{Ec} = 2, \mathrm{Pr} = 0.5$ and $R = 0.25$ (above right).

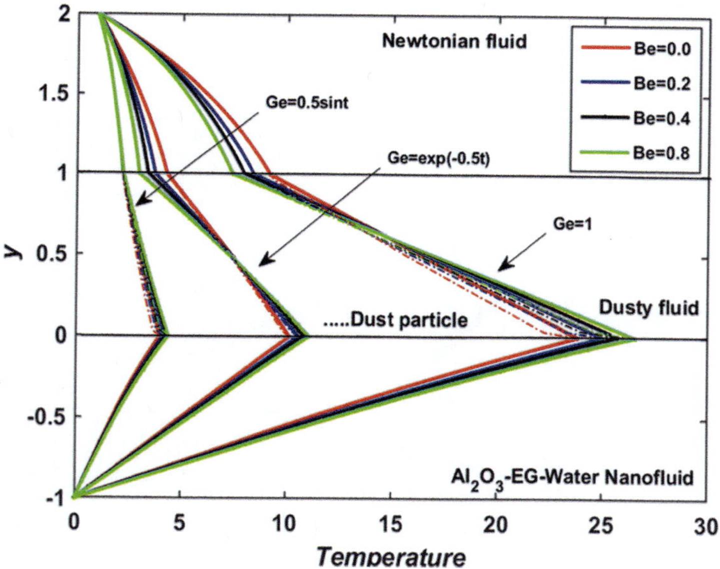

FIGURE 8.14 Temperature profiles of fluids computed using RBFPS with varying Be, when $t = 2, Ha^2 = 0.2, \Psi = 0.02, \mathrm{Re} = 2.5, Bi = 2, R_1^* = 0.268031159, R_2^* = 98.4865086, R_3^* = 0.2414$ $59895, R_4 = 0.5, R_5 = 0.5, R_6 = 100, R_7 = 0.5, R_8 = 0.5, R_9 = 0.5, R_{10} = 0.5, R_{11} = 0.5, R_{12} = 0.5,$ $R_{13} = 0.5, \mathrm{Ec} = 2, \mathrm{Pr} = 0.5$ and $R = 0.25$ (above left).

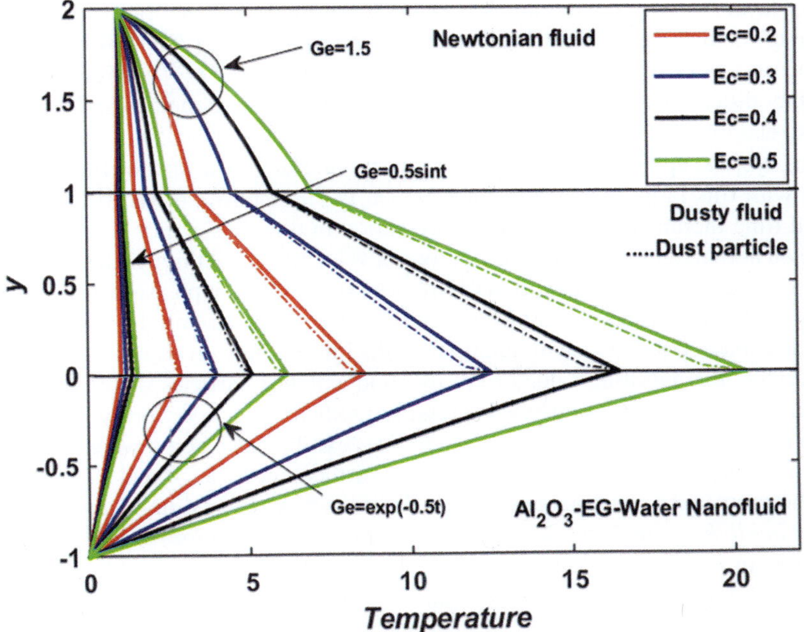

FIGURE 8.15 Temperature profiles of fluids computed using RBFPS with varying Ec when $t = 2, \mathrm{Ha}^2 = 0.2, \Psi = 0.02, \mathrm{Re} = 2.5, Bi = 2, Be = 0.2, R_1^* = 0.268031159, R_2^* = 98.4865086, R_3^* = 0.241459895, R_4 = 0.5, R_5 = 0.5, R_6 = 100, R_7 = 0.5, R_8 = 0.5, R_9 = 0.5, R_{10} = 0.5, R_{11} = 0.5, R_{12} = 0.5, R_{13} = 0.5, \mathrm{Pr} = 0.5$ and $R = 0.25$ (above right).

ratio $R_4 \left(= \rho_d / \rho_m \right)$ and viscosity ratio $R_5 \left(= \mu_d / \mu_m \right)$ of the dusty fluid relative to the nanofluid ethylene glycol water mixture and indeed also the Newtonian pure fluid. Since Newtonian fluid is less dense than dusty fluid and nanofluid, higher velocity magnitudes are computed in comparison to either Saffman dusty fluid or nanofluid, respectively. For the same reason, the zone 2 Saffman dusty fluid attains consistently greater velocity magnitudes at all locations in the duct relative to the zone 3 nanofluid. Furthermore, since the flow in all three zones is controlled by a *constant* pressure gradient, the velocity profile can reach a steady state. The dust particles in the dusty fluid produce a similar velocity distribution to the liquid since they are entrained and naturally are transported with the liquid. However, due to the presence of the Stoke drag effect, the velocity profile of the dust particles in the dusty fluid lags slightly behind the liquid velocity profile. This phenomenon has been confirmed in many previous studies including Zueco et al. [50], Bhatti et al. [51] and Chandrawat et al. [52], who deployed alternative numerical methods including electrothermal network simulation (PSPICE software), ANSYS FLUENT and cubic spline DQM. According to Figure 8.5, in the scenario where the triple-layer interfacial duct flow is mobilized by a *decaying pressure gradient*, the fluid velocities in all three zones and particle velocities in the middle zone exhibit an initial increase with time (up to $t = 0.3$) and all these profiles exhibit curvature. Furthermore, the peak velocity for the upper layer (zone 1-Newtonian) is near the lower duct wall and the peak velocity in the intermediate layer (zone 2-Saffman dusty) and base layer (zone 3-nanofluid) are at the respective upper interfaces. As time is progressed to $t = 10, 15, 20$, three distinct modifications arise. *Firstly*, the profiles assume linear topologies. *Secondly*, the peak velocity migrates away from the central zone in the upper zone 3 layer to exactly the base of the upper zone 3 (Newtonian/dusty interface). *Thirdly*, the flow is decelerated with this large increase in time, which is the opposite of the behaviour with small time increments. Clearly the progression in time has a very significant influence on both magnitudes and distributions of velocities in all three zones. The impact of decaying pressure gradient is very different from constant pressure gradient (studied in earlier plots) and this confirms that the nature of the imposed pressure gradient can be exploited judiciously to manipulate the internal flow dynamics in the duct. In Figure 8.6, the pulsating velocities of the fluid and dust particles within the duct are induced by a *periodic* (sinusoidal) pressure gradient. It is noteworthy that a steady state is not reached, and as such, the velocities exhibit an increase when $0 < t \leq \pi/2$, a reduction when $\pi/2 < t \leq 2\pi$, and a further increase when $2\pi < t \leq 5\pi/2$. Furthermore, very prominent negative velocities (flow reversal in the duct) are computed *in all three zones* for the time case, $t = 2\pi$ and partially for $t = 3\pi/2$ in the upper Newtonian zone and the intermediate dusty zone. However again for the lower nanofluid zone, for $t = 3\pi/2$ consistently negative velocity is computed. For all other times, positive velocity is maintained throughout the duct in all three zones. Maximum velocity is computed at the Newtonian/dusty interface for the case $t = \pi/2$ and the minimum velocity also appears at this location, although the former has a significantly larger magnitude. A much lower range of velocities is observed in the lower zone 3 (nanofluid) relative to the other two zones. Figure 8.7 demonstrates that the fluids present in the three zones all exhibit parabolic velocity profiles. The velocity profile of dust particles in dusty fluid follows a similar pattern to the fluid in that zone. While the profiles increase from zero at the top plate (no-slip) through the upper Newtonian zone and peak at the Newtonian/dusty interface, the profiles in the dusty zone 2 ascend from non-zero at the nanofluid/dusty fluid interface initially much more sharply but as the constant pressure gradient is boosted the rate of ascent is reduced. This behaviour is attributable to the elevation in volumetric flow rate with larger pressure gradient increases and may eventually reach a steady state. At first glance, the velocity profiles of the dusty fluid and the dust particles in zone 2 of the duct appear to be identical; however, the deviation is amplified as the pressure gradient rises. Peak velocity is computed just above the Newtonian/dusty interface in zone 3 (Newtonian) for maximum pressure gradient ($Ge = 20$). Much greater magnitudes of velocity arise in zone 3 (Newtonian) and zone 2 (dusty) relative to the lower zone 3 (nanofluid). No backflow is computed anywhere in the duct at any value of Ge (pressure gradient). Figures 8.8 and 8.9 show that the magnitude of the velocity profile increases with both Ion slip parameter (Bi) and Hall parameter (Be). The magnetic resistive force is decreased as Be and Bi are increased. For example for zone-*1* (*Nano-fluid:* $-1 \leq y \leq 0$), *the modified magnetic body force term is*

$$-\frac{\mathrm{Ha}^2}{\mathrm{Re}}\left(\frac{1 + BiBe}{(1 + BiBe)^2 + \mathrm{Re}^2 \dfrac{\Psi}{R_1^*} + (1 - \Psi)} \right) u_{nf}, \quad \text{in the momentum equation.}$$

Similar terms feature in the respective momentum equations for *Zone-II (Dusty fluid: $0 \leq y \leq 1$), viz* $-\left(\dfrac{1+BiBe}{\left(1+BiBe\right)^2+Be^2}\right)\dfrac{\mathrm{Ha}^2}{\mathrm{Re}\,R_4}u_d$ *and Zone-III (Newtonian fluid: $1 \leq y \leq 2$), viz* $-\dfrac{\mathrm{Ha}^2}{\mathrm{Re}}\left(\dfrac{1+BiBe}{\left(1+BiBe\right)^2+Be^2}\right)\left(\dfrac{1}{R_4 R_{10}}\right)u_n.$ Clearly the quadratic *BiBe* term in the denominator will dominate the term producing an effective depreciation in velocity. This will manifest in an acceleration in the flow in each of the three immiscible layers. Velocity magnitude will therefore be boosted for each fluid in all the zones (1–3) and due to entrainment in the dusty fluid in zone 2, dust particle velocity will also be elevated. The volume flow rate in lower zone is less than the middle and upper zone fluid. This is due to the presence of nanoparticles of Al_2O_3 in the lowest nanofluid zone which as elaborated earlier modifies the viscosity and opposes momentum development. In both plots peak velocity is computed for the Newtonian fluid (top zone) for *Be = 0.4 and Bi = 4*. The inclusion of ion slip and Hall current effects clearly contributes markedly to a modification in velocity distributions in electromagnetic duct flow. More precise simulations of nano-modified batteries and fuel cells can therefore be achieved with inclusion of these important magnetic effects. Although we have neglected magnetic induction, Maxwell displacement and oblique magnetic field effects [53–55] in this study, the model developed is quite comprehensive and provides a solid foundation for future refinements. Figure 8.10 has been plotted to compare the temperature heat profiles of the three fluids namely nanofluid (CuO, Al_2O_2, Ag, TiO_2) in zone 1, Saffman dusty in zone 2 and Newtonian fluids in zone 3 are considered in horizontal duct using the RBF-PS scheme for constant pressure gradient. The aim of this graph is to identity the efficient nanofluid mixture in terms of heat transfer. It is observed that mixture of Al_2O_2- ethylene glycol water achieves the best thermal enhancement relative to CuO, or Ag or TiO_2 nanoparticles combined with ethylene glycol-water. A parabolic decrement in temperature is present in zone 3 (Newtonian) whereas a linear decay is computed in the intermediate layer (dusty zone 2). In the lower nanofluid layer (zone 1), a linear scent in temperature is observed. Figure 8.11 visualizes the effect of Hartmann magnetic parameter squared, $\mathrm{Ha}^2\left(=\sigma B_0^2 h^2 / \mu_m\right)$ on the temperature profile of the fluids present in all three zones. It is observed that the temperature of fluids in all the zones along with the temperature of dusty particles decreases with the increase in Ha^2. This is produced due to the retarding nature of the magnetic field strength and the viscous dissipation effect. Unlike classical magnetohydrodynamic boundary layer flow which generates a heating effect in the fluid, in this duct flow the contrary effect is computed. This is also associated with the Hall cross flow effect which modifies the action of the static magnetic field. For the sinusoidal pressure gradient case, strongly parabolic profiles are observed in dusty zone 2 and nanofluid zone 1. A more gradual parabolic decay is computed in zone 3 (Newtonian). For the decaying and constant pressure gradient cases, consistently linear descents are computed in zone 3 (Newtonian) and zone 2 (dusty) whereas a linear growth is observed in the nanofluid zone 1. In zone 2 dust particle phase velocity is consistently lower than the fluid phase velocity at all values of Hartmann number. Maximum temperature arises at the dusty zone 2/nanofluid zone 3 interface. Figure 8.12 has been plotted to see the effect of the volume fraction parameter of Al_2O_3 nanoparticles in zone 1 on the temperature distributions of the three fluids in all three zones of the duct. It is observed that temperature magnitudes are boosted with an increase in volume fraction for both zone 1 (nanofluid) and zone 2 dusty fluid. However, the reverse trend is observed in zone 3 (Newtonian fluid). Maximum temperature is seen at the dusty/nanofluid interface again for the case of a constant pressure gradient. A parabolic decrement in temperature is seen in zone 3 (Newtonian) whereas once again sharp linear topologies are observed in zone 2 (dusty) and zone 1 (nanofluid). Minimum temperatures in all three layers correspond to the periodic (sinusoidal) pressure gradient and vanishing nanoparticle case ($\phi = 0\%$) indicating that pulsatile oscillations and absence of nanoparticle thermal conduction both damp heat transfer significantly in the duct. Figure 8.13 has been plotted to study the effect of ion-slip parameter (Bi) on the temperature profile of the fluids in all three zones of the horizontal duct. The ion-slip parameter (Bi) can have a significant impact on the transport properties of a fluid, including thermal conductivity and viscosity. With slip between the ions and the dust particles and nanoparticles, thermal conduction is encouraged. This results in a boost in temperature profile magnitude with elevation in ion-slip parameter in the nanofluid zone 1 and dusty zone 2. However, the converse trend is computed in the upper layer (Newtonian zone 3) since neither dusty nor nanoparticles are present. Again, parabolic profiles are computed for all pressure gradient cases in zone 3 (Newtonian) whereas linear profiles are present in zone 2 (dusty) and zone 2 (nanofluid). There is a clear temperature balance for dusty/nanofluids at the zone 2/zone 1 interface for both decaying pressure gradient as well as constant pressure gradient. However, there is a slight disparity for the periodic pressure gradient case. Figure 8.14 visualizes the influence of the Hall parameter (Be) on the temperature profiles of all three immiscible layers in the duct. All three layers are electrically conducting liquids and can generate magnetic field structures, such as magnetic reconnection layers, which can cause localized heating or cooling of the fluid. One can observe that temperatures are hiked up with stronger Hall current Be for nano fluid in zone 1 and decreases for Newtonian fluid in zone 3. These trends are sustained for all three pressure gradients. However, in the intermediate dusty zone 2, for the constant and decaying pressure gradient cases, only in the lower half space does an increase in Hall parameter elevate temperatures; in the upper half space the opposite pattern is observed. For the periodic pressure gradient case, however there is a consistent enhancement in temperature throughout the depth of the intermediate zone 2 with increasing Hall current parameter. Dusty particle temperatures also follow this trend. Peak temperature arises again the dusty/nanofluid interface. The cross flow generated by the Hall parameter has a different influence in the core zone of the duct (intermediate layer 2) than in the peripheral layers 1 and 3. It modifies the effective electrical conductivity and warps the influence of the Lorentzian magnetic retarding force very prominently in the intermediate layer. Figure 8.15 depicts the impact of the Eckert number (Ec) on the temperature distributions. The Eckert number is a dimensionless parameter that quantifies the conversion of kinetic energy into thermal energy (due to internal friction) relative to the enthalpy difference across the channel. The results demonstrate that an elevation in the Eckert number leads to a rise in the thermal energy of all three fluids and also the dust particles as a consequence of increased mechanical energy conversion to heat, thereby augmenting the energy level of the system. Constant pressure gradient produces maximum temperature, and sinusoidal (periodic) pressure gradient generates minimum temperatures, for all three immiscible layers.

8.6 CONCLUSIONS

A mathematical and computational study to examine the potential of using CuO, Al_2O_3, Ag, and TiO_2 nanoparticles to improve heat transfer in hydromagnetic fully developed horizontal duct flow arising in e.g. nano-battery configurations and electromagnetic duct materials processing, via elevation in thermal conductivity. The duct comprises three layers - *nanofluid, Saffman fluid, and Newtonian viscous fluid.* We examine the influence of the lower nanofluid on transport phenomena within the intermediate Saffman dusty fluid and upper Newtonian fluid strata, with *steady, oscillating, and decaying* pressure gradients. Lorentz magnetic body force, ion slip, and Hall current effects are incorporated. A 2-D model is formulated using coupled partial differential equations with appropriate wall and interfacial boundary conditions. The nonlinear boundary value problem is solved numerically with a meshfree radial basis function pseudo-spectral method (RBFPS). In the lower nanofluid stratum, Al_2O_3 nanoparticles are selected with an EG-water base fluid. A parametric study of the impact of key parameters on thermo-fluid characteristics is conducted. Validation with a differential quadrature method (DQM) is included. The RBFPS numerical solutions have shown that:

I. The volume fraction of Al_2O_3 nanoparticles although only present in the lower zone 1 of the duct significantly affects the energy distributions and momentum (velocity) characteristics of the intermediate zone dusty fluid and upper zone Newtonian fluid. Temperature is positively influenced whereas velocities are reduced (flow deceleration) with higher values of zone 1 nanofluid fraction in all the three immiscible zones.

II. The Hartmann magnetic number squared (Ha^2) has a significant impact on the temperature and velocity profiles in all three layers of the fuel cell, allowing for a systematic scrutiny of the influence of an external static magnetic field. As Ha^2 increases, the magnetic field strength intensifies, leading to an increase in the electromagnetic Lorentzian drag force. Owing to viscous dissipation terms in the heat conservation equations for all three zones, this results in a decrease in the temperature in all three zones, as well as the classical depletion in velocities.

III. At low values of Eckert number, the kinetic energy of the fluid dominates over the thermal energy, and the flow is considered to be highly energetic. In this regime, the fluid experiences large velocity gradients, and the energy (heat) transfer in the fluid is primarily driven by convection. The temperature profile of the fluid in this regime is characterized by steep temperature gradients near the boundaries, and the heat transfer rate is high. Higher Eckert number has a profound influence on temperature distributions in all three immiscible layers.

The present investigation has aimed to reveal some intricate fluid dynamic characteristics associated with triple-layer immiscible nano-battery and magnetic fuel cell transport phenomena. Excellent accuracy has been achieved with a hybrid numerical method (meshfree radial basis function pseudo-spectral method- (RBFPS)). Attention has however been confined essentially to momentum and heat transfer. Future investigations may address wall material property effects in MHD channels [56]. Efforts in these directions are underway and will be communicated imminently.

REFERENCES

1. Q. Zhang and J. Liu, "Nano liquid metal as an emerging functional material in energy management, conversion and storage," *Nano Energy*, 2013, doi: 10.1016/j. nanoen.2013.03.002.
2. S. H. Park and Y. S. Song, "Carbon nanofluid flow based biophotovoltaic cell," *Nano Energy*, 2021, doi: 10.1016/j.nanoen.2020.105624.
3. O. Mahian, A. Kianifar, S. Z. Heris, D. Wen, A. Z. Sahin, and S. Wongwises, "Nanofluids effects on the evaporation rate in a solar still equipped with a heat exchanger," *Nano Energy*, 2017, doi: 10.1016/j.nanoen.2017.04.025.
4. R. K. Cheedarala, M. Shahriar, J. H. Ahn, J. Y. Hwang, and K. K. Ahn, "Harvesting liquid stream energy from unsteady peristaltic flow induced pulsatile Flow-TENG (PF-TENG) using slipping polymeric surface inside elastomeric tubing," *Nano Energy*, 2019, doi: 10.1016/j.nanoen.2019.104017.
5. A. Salari, A. Kazemian, T. Ma, A. Hakkaki-Fard, and J. Peng, "Nanofluid based photovoltaic thermal systems integrated with phase change materials: Numerical simulation and thermodynamic analysis," *Energy Convers. Manag.*, vol. 205, 2020, doi: 10.1016/j. enconman.2019.112384.
6. M. Eisapour, A. H. Eisapour, M. J. Hosseini, and P. Talebizadehsardari, "Exergy and energy analysis of wavy tubes photovoltaic-thermal systems using microencapsulated PCM nano-slurry coolant fluid," *Appl. Energy*, vol. 266, 2020, doi: 10.1016/j. apenergy.2020.114849.
7. H. Kaneez, M. A. Qureshi, S. O. Alharbi, T. Aziz, and M. Nawaz, "Role of hybrid nano-structures and dust particles on transportation of heat energy in fluid with memory effects," *Ain Shams Eng. J.*, vol. 12, no. 2, pp. 2171–2180, 2021, doi: 10.1016/j.asej.2020.12.005.
8. N. Tonekaboni, H. Salarian, M. E. Nimvari, and J. Khaleghinia, "Energy and exergy analysis of an enhanced solar CCHP system with a collector embedded by porous media and nano fluid," *J. Therm. Eng.*, vol. 7, no. 6, pp. 1489–1505, 2021, doi: 10.18186/ thermal.990897.
9. D. G. Subhedar, Y. S. Patel, B. M. Ramani, and G. S. Patange, "An experimental investigation on the effect of Al2O3/ cutting oil based nano coolant for Minimum Quantity Lubrication drilling of SS 304," *Clean. Eng. Technol.*, vol. 3, 2021, doi: 10.1016/j. clet.2021.100104.
10. X. Zhang, J. Niu, and J. yong Wu, "Evaluation and manipulation of the key emulsification factors toward highly stable PCM-water nano-emulsions for thermal energy storage," *Sol. Energy Mater. Sol. Cells*, vol. 219, 2021, doi: 10.1016/j.solmat.2020.110820.
11. Y. Krishna, M. Faizal, R. Saidur, K. C. Ng, and N. Aslfattahi, "State-of-the-art heat transfer fluids for parabolic trough collector," *Int. J. Heat Mass Transf.*, vol. 152. 2020, doi: 10.1016/j.ijheatmasstransfer.2020.119541.
12. L. U. Gui, L. I. N. DianJi, and W. A. N. G. XiaoDong, "Progress of electrowetting applications in micro-nano energy conversion and utilization systems," *Kexue Tongbao/Chinese Science Bulletin*, vol. 62, no. 8. pp. 799–811, 2017, doi: 10.1360/N972016-00073.

13. P. G. Saffman, "On the stability of laminar flow of a dusty gas," *J. Fluid Mech.*, vol. 13, no. 1, pp. 120–128, 1962, doi: 10.1017/S0022112062000555.

14. M. Gnaneswara Reddy and M. Ferdows, "Species and thermal radiation on micropolar hydromagnetic dusty fluid flow across a paraboloid revolution," *J. Therm. Anal. Calorim.*, vol. 143, no. 5. pp. 3699–3717, 2021, doi: 10.1007/s10973-020-09254-1.

15. M. Gnaneswara Reddy, M. V. V. N. L. Sudha Rani, K. Ganesh Kumar, B. C. Prasannakumar, and H. J. Lokesh, "Hybrid dusty fluid flow through a Cattaneo-Christov heat flux model," *Phys. A Stat. Mech. Appl.*, vol. 551, 2020, doi: 10.1016/j.physa.2019.123975.

16. G. Ali, F. Ali, A. Khan, A. H. Ganie, and I. Khan, "A generalized magnetohydrodynamic two-phase free convection flow of dusty Casson fluid between parallel plates," *Case Stud. Therm. Eng.*, 2022, doi: 10.1016/j.csite.2021.101657.

17. R. K. Chandrawat, V. Joshi, O. Anwar Bég and D. Tripathi, "Computation of unsteady generalized Couette flow and heat transfer in immiscible dusty and non-dusty fluids with viscous heating and wall suction effects using a modified cubic B-spine differential quadrature method," *Heat Transf.*, vol. 51, no. 1, pp. 99–139, 2022, doi: 10.1002/htj.22299.

18. B. Mahanthesh, B. J. Gireesha, B. C. PrasannaKumara, and N. S. Shashikumar, "Marangoni convection radiative flow of dusty nanoliquid with exponential space dependent heat source," *Nucl. Eng. Technol.*, vol. 49, no. 8, pp. 1660–1668, 2017, doi: 10.1016/j.net.2017.08.015.

19. R. Muthuraj, K. Nirmala, and S. Srinivas, "Influences of chemical reaction and wall properties on MHD peristaltic transport of a dusty fluid with heat and mass transfer," *Alexandria Eng. J.*, vol. 55, no. 1, pp. 597–611, 2016, doi: 10.1016/j.aej.2016.01.013.

20. G. Kalpana, K. R. Madhura, and R. B. Kudenatti, "Impact of temperature-dependant viscosity and thermal conductivity on MHD boundary layer flow of two-phase dusty fluid through permeable medium," *Eng. Sci. Technol. Int. J.*, 2019, doi: 10.1016/j.jestch.2018.10.009.

21. Y. Abd Elmaboud, S. I. Abdelsalam, K. S. Mekheimer, and K. Vafai, "Electromagnetic flow for two-layer immiscible fluids," *Eng. Sci. Technol. Int. J.*, vol. 22, no. 1, pp. 237–248, 2019, doi: 10.1016/j.jestch.2018.07.018.

22. X. Chen and Y. Jian, "Entropy generation minimization analysis of two immiscible fluids," *Int. J. Therm. Sci.*, vol. 171, 2022, doi: 10.1016/j.ijthermalsci.2021.107210.

23. M. Salari, E. H. Malekshah, and M. H. Malekshah, "Natural convection in a rectangular enclosure filled by two immiscible fluids of air and Al2O3-water nanofluid heated partially from side walls," *Alexandria Eng. J.*, vol. 57, no. 3, pp. 1401–1412, 2018, doi: 10.1016/j.aej.2017.07.004.

24. B. Dong, Y. Y. Yan, W. Li, and Y. Song, "Lattice Boltzmann simulation of viscous fingering phenomenon of immiscible fluids displacement in a channel," *Comput. Fluids*, vol. 39, no. 5, pp. 768–779, 2010, doi: 10.1016/j.compfluid.2009.12.005.

25. R. K. Chandrawat and V. Joshi, "Numerical solution of the time-dependent flow of immiscible fluids with fuzzy boundary conditions," *Int. J. Math. Eng. Manag. Sci.*, vol. 6, no. 5, pp. 1315–1330, 2021, doi: 10.33889/ijmems.2021.6.5.079.

26. G. J. Wang, A. Damone, F. Benfenati, P. Poesio, G. P. Beretta, and N. G. Hadjiconstantinou, "Physics of nanoscale immiscible fluid displacement," *Phys. Rev. Fluids*, vol. 4, no. 12, 2019, doi: 10.1103/PhysRevFluids.4.124203.

27. M. Firkowski, K. Urbanowicz, and H. F. Duan, "Simulation of unsteady flow in viscoelastic pipes," *SN Appl. Sci.*, 2019, doi: 10.1007/s42452-019-0524-2.

28. M. I. Asjad, M. H. Butt, M. A. Sadiq, M. D. Ikram, and F. Jarad, "Unsteady Casson fluid flow over a vertical surface with fractional bioconvection," *AIMS Math.*, vol. 7, no. 5, pp. 8112–8126, 2022, doi: 10.3934/math.2022451.

29. Z. M. Mburu, S. Mondal, and P. Sibanda, "Numerical study on combined thermal radiation and magnetic field effects on entropy generation in unsteady fluid flow past an inclined cylinder," *J. Comput. Des. Eng.*, vol. 8, no. 1, pp. 149–169, 2021, doi: 10.1093/jcde/qwaa068.

30. A. S. Dogonchi, T. Tayebi, N. Karimi, A. J. Chamkha, and H. Alhumade, "Thermal-natural convection and entropy production behavior of hybrid nanoliquid flow under the effects of magnetic field through a porous wavy cavity embodies three circular cylinders," *J. Taiwan Inst. Chem. Eng.*, vol. 124, pp. 162–173, 2021, doi: 10.1016/j.jtice.2021.04.033.

31. M. Nikodijevic, Z. Stamenkovic, J. Petrovic, and M. Kocic, "Unsteady fluid flow and heat transfer through a porous medium in a horizontal channel with an inclined magnetic field," *Trans. Famena*, vol. 44, no. 4, 2021, doi: 10.21278/TOF.444014420.

32. I. I. Kim and K. D. Kihm, "Nano sensing and energy conversion using surface plasmon resonance (SPR)," *Materials*, vol. 8, no. 7. pp. 4332–4343, 2015, doi: 10.3390/ma8074332.

33. G. Zheng, M. Chen, J. Yin, H. Zhang, X. Liang, and J. Zhang, "Metal organic frameworks derived nano materials for energy storage application," *Int. J. Electrochem. Sci.*, vol. 14, no. 3. pp. 2345–2362, 2019, doi: 10.20964/2019.03.28.

34. A. J. Chamkha, "Solutions for fluid-particle flow and heat transfer in a porous channel," *Int. J. Eng. Sci.*, vol. 34, no. 12, pp. 1423–1439, 1996, doi: 10.1016/0020-7225(96)00036-5.

35. K. L. Hsiao, "Stagnation electrical MHD nanofluid mixed convection with slip boundary on a stretching sheet," *Appl. Therm. Eng.*, vol. 98, pp. 850–861, 2016, doi: 10.1016/j.applthermaleng.2015.12.138.

36. C. Barreneche, M. Martín, J. Calvo-de la Rosa, M. Majó, and A. I. Fernández, "Own-synthetize nanoparticles to develop nano-enhanced phase change materials (NEPCM) to improve the energy efficiency in buildings," *Molecules*, vol. 24, no. 7, 2019, doi: 10.3390/molecules24071232.

37. A. Panepinto and R. Snyders, "Recent advances in the development of nano-sculpted films by magnetron sputtering for energy-related applications," *Nanomaterials*, vol. 10, no. 10, pp. 1–27, 2020, doi: 10.3390/nano10102039.

38. E. J. Kansa, "Multiquadrics-A scattered data approximation scheme with applications to computational fluid-dynamics-II solutions to parabolic, hyperbolic and elliptic partial differential equations," *Comput. Math. Appl.*, 1990, doi: 10.1016/0898-1221(90)90271-K.

39. M. D. Buhmann, *Radial Basis Functions, Theory and Implementations*. Cambridge University Press, 2003.

40. B. Fornberg and N. Flyer, "Solving PDEs with radial basis functions," *Acta Numer.*, 2015, doi: 10.1017/s0962492914000130.

41. G. E. Fasshauer, *Meshfree Approximation Methods with MATLAB*, Wiley, New York. 2007.

42. A. J. M. Ferreira and G. E. Fasshauer, "An RBF-pseudospectral approach for the static and vibration analysis of composite plates using a higher-order theory," *Int. J. Comput. Methods Eng. Sci. Mech.*, 2007, doi: 10.1080/15502280701471632.

43. I. Waini, A. Ishak, and I. Pop, "Unsteady flow and heat transfer past a stretching/shrinking sheet in a hybrid nanofluid," *Int. J. Heat Mass Transf.*, 2019, doi: 10.1016/j.ijheatmasstransfer.2019.02.101.

44. S. S. Ghadikolaei, K. Hosseinzadeh, M. Hatami, and D. D. Ganji, "MHD boundary layer analysis for micropolar dusty fluid containing Hybrid nanoparticles (Cu-Al2O3) over a porous medium," *J. Mol. Liq.*, vol. 268, pp. 813–823, 2018, doi: 10.1016/j.molliq.2018.07.105.

45. H. A. Attia, W. Abbas, and M. A. M. Abdeen, "Ion slip effect on unsteady Couette flow of a dusty fluid in the presence of uniform suction and injection with heat transfer," *J. Brazilian Soc. Mech. Sci. Eng.*, vol. 38, no. 8, pp. 2381–2391, 2016, doi: 10.1007/s40430-015-0311-y.

46. H. A. Attia and K. M. Ewis, "Magnetohydrodynamic flow of continuous dusty particles and non-Newtonian Darcy fluids between parallel plates," *Adv. Mech. Eng.*, vol. 11, no. 6, 2019, doi: 10.1177/1687814019857349.

47. M. Devakar and A. Raje, "A study on the unsteady flow of two immiscible micropolar and Newtonian fluids through a horizontal channel: A numerical approach," *Eur. Phys. J. Plus*, vol. 133, no. 5, 2018, doi: 10.1140/epjp/i2018-12011-5.

48. M. Devakar and A. Raje, "A magnetohydrodynamic time dependent model of immiscible newtonian and micropolar fluids through a porous channel: A numerical approach," *J. Appl. Fluid Mech.*, vol. 12, no. 2, pp. 603–615, 2019, doi: 10.29252/jafm.12.02.28548.

49. E. Abu-Nada and A. J. Chamkha, "Effect of nanofluid variable properties on natural convection in enclosures filled with a CuO-EG-Water nanofluid," *Int. J. Therm. Sci.*, 2010, doi: 10.1016/j.ijthermalsci.2010.07.006.

50. J. Zueco, P. Eguía, E. Granada, J. L. Míguez, and O. Anwar Bég, "An electrical network for the numerical solution of transient MHD Couette flow of a dusty fluid: Effects of variable properties and Hall current," *Int. Comm. Heat Mass Transf.*, vol. 37, 2010, pp. 1432–1439.

51. M. M. Bhatti, A. Zeeshan, N. Ijaz, O. Anwar Bég, and A. Kadir, "Mathematical modelling of nonlinear thermal radiation effects on EMHD peristaltic pumping of viscoelastic dusty fluid through a porous medium channel," *Eng. Sci. Technol.*, vol. 20, no. 3, 2017, pp. 1129–1139.

52. R. K. Chandrawat, V. Joshi, and O. Anwar Bég, "Ion slip and Hall effects on generalized time-dependent hydromagnetic Couette flow of immiscible micropolar and dusty micropolar fluids with heat transfer and dissipation: A numerical study," *J. Nanofluids*, vol. 10, 2021, pp. 431–446.

53. S. K. Ghosh, O. Anwar Bég, J. Zueco, and Prasad, V. R., "Transient hydromagnetic flow in a rotating channel permeated by an inclined magnetic field with magnetic induction and Maxwell displacement current effects," *ZAMP: J. Appl. Math. Phys.*, 61, 2010, pp. 147–169.

54. M. J. Uddin, M.N. Kabir, O. Anwar Bég, and Y. Alginahi, "Chebyshev collocation computation of magneto-bioconvection nanofluid flow over a wedge with multiple slips and magnetic induction," *Proc. IMechE: Part N-J. Nanomater. Nanoeng. Nanosyst.*, vol. 232, 2018, pp. 109–122.

55. S. Akter, M. Ferdows, T. A. Bég, O. Anwar Bég, A. Kadir, and S. Sun, "Spectral relaxation computation of electroconductive nanofluid convection flow from a moving surface with radiative flux and magnetic induction," *J. Comput. Des. Eng.*, vol. 8, no. 4, 2021, pp. 1158–1171.

56. V. K. Rohatgi, "High temperature materials for magnetohydrodynamic channels," *Bull. Mater. Sci.*, vol. 6, 1984, pp. 71–82.

9 CFD Modelling of Combustion and NOX Production in a Propulsion Duct with Aspect Ratio Effects

O. Anwar Bég, Theo Morand,
Tasveer A. Bég, S. Kuharat, B. Vasu, Henry
J. Leonard, and Martin L. Burby

9.1 INTRODUCTION

Combustion flows are fundamental to aerospace propulsion [1]. Many complex phenomena arise in combustion including heat transfer, mass transfer, turbulence, flame propagation (weak flames, flames with repetitive extinction and ignition, and stable flames), emissions, etc. In recent decades, as a compliment to experimental studies which feature for example laser diagnostics used for flame studies at high resolution in time and space [2], there has been a significant development in computational-based methods. Many excellent numerical studies of combustion flows have been reported, often for premixed systems in channel configurations. Such flows are invariably turbulent and also involve chemical reactions and multi-species transport. Ghoneim et al. [3] deployed a random vortex method (r.v.m.) to simulate turbulent flow associated with combustion, tracing and capturing the action of elementary turbulent eddies and their cumulative effects. They also used a flame propagation algorithm, and volume sources modelling and considered flow in a lean propane-air mixture within a combustion tunnel (where the flame is stabilized by a back-facing step) and obtained very good agreement with high speed Schlieren photography. Kang et al. [4] utilized the open-source framework OpenFOAM finite element software for simulating time-accurate, low-Mach number reacting combustion flow with small-scale flames. They successfully computed the structure of diffusion flame street and deployed a conjugate heat transfer model with special solid-wall heat conduction features which more accurately analyses axial thermal conductivities. Very cost-effective capabilities were also demonstrated, and superior wave-damping abilities also achieved during early stages of the simulations. Yamamoto et al. [5] used a lattice Boltzmann method (LBM) to simulate computationally both 2D and 3D flame structures in turbulent combustion in simple flow geometries (channels). They accurately

DOI: 10.1201/9781003473749-11

computed the laminar burning velocity is obtained and examined the interaction of unsteady flames with vortices, including refined chemistry for a counter-flow flame. Kang et al. [6] deployed a (DRM-19) chemical reaction model and a convergence index (GCI) for refining the computational grid, to compute the premixed methane/air combustion in narrow channels. They considered a range of ignition strategies to initiate the flame and obtained good correlation with other experimental and numerical studies. Moreau [7] used the Star-CD finite volume code to study three different flame configurations in premixed combustion turbulent flow. He used a self-similar turbulent flame model to characterize the turbulent flame velocity and included a stretch factor, for controlling the flame behaviour close to the flame holder and near the walls. Usman Allauddin [8] employed both a Large Eddy Simulation (LES) method and standard Reynolds-averaged Navier-Stokes (RANS) approaches to simulate turbulent premixed combustion. To overcome the problem of flame thickness which is generally too narrow for resolution on the typical LES filter sizes, he implemented a model for subgrid scale flame wrinkling and a Flame Surface Density (FSD) method to simulate the filtered LES reaction rate. He also computed the flame shape and thickness evolution and the effects of high pressure and the Lewis number (Le = thermal diffusivity/mass diffusivity) on combustion characteristics. Gruber et al. [9] used direct numerical simulations (DNS) to study the spontaneous ignition of hydrogen flames at laminar, turbulent, adiabatic and non-adiabatic conditions, for both stable and unstable flames. They demonstrated that for hydrogen reheat combustion, compressibility effects are critical for flame stability and that unstable ignition and combustion are consistently encountered for reactant temperatures close to the mixture characteristic crossover temperature. They also evaluated the instantaneous fuel consumption rate within the reaction front at different temperatures, pressures, and turbulence intensity levels. Ayoobi and Schoegl [10] conducted numerical studies on premixed flames in air/methane laminar combustion propagating within small channels to elucidate the difference from flame propagation at conventional scales. They studied the mechanisms contributing to symmetry breaking and limit cycle behaviour that are fundamental to these combustion modes for different equivalence ratios (0.53 and 0.7) and channel widths (2 mm and 5 mm) for a constant channel length and bulk inlet velocity, with refined chemical reaction models. They observed that different flame front topology is present with different channel widths. For the 5 mm channel, although flame fronts commenced symmetrically, the symmetry vanished very quickly following ignition. However, in the 2 mm channel, both equivalence ratios produce flames that stabilize with symmetric flame fronts after propagating upstream. They identified that the absence of instabilities and asymmetries for the 2 mm case is connected to inadequate wall separation, which is similar in magnitude to the flame thickness. An important consideration in modern propulsion systems has been the need for reducing pollution, i.e., nitrogen oxides (NOx) emission reduction. Significant studies in this area have been conducted at the NASA Glenn Research Centre, USA on flame tube designs and mitigation of NOX production to curtail damage to the environment [11,12]. Other key works have been presented for various gas turbine propulsion designs by Tacina et al. [13] (who introduced the lean direct wall injection technique), Leonard and Stegmaier [14] (who developed the "autoderivative" approach"), Sattelmayer et al. [15] (who examined lean premixed combustors). Marek et al. [16] have also scrutinized the use of lean direct injection techniques for suppressing NOX emissions in hydrogen combustion. In many of these studies it has been shown via computational simulations of combustion in channel geometries that stationary, non-stationary, or asymmetric modes which depend on properties of the incoming reactant flow as well as channel geometry and wall temperatures, have a significant effect on NOX emissions (mass fractions are strongly influenced).

The strong trend in producing environmentally more friendly combustion propulsion systems [17] has provided strong motivation for the current work in which we examine the effects of channel aspect ratio and inlet methane velocity on combustion characteristics in 2-D channel geometries. Appropriate stochiometric relations are adopted and mass fractions for water, carbon dioxide, oxygen (oxidiser) and methane (fuel) presented to compute pollutant distributions. ANSYS FLUENT CFD software is used via the pressure-based solver (incompressible flow is considered). A k-epsilon turbulence model is deployed. Mesh independence is included. Extensive modifications in transport characteristics (temperature, pressure, vorticity, mass fraction) are identified with a change in geometry and inlet methane velocity under temperatures exceeding 2,000 K which are appropriate for rocket designs. In particular, the use of contour plot visualization is shown to provide an excellent, low-cost facility for appraising NOX reduction in turbulent combustion without the need for expensive laboratory work with laser diagnostic systems. The novelties are the inclusion of aspect ratio, NOX production computation and full visualization with all stages of the simulation included, features which is often neglected in other studies.

9.2 NOX COMBUSTION

Every propulsion system, whether solid, liquid, hybrid, nuclear or electric, involves combustion. The mixture of propellants (oxidizer and fuel) generates this combustion and thus allows to create the necessary thrust and the flame that is observed at the nozzle exit. The combustion will produce the required thrust of the engine and depends on the mixture used, also on the pressure inside the combustion chamber and of course on the mass flow rate. To optimize designs, therefore it is essential to study transport combustion characteristics with flame propagation in a duct. At the exit of the nozzle, the flow is almost turbulence free, the flow (the flame) is laminar. However, under the engines, the flow becomes chaotic, the atmosphere mixes with the flame and the flow becomes turbulent. In the case of a rocket duct propulsion system, this effect is even more pronounced because the gases from the generator must be managed, as some of them must be evacuated and may be the cause of this pollution. Combustion chemistry is central to NOX production. The family of nitrogen oxide pollutants generated in premixed combustion is relatively large and included the following species [17]:

$$N_2O, NO, N_2O_2, N_2O_3, N_2O_4, \text{ and } N_2O_5 \tag{9.1}$$

The gaseous species are usually:

$$N_2O, NO, NO_2, N_2O_4, \text{ and } N_2O_2 \tag{9.2}$$

However, the abundant ones are *NO*, NO_2 which are called NOx. The NO-fuel is formed from the organic nitrogen species contained in the fuel. This mechanism is very present when the fuel contains nitrogen (80% of total NO) and is absent in natural gas. Nitrous oxide-early has a rapid reaction taking place very early in the flame [18]. Production is higher in rich combustion. In lean burn conditions, the amount of nitrogen oxide produced becomes negligible compared to the total NO production. The last formation of nitrogen oxide is a so-called thermal production which has slow reactions. It is sensitive to residence time, temperature and oxygen concentration. In a rocket engine, the combustion is done by the propellants which will be mixed (on the one hand the oxidizer and on the other hand the fuel). Engineering research has generally focused on establishing the best mixture in order to have the best possible performance. Indeed, these propellants must be handled with care and with the right percentage in terms of ratio. It is useful therefore to consider the chemistry in order to understand the phenomena and predict the products of the reaction.

9.2.1 Combustion Theory

The combustion is the result of a continuous chemical reaction between a fuel and an oxidant. Under the effect of an energy source, the propellants will react together, and the aim is therefore to maintain this reaction in order to obtain the products of combustion, which are burnt gases (other chemical products) and heat. Due to this physical phenomenon, a flame is produced, and one achieves the necessary thrust for a rocket. Since the atmospheric pressure varies greatly during flight, the engines of the first stages are only pressure-adapted for a medium altitude [19]. At take-off, the engines are not pressure-adapted, which is why the engine outlet pressure is lower than the ambient pressure. At high altitudes, however, their pressure is much higher than the air pressure, so the flame overflows a lot to the sides and is pushed downwards as well as to the sides.

9.2.1.1 Mass Fraction

In combustion species transfer, the key characteristic is *mass fraction*. This is defined as:

$$Y_k = \frac{m_k}{m} \tag{9.3}$$

Here k represents the species used, m_k is the mass of the species k and m represents the total mass of the mixture. For the total species produced, we have:

$$\sum_{k=1}^{N} Y_k = 1 \tag{9.4}$$

In a chemical reaction, and in our case, the two species (fuel and oxidizer) mix to create a reaction product which then depends on the starting mixture. The goal is to know, the different ratio, in order to know the best compromise for the mixture's composition which will have an important impact of the quality of the combustion, of the emission o pollutant and of the flame dynamics. In our case study, we consider that the propellant is not mixed together before entering the gas generator or the combustion chamber but introduce separately via the injectors. Moreover, we will make the assumption that we have a stoichiometric mixture, which will change primarily the equivalence ratio. A stoichiometric mixture is a very efficient mixture because all the reactants are completely consumed in the final state. In reality, there are often some reactants left in the initial state that are not consumed, which is why we make the assumption. According to Turns [17], stoichiometric takes the definition:

$$S = \left(\frac{Y_O}{Y_F}\right)_{st} = \frac{v'_O w_O}{v'_F w_F} \tag{9.5}$$

Here v'_O and v'_F the coefficient corresponding to the fuel and oxidizer, while w_O and w_F are the atomic weight (in kg/mole) of the fuel and oxidizer. This stoichiometric ratio allows one to calculate how much mass of oxidizer should we have to burn fuel. Let us introduce some reactions to perform calculations in order to obtain the stoichiometric ratio for different case examined in the ANSYS simulations, described later, based on Equation (9.5).

Table 9.1 documents the stoichiometric ratio S for different combustion reactions in air. Different gases have been used, starting first with methane, then propane and also hydrogen, all of which are common rocket fuels. The ratio clearly changes. According to the species used, the mass of the oxidizer will not be the same and is highest when there is a reaction with the hydrogen. For the first combustion, the methane reacts with the oxygen in the air and then produces carbon dioxide.

9.2.1.2 Equivalence Ratio

In this study, the equivalence ratio is equal to 1 due to the stoichiometry:

$$\phi = S\left(\frac{Y_F}{Y_O}\right) = 1 \tag{9.6}$$

Without the hypothesis of the stoichiometric mixture, when $\phi > 1$, the fuel is in excess, and when $\phi < 1$, the oxidizer is in excess.

TABLE 9.1

Stoichiometric Calculations for Different Chemical Element Reactions

Reaction	S
$CH_4 + 2(O_2 + 3.76N_2) \rightarrow CO_2 + 2H_2O + 7.52N_2$	$2 \times \dfrac{32}{16} = 4$
$C_3H_8 + 5(O_2 + 3.76N_2) \rightarrow 3CO_2 + 4H_2O + 18.8N_2$	$\dfrac{5 \times 2 \times 16}{(12 \times 3) + 8} = 3.63$
$2C_8H_{18} + 25(O_2 + 3.76N_2) \rightarrow 16CO_2 + 18H_2O + 94N_2$	$\dfrac{25 \times 2 \times 16}{(2 \times 18 + 12 \times 8 \times 2)} = 3.51$
$2H_2 + (O_2 + 3.76N_2) \rightarrow 2H_2O + 3.76N_2$	$\dfrac{2 \times 16}{4} = 8$

9.2.1.3 Fuel Mass Fraction and Ternary Diagram

It is now possible to calculate the final formula of the *fuel mass fraction*:

$$2H_2 + (O_2 + 3.76N_2) \rightarrow 2H_2O + 3.76N_2 \tag{9.7}$$

$$\frac{Y_{N_2}}{Y_{O_2}} = \frac{aW_{N_2}}{W_{O_2}} \tag{9.8}$$

It follows that:

$$Y_F = 1 - Y_{N_2} - \frac{SY_F}{\phi}$$

$$Y_F\left(1 + \frac{S}{\phi}\right) = 1 - Y_{N_2}$$

$$Y_F\left(1 + \frac{S}{\phi}\right) = 1 - Y_{O_2}\left(\frac{aW_{N_2}}{W_{O_2}}\right) = 1 - S\frac{Y_{H_2}}{\phi}\left(\frac{aW_{N_2}}{W_{O_2}}\right) \tag{9.9}$$

This leads to:

$$Y_F = \frac{1}{1 + \dfrac{S}{\phi}\left(1 + \dfrac{aW_{N_2}}{W_{O_2}}\right)} \tag{9.10}$$

This formulation is implemented in ANSYS FLUENT combustion solver [20] and elaborated later.

A ternary (flammability) diagram is widely used, particularly in rocket thermo-chemistry, to represent the flammability of certain gases in air. It is composed of three axes and therefore we can use it to have precision on the reaction of methane in air. Figure 9.1 Methane ternary flammability diagram. The point at the bottom left is pure oxygen, the point at the top of the triangle is pure methane and the point at the bottom right is pure nitrogen (since there is no oxygen or fuel). The blue line intersects the bottom of the triangle at a point that corresponds to pure air (79% nitrogen and 21% oxygen). The blue line corresponds to the mixtures that can be made between pure air and pure methane. The zone of flammable mixture divides the diagram into region where the limits is important in order to know the points of flammability. We can see that this region is extensive and demonstrates that a large range of mixtures between these three species is possible in order to obtain the in flammability, hence the combustion. Indeed, the UEL and LEL point corresponds to the lower and upper limit of explosion in air. The green arrows show also the limits of methane quantity mixed with oxygen (without nitrogen) to have the in flammability. Finally, the red arrow depicts the minimal oxygen quantity to have the in flammability; below this value of 12% it is not possible to have the combustion process.

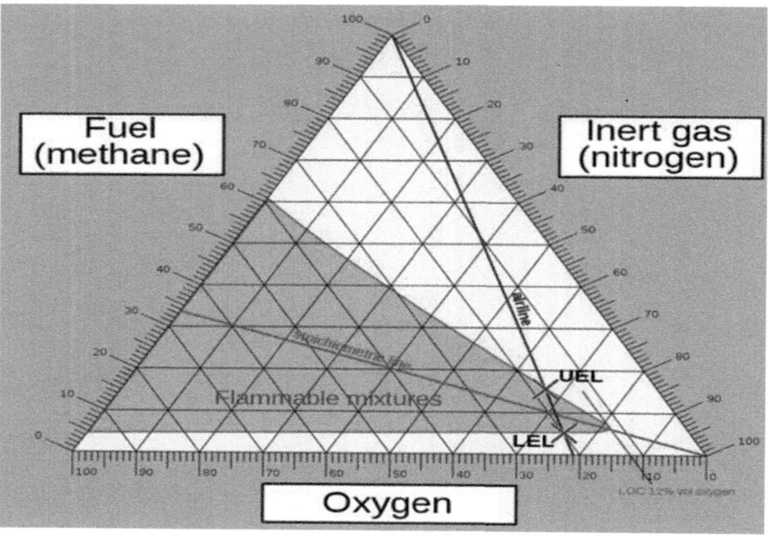

FIGURE 9.1   Methane ternary flammability diagram.

9.2.1.4 Chemical Components in Propellant

There are different criteria in order to choose the propellants. Sometimes the choice is very limited due to a specific mission requires some really precise characteristic, such as high thrust for example which mean that just a few propellants can overcome these constraints. Oxygen, hydrogen, and methane characteristics are given below as these are considered in the present ANSYS FLUENT simulations (Table 9.2).

Important criteria are (i) *density of the propellant*, which allows a scaling down of the size of the tank (for example high density for liquid state), (ii) *energy value* associated with a low molar mass value of the products, (iii) *high thermal capacity* C_p (J/K) in order to allow a good cooling of the combustion system (iv) appropriate thermal conductivity λ (W/mK) (v) sufficient critical temperature and (vi) optimized critical pressure. Other considerations are low toxicity, good thermal stability especially for the cooling system, and also low supply cost.

TABLE 9.2

Critical Conditions of Fuel and Oxidizer

	$T_{critical}$ (K)	$P_{critical}$ (bar)
O_2	154.5	50.43
H_2	33.2	12.98
CH_4	190.5	45.96

9.3 HEAT TRANSFER

The pressurization and the combustion of the propellants produces heat as well as energy to produce the necessary thrust. Depending on the type of propulsion system used, this phenomenon is always present since without energy and heat the thrust cannot be produced. Heat transfer must be simulated therefore in both rocket propulsion and jet engine gas turbine propulsion systems. Heat transfer is a thermodynamic process between two bodies with different temperatures. It is an exchange of energy between two systems that depends on the nature of the transformation, the state law and the principles of thermodynamics. When two bodies have the same temperature, they are in equilibrium. *Thermal conduction* is based on the classical *Fourier law of heat conduction*. In a 3-D (x,y,z) coordinate system, which neglects thermal relaxation and takes the parabolic form:

$$\rho c_p \frac{\partial T}{\partial t} = \frac{\partial}{\partial x}\left(K_x \frac{\partial T}{\partial x}\right) + \frac{\partial}{\partial y}\left(K_y \frac{\partial T}{\partial y}\right) + \frac{\partial}{\partial z}\left(K_z \frac{\partial T}{\partial z}\right) + q_v \qquad (9.11)$$

Here T is temperature (K) '$K_{x,y,z}$' is the thermal conductivity of body material in the x, y and z directions respectively (W·(m K)) (when $K_x = K_y = K_z$ the material is assumed isotropic), q_v is the rate at which energy is generated per unit volume (W/m^3), ρ is density (kg/m^3), c_p is specific heat capacity (J/kg K), '$\partial T/\partial x$' is the x-direction temperature gradient (K/m), 'T/y' is the y-direction temperature gradient (K/m), and '$\partial T/\partial z$' is the x-direction temperature gradient (K/m). Thermal conductivity 'k' provides an indication of the rate at which heat energy is transferred through a medium by the thermal conduction process. Inherent to Fourier's equation is the assumption that steady state heat conduction takes place, bounding surfaces are *isothermal* in character, i.e., constant and uniform temperatures are maintained at the rocket chamber (channel) interior surface and there is no internal heat generation. Since heat transfer in turbulent combusting gas flow in a channel is considered, the energy conservation includes also convective heat transfer. The appropriate formulations are discussed next, as implemented in ANSYS FLUENT [20].

9.4 ANSYS SIMULATION- GOVERNING EQUATIONS AND GEOMETRY

9.4.1 GOVERNING EQUATIONS

CFD (computational fluid dynamic) modelling of species transport, combustion, and NOx simulations of 2-D steady, incompressible, turbulent heat and species transfer in a channel are conducted. The fundamental governing general conservations equations [20] in ANSYS FLUENT in vectorial notation are the mass, momentum, energy (heat), and species (NOX) for 3D unsteady, viscous, incompressible and laminar flow which are as follows. The turbulence aspects are introduced later, and the equations are considered only for 2-D steady combustion.

Mass conservation

$$\frac{\partial \rho}{\partial t} + \nabla \cdot \left(\rho \vec{v} \right) = S_m \tag{9.12}$$

Momentum conservation

$$\frac{\partial}{\partial t} \left(\rho \vec{v} \right) + \nabla \cdot \left(\rho \vec{v} \vec{v} \right) = -\nabla p + \nabla \cdot \left(\overline{\tau} \right) + \rho \vec{g} + \vec{F} \tag{9.13}$$

Energy conservation

$$\frac{\partial}{\partial t} \left(\rho E \right) + \nabla \cdot \left(\vec{v} \left(\rho E + p \right) \right) = \nabla \cdot \left(k_{\text{eff}} \nabla T - \sum_{j} h_j \vec{J}_j + \left(\overline{\tau}_{\text{eff}} \cdot \vec{v} \right) \right) + S_h \tag{9.14}$$

Species conservation

$$\frac{\partial}{\partial t} \left(\rho Y_i \right) + \nabla \cdot \left(\rho \vec{v} Y_i \right) = -\nabla \cdot \vec{J}_i + R_i + S_i \tag{9.15}$$

Here the usual notation is deployed for velocity vector, pressure, body force, etc. [20]. Also, k_{eff} is the effective thermal conductivity and $\overline{\tau}_{\text{eff}}$ the effective stress tensor. The turbulence is accounted for with the use of these to parameters. When solving the conservation equation(s) for chemical species, **ANSYS FLUENT** predicts the *local mass fraction* of each species, Y_i, through the solution of a convection-diffusion equation for the *i*th species. In Equation (9.15), R_i is the net rate of production of species *i* by chemical reaction and S_i is the rate of creation by addition from the dispersed phase plus any user-defined sources. An equation of this form will be solved for $N-1$ species where N is the total number of fluid phase chemical species present in the system. Since the mass fraction of the species must sum to unity, the Nth mass fraction is determined as one minus the sum of the $N-1$ solved mass fractions. To minimize numerical error, the Nth species should be selected as that species with the overall largest mass fraction when the oxidizer is air.

9.4.2 Turbulence Model

The pressure-based SIMPLE algorithm is deployed in ANSYS which computes the velocity field via the momentum equation. It has good convergence properties. Since there is no time dependence in the simulation, a "steady" simulation is selected in the ANSYS workbench modelling. The velocity formulation is absolute since the flow in most of the domain is not rotating. The *k-epsilon model* is one of the most common turbulence models. It is a two equation model which means that it includes two transport equations to represent the turbulence of the flow, as it is important to perform a study that is as close to reality as possible. The first transported variable is turbulent kinetic energy, k. The second transported variable in this case is the turbulent dissipation ε. It is the second variable that determines the scale of the turbulence, whereas the first variable, k, determines the energy in the turbulence. The appropriate equations are:

Turbulence energy (k):

$$\frac{\partial}{\partial t}(\rho k)+\frac{\partial}{\partial x_i}(\rho k u_i)=\frac{\partial}{\partial x_j}\left[\left(\mu+\frac{\mu_t}{\sigma_k}\right)\frac{\partial k}{\partial x_j}\right]+G_k+G_b-\rho\varepsilon-Y_M+S_k \quad (9.16)$$

Energy dissipation (epsilon):

$$\frac{\partial}{\partial t}(\rho\varepsilon)+\frac{\partial}{\partial x_i}(\rho\varepsilon u_i)=\frac{\partial}{\partial x_j}\left[\left(\mu+\frac{\mu_t}{\sigma_f}\right)\frac{\partial\varepsilon}{\partial x_j}\right]+C_{1\varepsilon}\frac{\varepsilon}{k}(G_k+C_{3\varepsilon}G_b)-C_{2\varepsilon}\rho\frac{\varepsilon^2}{k}+S_\varepsilon$$

$$(9.17)$$

In these equations, G_k represents the generation of turbulence kinetic energy due to the mean velocity gradients. G_b is the generation of turbulence kinetic energy due to buoyancy, Y_M represents the contribution of the fluctuating dilatation in compressible turbulence to the overall dissipation rate. C_1, $C_{2\varepsilon}$, $C_{3\varepsilon}$ are constants. σ_k, and σ_ε are the turbulent Prandtl numbers for the kinetic energy and for the dissipation and finally, S_k, S_ε are user-defined source terms (standard k-epsilon model, Ansys fluent theory guide [21]). The turbulent viscosity or eddy viscosity, μ_t, is computed by combining k and ε as follows. Here we have:

$$\mu_t=\rho C_\mu\frac{k^2}{\varepsilon} \text{ (Turbulent viscosity)} \quad (9.18)$$

$$P_k=-\overline{\rho u_i' u_j'}\frac{\partial u_j}{\partial x_i} \text{ (k production)} \quad (9.19)$$

$$S\equiv\sqrt{2S_{ij}S_{ij}} \text{ (modulus of the mean rate of strain tensor)} \quad (9.20)$$

The k-epsilon model is used in ANSYS FLUENT by selecting: *k-epsilon under model* → *then Standard* → *with a Standard Wall Functions* → *OK*. The standard $k-\varepsilon$ model [22] is a semi-empirical model based on model transport equations for the turbulence kinetic energy (k) and its dissipation rate (ε). It is assumed that the flow is fully turbulent, and the effects of molecular viscosity are negligible. The standard $k-\varepsilon$ model is therefore valid only for fully turbulent flows. Here C_μ is a constant. The model constants $C_{1\varepsilon}$, $C_{2\varepsilon}$, C_μ, σ_k, and σ_ε have the following default values: $C_{1\varepsilon}=1.44, C_{2\varepsilon}=1.92, C_\mu=0.09, \sigma_k=1.0, \sigma_\varepsilon=1.3$. These default values have been determined from experiments with air and water for fundamental turbulent shear flows including homogeneous shear flows and decaying isotropic grid turbulence. They are appropriate for gas turbine simulations. The time-dependent terms (in grey) are ignored in Equations (9.16) and (9.17) for the simulations since only steady state turbulent combustion is examined.

9.4.3 ANSYS Geometric Channel Model

Figure 9.2 shows the dimension of the combustion channel with the air inlet and Methane inlet dimensions highlighted (not to scale). The inlet dimension for the air and methane will stay the same all along the different analysis but it is the

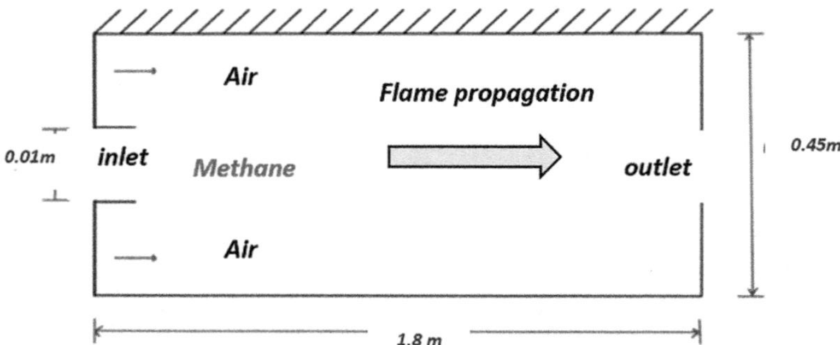

FIGURE 9.2 Aspect ratio 4 (length/depth) with combustion channel dimensions and inlets (not to scale).

length which will be changed when duct *aspect ratio* is modified in the simulations described later.

9.4.4 BOUNDARY CONDITIONS AND ANSYS SPECIFICATIONS

As it is a combustion analysis, the species model must be defined by doing the following: below viscous model, choose Species → the model selected as "species transport" → with volumetric reactions → under options, *Inlet diffusion and Diffusion Energy Source* has been defined, the mixture material is composed by methane-air with Eddy-Dissipation for the Turbulence-Chemistry Interaction. A final model was set in order to be able to obtain and plot the results of NOx. Indeed, the species model was set previously, so here, the NOx needs to be turned on. The *thermal NOx and the Prompt NOx* was selected. The rocket fuel is of course the methane, and the model was set as *partial equilibrium.* For the turbulence interaction mode, temperature was selected, and the type was set to *beta.* Temperature variance and T_{max} option was selected, to be transported as a global-Tmax. The fuel carbon number was set to 1 with an equivalence ratio of 0.76.

The *inlet boundary conditions* were set following this step:

Setup → Boundary Conditions, double clicking on the air-inlet previously define on the mesh section. After that a window of velocity inlet appears, magnitude normal to boundary was selected in the velocity specification method, and the velocity of the air is set to 0.5 m/s. In this same window, other parameters need to be taken into account such as the turbulence. Indeed, under the specification method, Intensity and Hydraulic Diameter was selected, the Turbulent Intensity is at 10% and the Hydraulic diameter was selected to be 0.44 m. 300 K was entered in the thermal window for the external air temperature. Another Inlet Boundary condition was set for the methane. It uses the same velocity specification method as air but instead of 0.5 m/s the methane will enter into the channel with a high injection velocity of 80 m/s. The turbulence specification method is the same, the turbulent intensity is still 10% but the hydraulic diameter is 0.01 m. As in this inlet, there is only methane as species, the species mass fraction was set to 1 for CH_4.

The *outlet boundary conditions* were set by clicking on the outlet appearing on the list, the gauge pressure has been let to 0 Pa, the backflow reference frame is absolute. For the turbulence, the intensity is still at 10% for a backflow hydraulic diameter of 0.45 m (the sum of the inlet Hydraulic diameter). After that, there are no other conditions to set for the output because it is the fluid that will give the values after the iterations and once the calculations have started.

The *last boundary conditions* have been set for the *top and base horizontal duct walls*. The wall motion was set to stationary, with no slip shear condition, and the wall roughness parameters, i.e., velocity inlet parameters, methane: pressure outlet parameters, are standard. Furthermore, the temperature on the wall will be constant at 300 K. The combustion temperatures are much higher and exceed 2,000 K, as shown in the simulations.

After that, the solution was initialized with a hybrid initialization. The simulation was run for 500 iterations in order to obtain the convergence. With all the previous parameters set on fluent, a range of results will be obtained from the temperature, the pressure to the pollutant. Fluent allows visualization of the combustion flow characteristics in terms of contour plots for *temperature, pressure, vorticity, mass fraction* etc.

9.4.5 ANSYS FINITE VOLUME MESH DESIGN FOR CHANNEL (DUCT)

The mesh design (grid) has been produced using 2-D quadrilateral elements, with an element size of 0.0075 m. The following step (described earlier) is the specification of the boundary condition, air inlet, methane inlet, pressure outlet and wall constant temperature. Afterwards, the appropriate Ansys solver has been – selected, i.e., *FLUENT pressure-based, k-epsilon model, steady state analysis*. Once all these steps have been completed, the calculation can be run, initialization of the setup is required and then the number of iterations must be chosen. After that all the results can be obtain (static temperature, static pressure, vorticity magnitude, species mass fraction, mass fraction of pollutant) thanks to the contours and graphs in order to compare them and understand the behaviour to draw conclusions (Figure 9.3).

In an iterative solution, the residual are the solution imbalances. It represents the error between each iteration of the continuity equations. The residual curves are drawn and move forward during the calculation. As it is represented in the following figure, there are thirteen curves from the *continuity curve to the species curves* and also the curves which represent the pollutant (species). At the beginning of the iterations, we can see that there are some oscillations, which is gradually decreasing. We had started the calculation with 500 iterations but after the 450th iteration the solution stopped. At this level, we can say that convergence has been reached because the residuals are of the order of 1e-10 (Figure 9.4).

FIGURE 9.3 Final mesh domain.

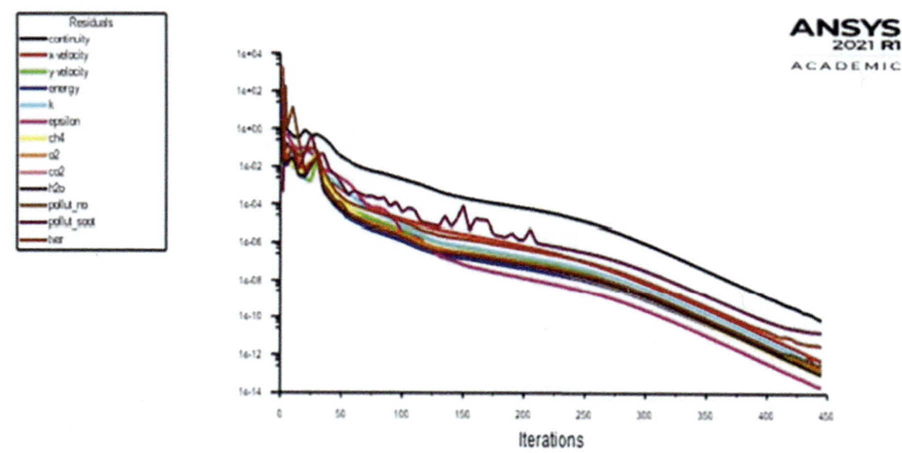

FIGURE 9.4   Residual iteration count for convergence.

9.5 GRID CONVERGENCE STUDY

The objective of a grid convergence study is to use different mesh refinements to demonstrate that the solution is grid independent. Indeed, this allows a compromise between the accuracy of the results and the cost of calculations which can be really high and resource intensive. It is also to show that even with a more refined mesh, the results eventually will become the same. The analysis has been run for different mesh designs and the contour plots focus on the same variable in order to compare the results for the different mesh. The static temperature was selected. In order to do this, three different meshes were adopted- a coarse mesh with an element size of 0.01 m, a medium mesh with a smaller size of element which is 0.0075 m and a high-density mesh with 0.005-m element size. With the coarse mesh, the structure is composed by 4,344 nodes for a total number of elements of 4,140. The geometry with the medium mesh is composed of 7,347 nodes for 7,080 elements. To finish, the maximum density mesh has 16,514 nodes and 16,110 elements. Figure 9.5 shows the computed temperature distribution in the channel the mesh independence studies in Figure 9.6. To prove the grid independence, we will focus on two regions in the domain. The first one represented by the line 1 on the middle of the channel and the second region represented by the line 2 on the upper part. These two lines represent an important region because the first one depicts the middle of the channel, where the species are mixing together really quickly, and the second line is also interesting as it situated in the region where the flame does not form immediately. Hence there is a sudden change when the two species mix and create this flame. This is around 0.35/0.4 m along the channel from the inlet. The test case for mesh design is for a unity channel aspect ratio.

The high mesh density is adopted in all subsequent simulations described in the next section.

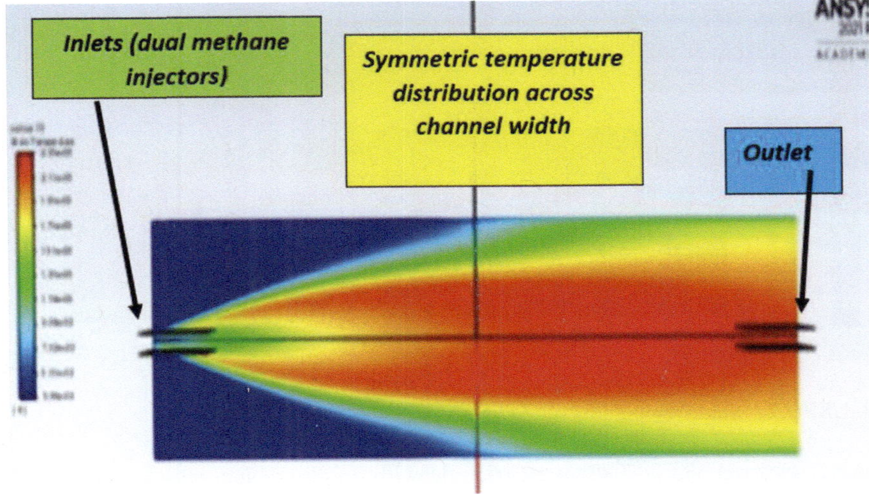

FIGURE 9.5 Static temperature in region 2 with highest mesh density (16,110 elements).

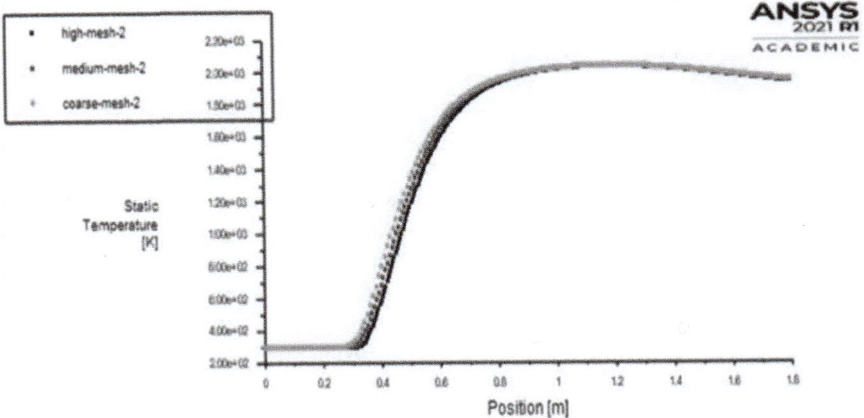

FIGURE 9.6 Comparison of static temperature in region 2 for different refinements in the mesh independence study.

9.6 SIMULATIONS FOR ASPECT RATIO AND METHANE INLET VELOCITY EFFECTS

9.6.1 Simulation I- Aspect Ratio 4 with Inlet Methane Velocity of 80 m/s

9.6.1.1 Static Temperature

For the *aspect ratio = 4 case*, plots are first described. The static temperature contour plot (Figure 9.7) allows us to represent the temperature of the flame inside the channel. As the following figure depict, we clearly see the shape of the flame. At the,

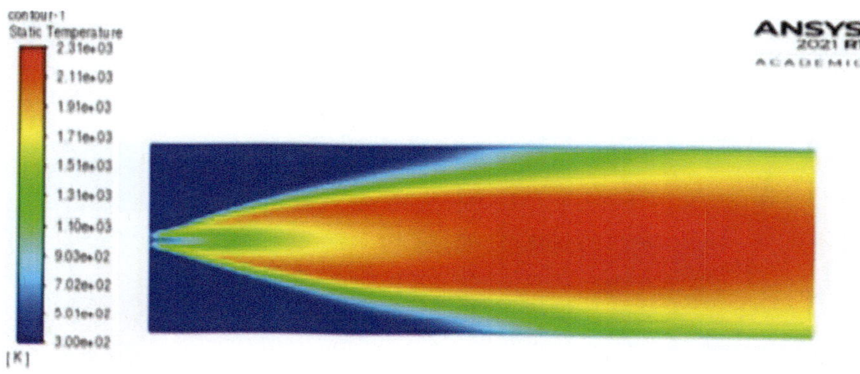

FIGURE 9.7   Static temperature.

there is two regions in a shape of triangle with a colour blue, which represent the initial temperature put on the ANSYS FLUENT parameters (300 K). The static temperature profile forms a cone. This phenomenon can be explained by the fact that the temperature begins to increase when the two specie (CH_4 and air) meet and mingle together. From the inlet, we can see that the temperature increases strongly in the centre to reach the maximum value around the three quarters of the channel with a value of 2,310 K which is consistent with the massive temperatures encountered in rocket combustion chambers [18,19]. Due to the boundary condition, there is no change in temperature on the top and bottom wall which stay at constant 300 K, represented by a blue line on the previous figure (the low temperature is enforced to sustain cooling and avoid material degradation of the duct boundaries). We can see that the shape of the flame fills the entire space of the channel. At the outlet, the temperature is almost maximal on the entire wall and there is just a lower temperature on the side within the range of 1,510 and 1,710 K. Temperature is generally computed smoothly along the length of the channel and also transversely, as witnessed by the symmetry. There is a significant modification in the entry section of the channel and then the temperature stabilises and gradually becomes almost identical over the whole area of the channel.

9.6.1.2   Static Pressure

Figure 9.8 shows the static pressure evolution through the rocket duct (channel). We can see that there is very low static pressure away from the inlet. Static pressure is maximal at the inlet in the middle core zone, around the methane injector. The value is 21.3 Pa and, otherwise, the contour plot is blue, which mean a very low static pressure and at some particular location, a negative static pressure (−1.68 Pa) due to the suction in the fluid field. It can be noted that the location of the high static pressure in red corresponds to the region on the static temperature plot where the temperature is invariant in the middle at the inlet (between the two methane injectors).

9.6.1.3   Vorticity Magnitude

Figure 9.9 depicts the vorticity magnitude along the channel. The vorticity corresponds to the rotational dynamics in the turbulent motion of the fluid (it is related to the difference in velocity gradients). In the case study, the highest vorticity magnitude

FIGURE 9.8   Static pressure.

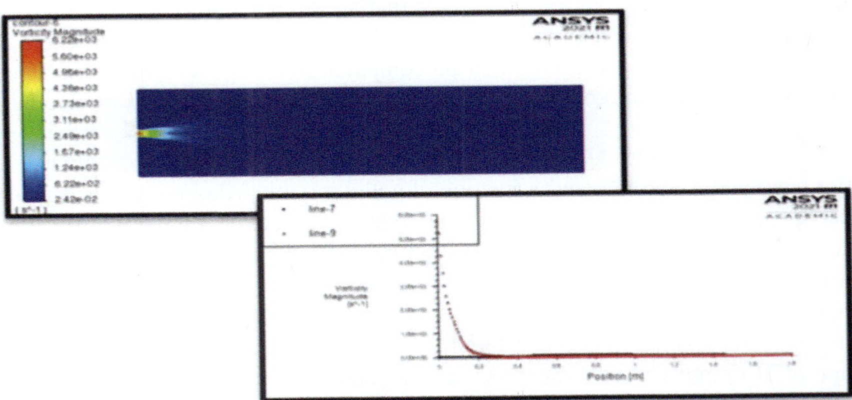

FIGURE 9.9   Vorticity magnitude contour plot with inset of vorticity distribution versus location along channel.

obtained as the value of 6.22×10^3 /s and the lowest magnitude of 2.42×10^{-2} /s. The highest value for the vorticity is shown at the methane inlet injectors. At the middle of the channel (core zone), the value decreases from the inlet, and at the upper and lower regions, the vorticity sustains the lowest value, since it is furthest from the disturbance at the inlet locations. It is on the centre of the channel where the evolution of the vorticity magnitude is the most important. Dually symmetric circulation zones are clearly computed about the centre line which are characteristic of rocket combustion flame dynamics [18].

9.6.1.4   Mass Fraction of Pollutant

The production of the NOx (the pollutant) originates from the chemical reaction of the species (methane and air) in order to create the ignition. It is important to take this phenomenon into account due to the undesirable effect on the environment and the pollution it can create depending on the quantity of the NOx produced. The reduction of the pollutant emission is one of the main challenges in the field of aerospace propulsion. Indeed, the new developments of rocket motor technology chiefly focus on this idea and to optimize thrust with simultaneously a very small environmental footprint. As depicted in Figure 9.10, the first part of the channel is not subject to pollutant production. It is the blue part on the figure with zero emission mass fraction. Gradually, the mass fraction of NOx begins to increase around the middle of the channel with a first value of around 0.0015. The pollutant mass fraction is constantly increasing to reach the maximum value of 0.00507 at the end of the channel. Inset with the contour plot is the evolution of mass fraction with channel length.

Inspection of Figure 9.10 reveals that mass fraction of pollutant also exhibits a symmetric behaviour about the middle axis of the channel. It is important to look at the concentration of pollutant on the surface of the channel (walls). That is, one needs to locate the highest concentration in red on the previous figure, look at which stage it commences and make a ratio in order to calculate the percentage of NOx in high concentration according to the total surface. In this first case, referring to the vertical line draw on the previous figure, we can see that the red flame begins at 1.13 m, hence, it corresponds to 37% of the wall surface (slightly less in the middle part where less NOx is present). This percentage will be compared thereafter with the other aspect ratio cases simulated in order to study the impact of the length on NOx production.

9.6.1.5 Species Mass Fraction

After having presented the mass fraction of pollutant, it is judicious also to visualize other species formed due to the turbulent combustion reaction. Figure 9.11 illustrates that inside the channel different species are present such as H_2O, CO_2, O_2 and CH_4. They arise in different quantities but with the same order of magnitude, which allows a comparison of their evolution on the following graph as a function of the length of the channel. Concerning the shape, it is relatively straightforward to see that the shape of H_2O and CO_2 are the same, and resemble the topology computed in the

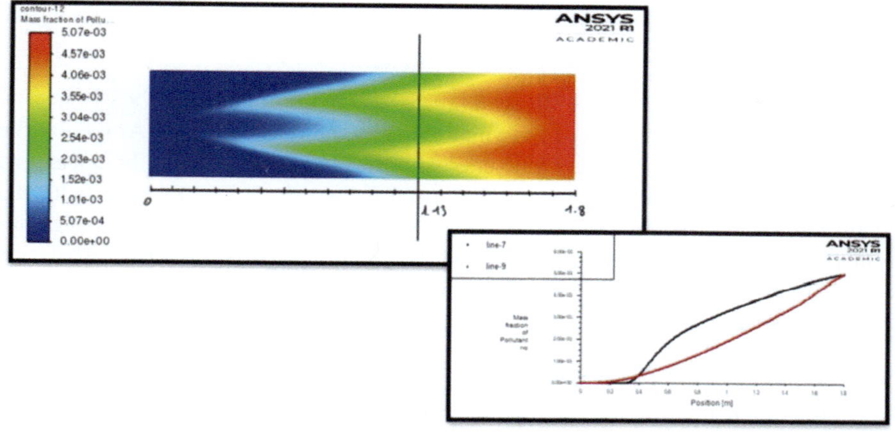

FIGURE 9.10 Mass fraction of pollutant contour plot and distribution along channel (insert).

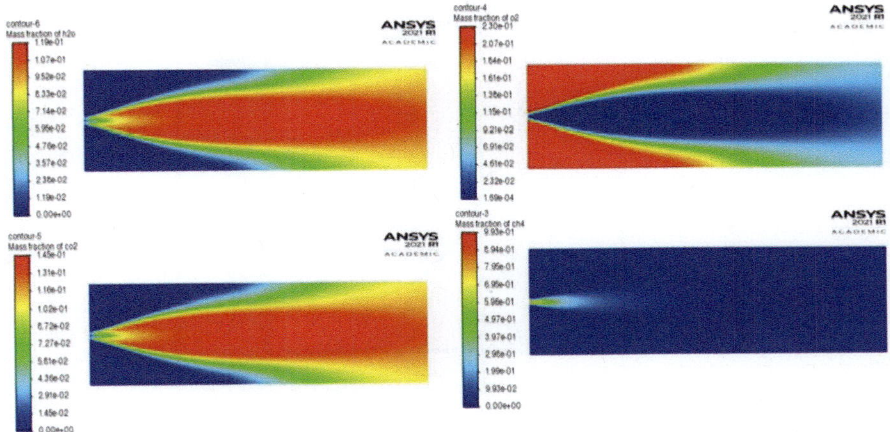

FIGURE 9.11   Species mass fraction of H_2O, CO_2, O_2 and CH_4.

static temperature plot, with the singular shape of a cone. On the entire surface, the highest mass fraction of H_2O obtained has the value of 0.119 while the highest value for the CO_2 mass fraction is 0.145. For these two species, the lowest value is 0 at the level of the air inlet and before the air mixes with the methane.

For the contour of the O_2, the behaviour deviates from H_2O and CO_2 cases and the maximum species mass fraction computed for O_2 is 0.23. At the middle, the cone has the lowest value of O_2 mass fraction, it could be considered as 0 (0.000169). The largest quantity of O_2 arises in the first half entry section of the channel. Finally, we find the CH_4 just after the injectors only exists in a very small portion owing to the methane being quickly transformed during the combustion reaction. The lowest value for the mass fraction of CH_4 is of course 0 when there is no methane in the channel following the completion of the burning process, and the highest CH_4 mass fraction has the value of 0.993 (Figure 9.12).

A specific region has been studied in order to show the evolution of the different combustion species. The previous graph gathers all the four curves which represent the evolution of the different species. This region has been defined previously by the line already created. The region has the following coordinates ($x=0$, $x_1=1.8$, $y=0.1125$ and $y_1=0.1125$), represented by the "line 7". We can see that on this upper location of the channel, the evolution is quite different depending on the species. It can be seen that there is no methane except in the middle zone of the duct between 0.6 and 1 m with a very low value. Then we can notice that O_2 is the most abundant species between 0 and 0.4 m into the channel with a stagnant level at the value of around 0.25 and thereafter from 0.4 m onwards, O_2 starts to decrease until 1.3 m and then finally slightly upsurges at the termination of the duct (outlet). CO_2 and H_2O behave in a similar fashion in this region. Indeed, they are not present until 0.375 m and then they increase significantly. From this growth, a small gap is created between the two species as can be seen with the red and green curves. Finally, at the height of 1 m in the channel, both species stabilize. The most represented species from 0.6 m in the channel is therefore CO_2. It is also noteworthy that for the pollutant

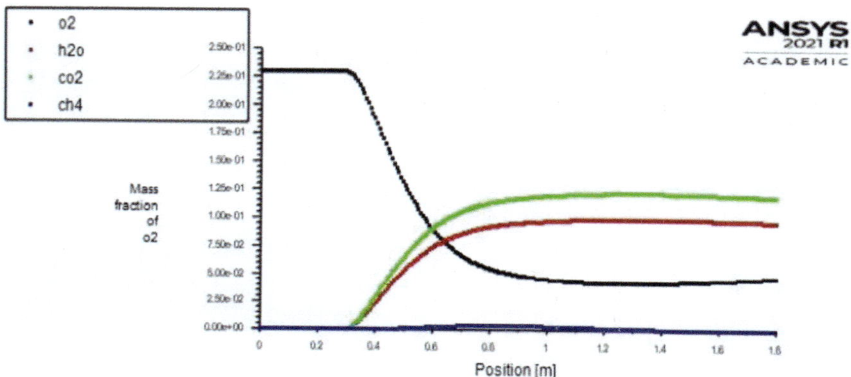

FIGURE 9.12   Comparison of the species evolution.

mass fraction the order of magnitude is 10^{-3} while for the CH_4, O_2, H_2O and CO_2, the order of magnitude is 10^{-1} so we can conclude that there much more species (CH_4, O_2, H_2O and CO_2) than NOx on the entire channel with the analysis performed.

9.6.2   SIMULATION II- ASPECT RATIO 4 WITH INLET METHANE VELOCITY OF $CH_4 = 20$ M/S

Next we consider again an aspect ratio of 4 with length again at 1.8 m and the height as 0.45 m (aspect ratio = 1.8/0.45 = 4). However, in this second set of computations, we alter the velocity of the methane inlet. It was set for the main analysis a methane velocity of 80 m/s (Section 9.6.1) and in this second case, is altered to a much lower value of *20 m/s* in order to be able to assess its impact for smaller rocket thrust designs. All the other parameters have been set to the initial setup. Three set of results will be compared with the previous analysis, the static temperature, the mass fraction of pollutant and the vorticity magnitude.

9.6.2.1   Static Temperature

Figure 9.13 depicts the static temperature contour with the new methane inlet velocity. the extreme value (lowest and highest) of the temperature does not evolve significantly from the previous case with inlet methane velocity of 80 m/s. However, there are some modifications in magnitudes and topology (described later). Indeed, the lowest value in blue is the same with a value of 300 K, while the highest static temperature has the value of 2,300 K, a difference of 10 K with the static temperature for the 80 m/s inlet methane velocity case. This change is not at all significant, and it can be concluded that the methane velocity does not have a huge impact on the static temperature distributions. *Similar high temperatures* are therefore produced even with a *75% reduction* in inlet methane velocity.

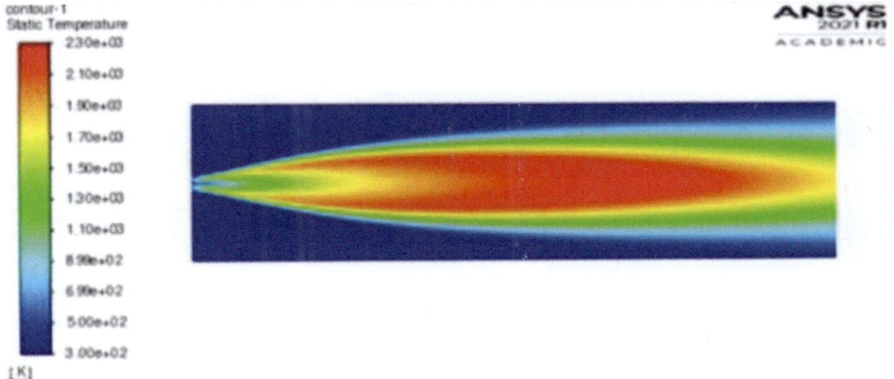

FIGURE 9.13   Static temperature with CH_4 velocity of 20 m/s.

However, the change in the methane velocity at the inlet has a direct impact on *the shape of the flame*. Here in this new analysis, it is much more compressed, thin and stretched. So, there is less surface area with a high static temperature (in red on the contour plot). The contour of the static temperature at the outlet is not high. At the base and upper sections of the duct, the outlet the temperature is still 300 K (a corridor is formed from the inlet to the outlet with the same temperature of 300 K in blue). However, there is a modification in the *growth of temperature*. With a velocity of 80 m/s for the methane, at a certain location in the channel, the temperature was in excess of 1,310 K (far above 300 K). Furthermore, the maximum static temperature at the outlet is reached at one point and not on the entire surface when methane velocity is reduced. For future study, it would appear to be logical to take into account for the fuel inlet velocity which can bring changes for the results.

9.6.2.2   Vorticity Magnitude

For the vorticity magnitude, there are minor changes concerning the behaviour. Again there is the same location for the maximum magnitude (at the methane injectors region), with a rapid decline of the magnitude as show on the graph for the middle line 9 and on the upper part (line 7 of the legend), a vorticity magnitude is computed of almost zero (Figure 9.14).

The biggest change is in the magnitudes, not the locations. Indeed, we go from a maximum value of 6.22×10^3 /s to 1.52×10^3 /s. Similarly, the range for the lowest magnitudes is from 2.42×10^{-2} /s to 1.40×10^{-3} /s. The reduction of the methane inlet velocity has also an impact on the vorticity magnitude, and this is attributable to the reduction in momentum at the channel inlet. This suppresses also turbulence effects and leads to a general deceleration in the combustion flow in the channel. Therefore, while temperatures are weakly affected, there is a much *stronger modification (depletion) in vorticity distribution* with a lower inlet methane velocity.

9.6.2.3   Mass Fraction of Pollutant

Last, we consider the mass fraction of pollutant. As noted earlier, the shape of the mass fraction of pollutant is different, hence the repartition of the NOx differs from the higher methane inlet velocity (80 m/s) case studied earlier. Figure 9.15 shows

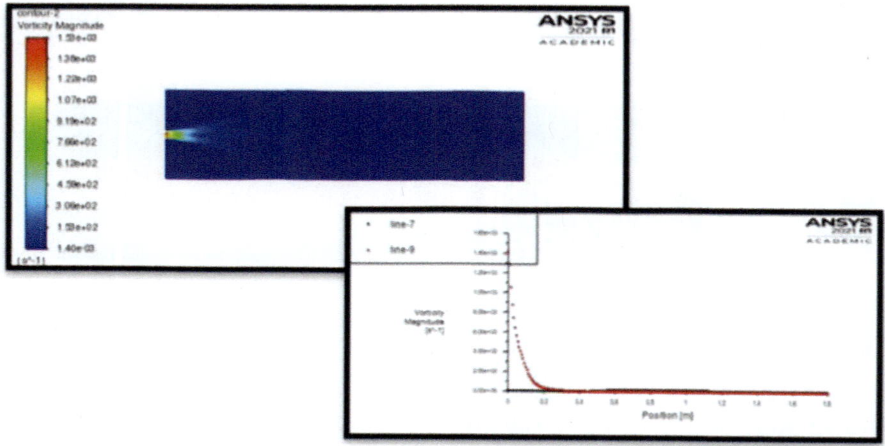

FIGURE 9.14   Vorticity magnitude with new CH$_4$ inlet velocity of 20 m/s.

that the lowest value is still 0 but the highest mass fraction is somewhat higher at 7.74×10^{-3} instead of 5.07×10^{-3}. The lower methane inlet velocity permits a more robust production of mass fraction of the pollutant and leads to elevation in the maximum value. This may be connected to the slower momentum in the flow which permits more intense chemical reactions in the combustion process. Also, the shape changes; the surface in which the NOx has it maximum value is smaller than the contour plot computed for the 80/s methane inlet velocity case. Therefore, a compromise is indicated. Either a greater value is produced but on a smaller surface of the channel or a lower maximal value for the pollutant which extends along a larger surface of the channel (duct) walls.

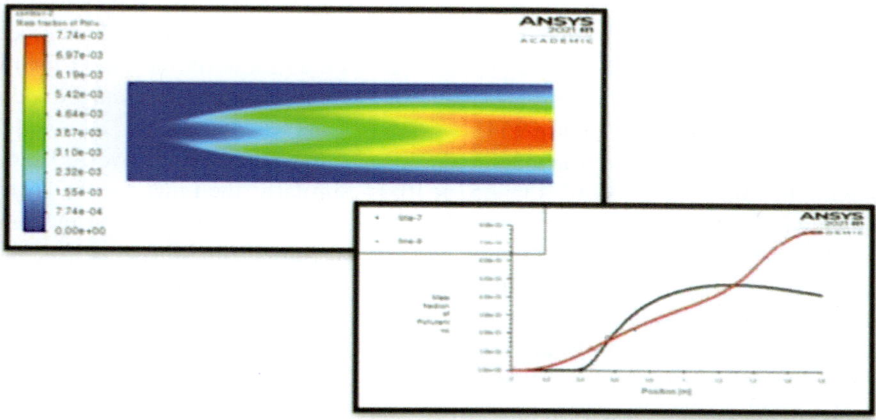

FIGURE 9.15   Mass fraction of pollutant contour plot and distribution for inlet methane velocity of 20 m/s.

9.6.3 Simulation III- Aspect Ratio 8 with Inlet Methane Velocity of 80 m/s

In this third simulation, we double the duct length to 3.6 m but maintain a height of 0.45 m. The aspect ratio is therefore now 3.6/0.45 = 8. The same inlet methane velocity as in *simulation I* is applied i. e. 80 m/s. As the geometry dimension has changed, a new analysis has been run in order to obtain the results corresponding to this new case study. Other residual curves were obtained and despite some oscillations, convergence is comfortably achieved in less iterations i. e. 350 iterations, rather than 450 iterations which were required for simulation I (aspect ratio of 4). The results are now described.

9.6.3.1 Static Temperature

As observed in Figure 9.16, the maximum value and the minimum value for static temperature are the same between the aspect ratio 4 (simulation I) and the new aspect ratio of 8 (300 K for the lowest value and 2,310 K for the highest one). However, as the length of the channel is twice as large, the shape and the behaviour of the flame is different. Indeed, the contour of the temperature is larger – this allows greater details to be captured at the back of the flame which was not possible in the first shorter length duct case (simulation I). We can see the formation of the flame at the beginning, following by a region where the static temperature is at its maximum value, and then, the temperature decreases and becomes more homogeneous around the outlet region (in yellow). When the flame reaches the bottom and the top wall, it fills the entire surface of the channel. A large part of the outlet has a temperature of around 1,710 K, while the outlet of the aspect ratio 4 (simulation I) had almost the maximum value. The flame has the space to spread, i.e., propagate more gradually in both the axial and transverse directions. The red part corresponds to *the laminar regime* from the exit of the nozzle while further on, the rather yellow part with a

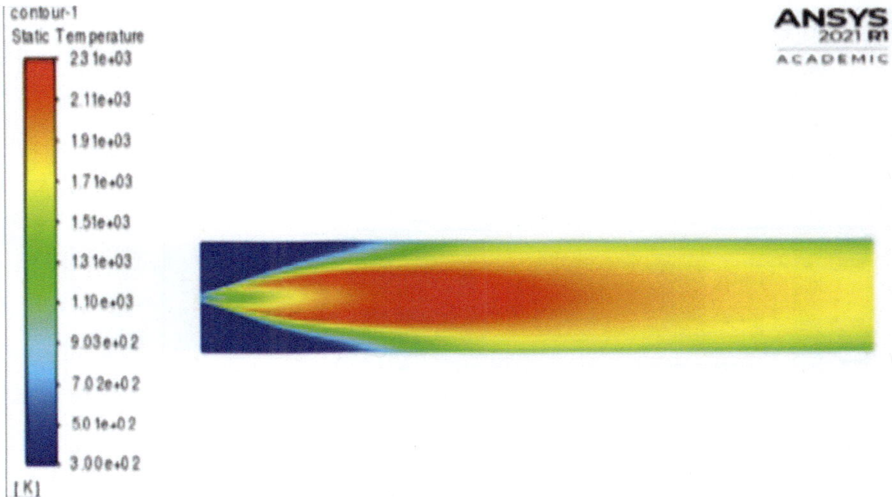

FIGURE 9.16 Contour of the static temperature with aspect ratio of 8.

lower temperature corresponds to the regime where the flow is *turbulent*. Increasing the length of the channel allows the flame to develop to its entirety and therefore allows more precise simulations of the combustion process.

9.6.3.2 Static Pressure

Figure 9.17 depicts the contour of the static pressure. The location of the highest static pressure is the same as in *simulation I*. The values obtained are relatively identical, indeed the highest static pressure has the value of 21.2 Pa instead of 21.3 Pa for the first analysis and we can also note that there is always the suction phenomenon in the fluid field with a minimal pressure of −1.76 Pa compared to the −1.68 Pa of the aspect ratio 4. Despite the increase in length, there is almost no impact on the static pressure. Indeed, as the dominant static pressure is just located at the level of the methane injectors, it does not exhibit any great sensitivity to a change in the length of the combustion duct (channel).

9.6.3.3 Vorticity Magnitude

As with the static pressure, no tangible modifications in vorticity magnitude are computed (Figure 9.18). The duct length, therefore, produces no substantial influence on the vorticity. Very similar behaviour is computed for the aspect ratio 8 case as witnessed earlier for the aspect ratio 4 (simulation I) case. There is an absence of vorticity on the upper half part of the channel and a decline at the middle as soon as the methane is injected. As a summary, the highest value is 6.22×10^3 /s instead of 6.23×10^3 /s and the minimum value is 2.34×10^{-2} /s instead of 2.42×10^{-2} /s.

9.6.3.4 Mass Fraction of Pollutant

It is interesting for this larger channel geometry, to analyse some of the NOx mass fractions. As presented in Figure 9.19, it can clearly see that it is different compared to the contour plot for *simulation I* (aspect ratio 4). As in simulation I, the inlet of the

FIGURE 9.17 Static pressures for aspect ratio 8.

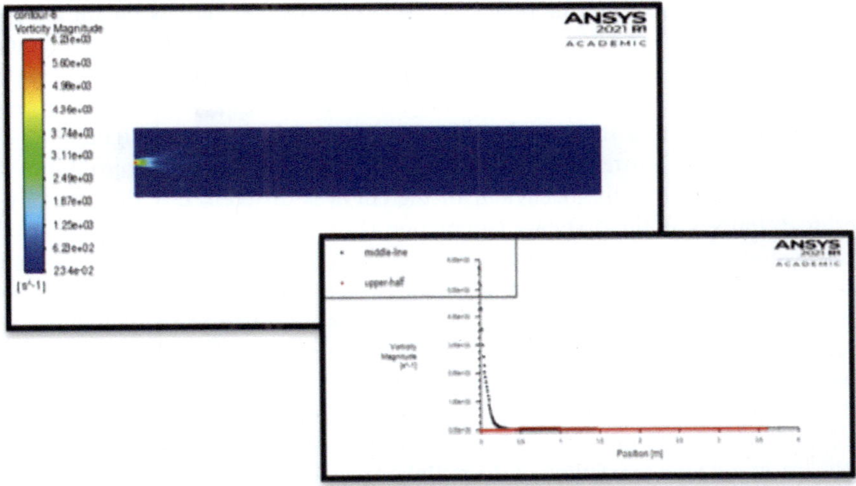

FIGURE 9.18 Vorticity magnitudes for aspect ratio 8.

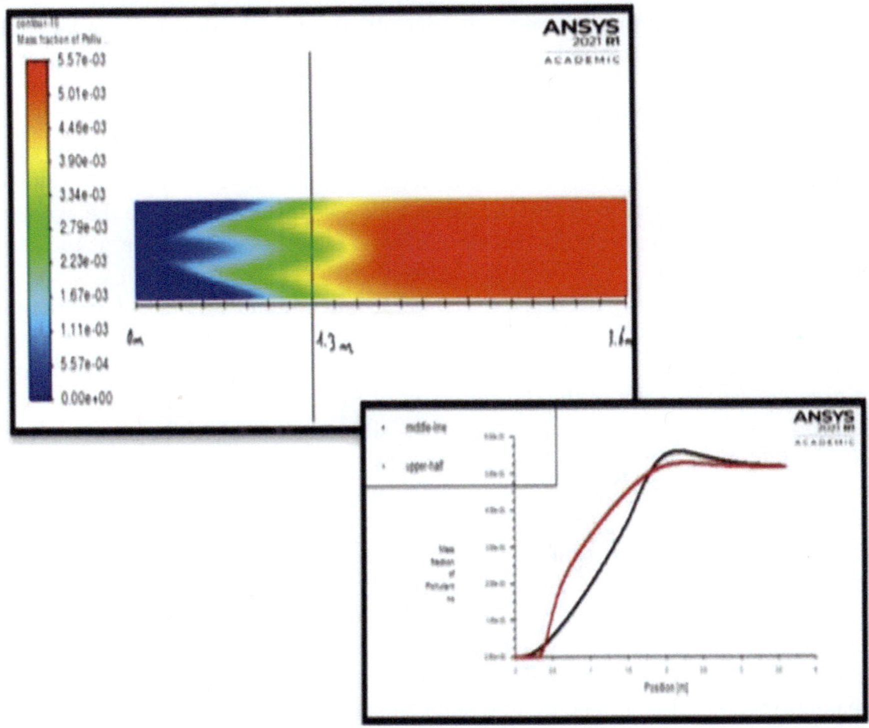

FIGURE 9.19 Mass fraction contours of pollutant.

channel is not subjected to the pollutant (NOX) production. The pollutant starts to form from 0.16 m into the duct with exactly the same shape as the contour computed in simulation I. However, the augmentation of the length of the channel has an impact on the mass fraction of the pollutant. Indeed, it is now possible to see a wider range of magnitudes and thus the formation of NOx along the flame. The highest value is equal to 5.57×10^{-3} *instead of* 5.07×10^{-3}. The lowest value is still zero. From the graph, we can see that the behaviour of the curve is different i. e. the topology is modified. Indeed, in the first case (simulation I) the slope of the curves was always positive since they were constantly growing. In this third simulation case, the higher aspect ratio produces a constant value for the two regions which appears around 3 m. Concerning the central (core) region, there is even a decline of the pollutant mass fraction after a peak at 2 m in the channel. From the contour plot, the red zone which represents the latter half section (downstream), the pollutant mass fraction magnitude is much higher compared to the first case with aspect ratio 4 (it starts to increase at 0.4 m instead of 0.1 m as observed in simulation I). Finally, as it was done previously, we studied the repartition of the pollutant according to the formation of the red flame (which means the highest value). In the inset figure, this phenomenon starts at 1.3 m. Hence, the red zone which covers the channel space represent 63% of the total channel domain. While with a smaller channel (*simulation I, i.e., aspect ratio of 4*), this percentage was 37%. This indicates that the development of NOx is linked to the length of the channel. Indeed, it has a significant impact due to the fact that the pollutant production is strongly increased especially towards the downstream locations and sustained towards the outlet.

9.6.3.5 Species Mass Fraction

For all the 4 species considered, i.e., H_2O, CO_2, O_2 and CH_4 the results are the same between the two aspect ratios of 8 (simulation III) and 4 (simulation I) in terms of magnitudes. It is logical to obtain the same values as these species are the products (results) of the chemical reaction between air and methane which remains the same for all analyses

So, the same quantity of species will be produced, and this is independent of the size of the domain (length of the channel). It is the form of the contours which change (Figure 9.20). The longer channel length and larger domain size allows for greater visualization of the development of these species over the domain and therefore more details of the structure are visible compared with a smaller channel length (simulation I) where they are not as clearly visible. The region studied on the following graph is the same as earlier with the coordinates ($x=0$, $x_1=3.6$, $y=0.1125$ and $y_1=0.1125$).

Figure 9.21 shows that from 0.5 m, the most represented species in term of mass fraction is carbon dioxide following by the water, then the oxygen is present in large numbers up to 0.5 m, and finally, we found the methane almost absent all along the channel (it is only present in very small concentrations between 0.5 and 1.5 m). On the insert graph, we can also see the curve evolution after 1.8 m and note that the behaviour become linear, each curve stabilises at a constant value. Only dramatic changes in species generation arise near the inlet zone and are stabilized quickly with progression along the channel (position).

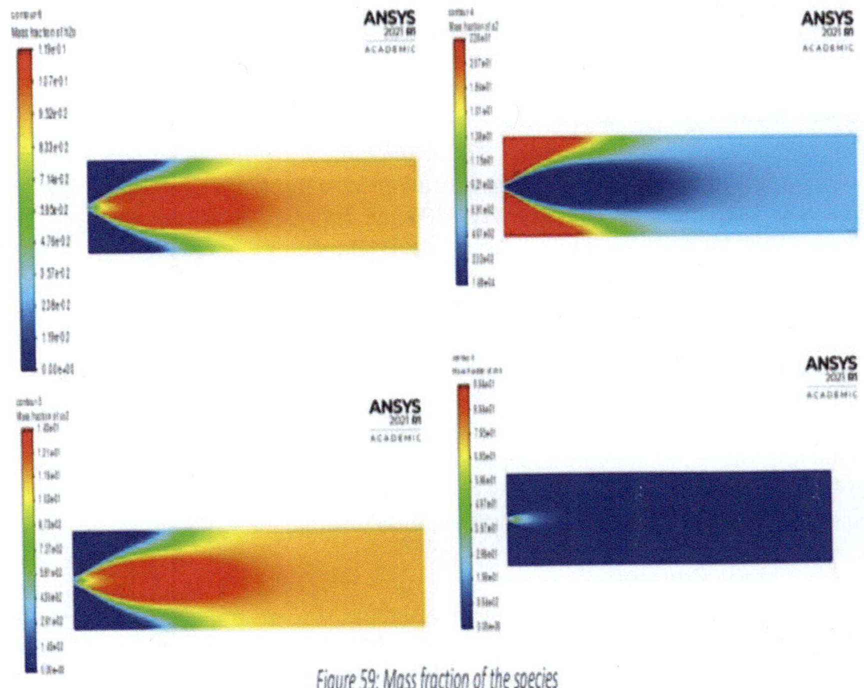

Figure 59: Mass fraction of the species

FIGURE 9.20   Mass fraction of species – aspect ratio 8.

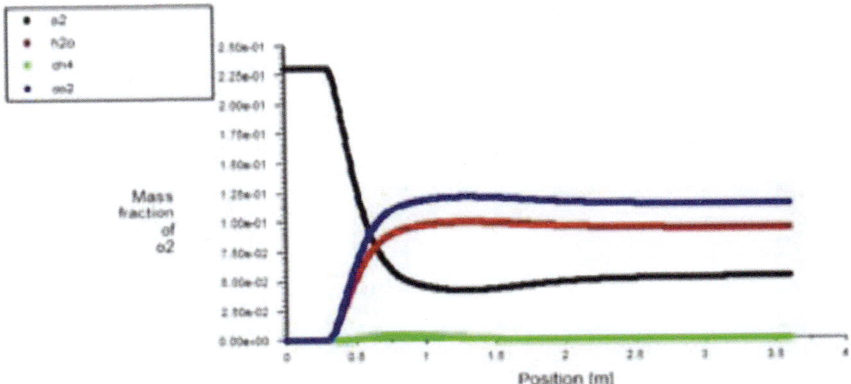

FIGURE 9.21   Species curves comparison.

9.6.4   Simulation IV- Aspect Ratio 12 with Inlet Methane Velocity of 20 m/s

In this last simulation, a third aspect ratio is studied. The length of the channel was increased from 3.6 to 5.4 m. Aspect ratio was therefore 5.4/045 = 12. The same analysis process as the previous part in this chapter will be developed below. Also, the methane inlet velocity is changed to 20 m/s. With this new aspect ratio, the shape of the channel has again changed and is a much slender design, which affects combustion flow characteristics.

9.6.4.1 Static Temperature

Static temperature values in Figure 9.22, are the same with all other previous analysis. It can be concluded that for each aspect ratio, the static temperature does not vary and always keep the same value between 300 and 2,310 K. The different length of the channel does not have any impact on the static temperature parameter. On the other hand, the contour of the static temperature has also the same behaviour for each aspect ratio. However as noted earlier for the aspect ratio of 8 (simulation III) case, the greater length of the duct again allows us to see the flame in its entirety, and even more detail is revealed with the new aspect ratio of 12. The shape of the flame is better captured and in the proximity of the end of the channel, the temperature decreases to around 1,510 K. At this location, the temperature is given by the upstream flame which starts to fade (the shape topology disappears, and a homogeneous temperature is observed across the width with a central yellow core and outer green cooler regions). The location of the maximum temperature is the same as in simulation III with aspect ratio 8.

9.6.4.2 Static Pressure

As with *simulations I–III*, static pressure contours (Figure 9.23) computed for this aspect ratio of 12 are similar. There is a very small difference in the values, but it is not significant (0.01 Pa for the highest value and 0.02 Pa for the minimum value). The hypothesis introduced earlier can be confirmed, which means that the change in the length of the channel has no tangible impact on the static pressure. At a certain point along the channel length, the entire area of the channel, the static pressure does not evolve at all, and this is very close again to the methane inlets.

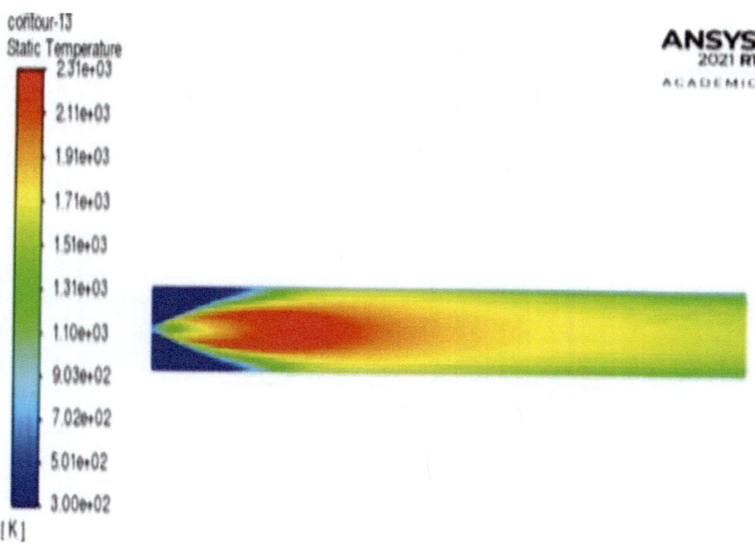

FIGURE 9.22 Static temperature for aspect ratio.

FIGURE 9.23   Static pressure of the aspect ratio 3 of 12 (longest duct length).

9.6.4.3 Vorticity Magnitude

Inspection of Figure 9.24 however shows that a change in the vorticity magnitude is noticeable when compared with simulation III (vorticity magnitude for aspect ratio 8 with higher inlet methane velocity of 80 m/s). Indeed, the maximum value reduces, as anticipated from 6.23×10^3 /s to 6.21×10^3 /s although for the minimum value it actually increases from 2.34×10^{-2} /s to 2.76×10^{-2} /s, and again this is due to momentum alteration with the lower inlet velocity prescribed. This changes the velocity evolution along the duct length. However, the location at where the vorticity magnitude is greatest and lowest is not displaced, demonstrating that a change in length of the channel from 3.6 to 5.4 m only weakly modifies the vorticity magnitude but not the locations of the lowest and greatest values of the vorticity magnitude.

9.6.4.4 Mass Fraction of Pollutant

Figure 9.25 displays the mass fraction of the pollutant inside the channel with aspect ratio 12. When comparing the value of the NOx for the aspect ratio 8, the greatest values has not changed. The minimum value in blue is still zero for all the analyses. The first most noticeable change when comparing it with the mass fraction of pollutant of the aspect ratio 8 (simulation III) and aspect ratio 4 (simulation I) is *the increase of the extent of the channel over which the pollutant is distributed*. The red area corresponding to the maximum mass fraction of the pollutant cover a considerably protracted section of the duct. By doing the calculation, we find that it covers 75% of the total area, which is an important change compared with the 63% for the aspect ratio 8 and the 37% for the aspect ratio of 8. Clearly the spatial distribution of pollutant generated is intimately associated with the length of the channel and the mass fraction of the pollutant is extended further along the duct. According to the length evolution, the pollutant also increases in term of value between the aspect

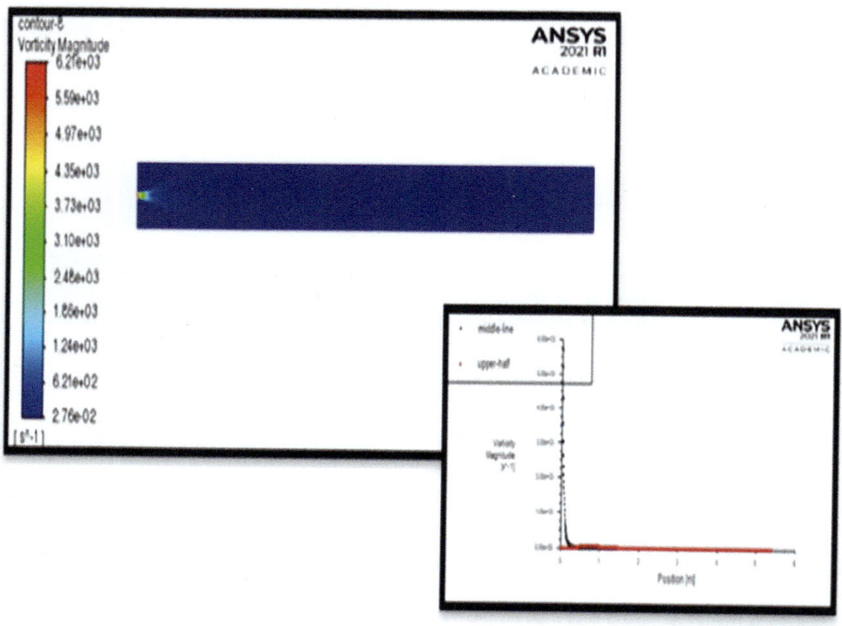

FIGURE 9.24   Vorticity magnitude with aspect ratio 12.

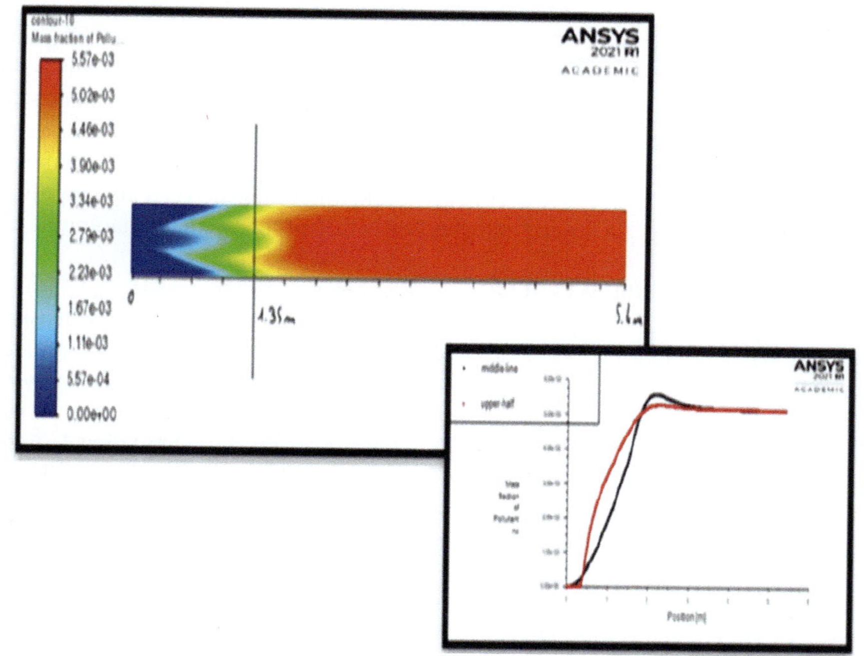

FIGURE 9.25   Mass fraction aspect ratio 12.

ratio 4, is higher for aspect ratio of 8 and achieves a maximum for the aspect ratio of 12 in terms of the duct zone extent over which it is distributed. The greater the length, the higher presence of NOx. The implication is that shorter ducts may be a viable design criterion for lowering NOx production [16] in combination with a lower methane injection velocity at the inlet.

9.6.4.5 Species Mass Fraction

Finally, Figure 9.26 depicts the mass fraction contours of the other species in the flame. The values are the same, no changes with the mass fraction of the species are. This case is an expansion of the aspect ratio of 8 case. Figure 9.27 shows the spatial evolution along the channel length (position) of species mass fractions. After a certain location, the species mass fraction attains a constant value i. e. a steady state behaviour is computed rapidly and no changes appear in the topology after a certain distance into the duct. Hence, it is not necessary to have a greater aspect ratio (i.e. a longer duct length) for the study of the mass fraction of the species. Homogenous distributions are produced as before and the mass fractions of all species rapidly converge to the constant plateau values which are sustained to the channel exit.

9.7 CONCLUSIONS

Detailed ANSYS FLUENT CFD simulations have been presented to examine the influence of *aspect ratio and inlet methane velocity* on static temperature, pressure, vorticity and species mass fraction for turbulent combustion in a 2-D channel

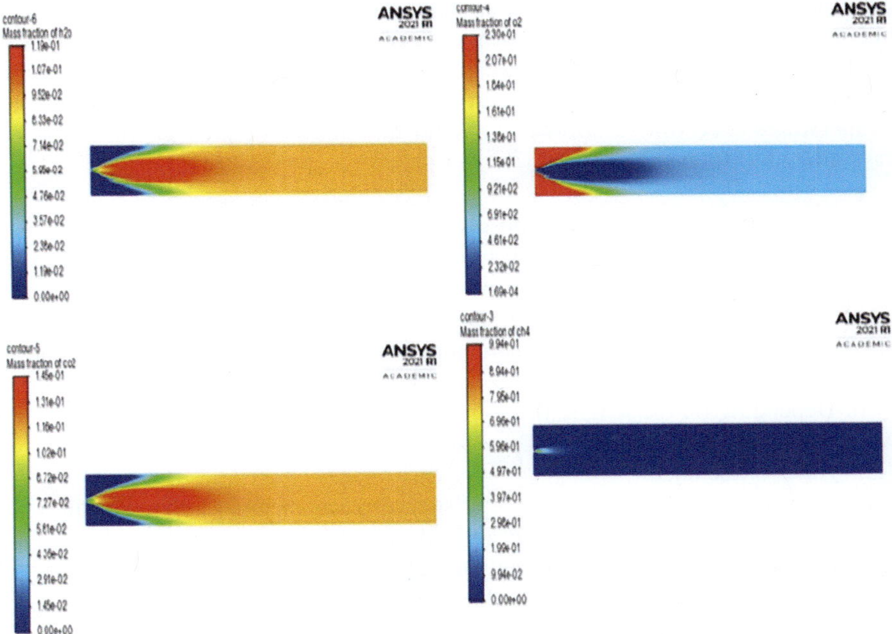

FIGURE 9.26 Species contours plot (aspect ratio of 12).

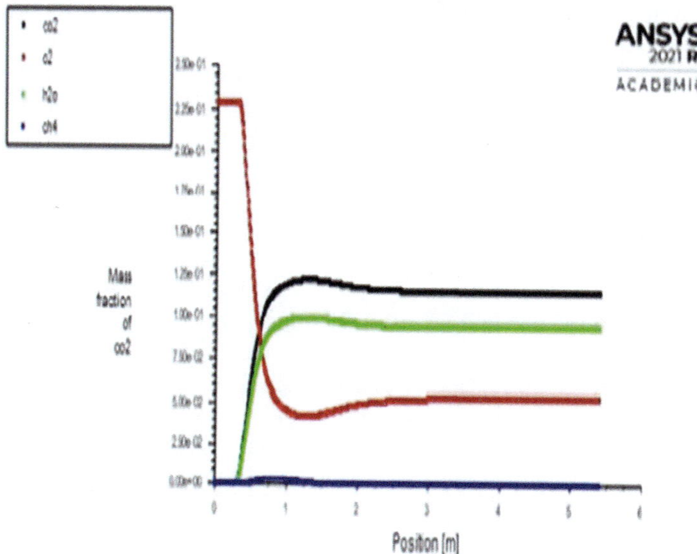

FIGURE 9.27   Species for aspect ratio of 12.

as a model of a propulsion system. Appropriate stochiometric relations have been adopted and mass fractions for water, carbon dioxide, oxygen (oxidiser) and methane (fuel) presented in addition to pollutant (NOX) distributions. A *k*-epsilon turbulence model was deployed. Mesh independence has been included. Extensive modifications in transport characteristics (temperature, pressure, vorticity, mass fraction) are identified with a change in geometry and inlet methane velocity under temperatures exceeding 2,000 K. The main findings of the present study may be crystallized as follows:

i. Significant variations in combustion flow characteristics are computed with a change in the duct length of the channel i. e. as aspect ratio is increased from 4 to 8 and then 12 (corresponding to an increase in duct length from 1.8 to 3.6 m and then 5.4 m) with the duct width sustained at 0.45 m throughout, the value and the contour obtained for the mass fraction of pollutant increases substantially.

ii. Depending on the length of the propulsion duct, the mass fraction of pollutant (NOX) is elevated and extends deeper along the duct covering larger zones of the duct.

iii. A substantial evolution in characteristics also accompanies a change in inlet methane velocity was changed. Indeed, from 80 to 20 m/s, the topology of contours of the flame are strongly modified. Thus, the methane inlet velocity is an important parameter which impacts the analysis and furthermore also affects the pollutant production (NOX).

iv. Pressure distribution is not altered significantly either with methane inlet velocity or duct length (aspect ratio).

    v. Vorticity magnitudes and locations of peak and trough values are changed with aspect ratio and methane inlet velocity indicating a strong sensitivity in the flow circulation within the duct to inlet conditions and also the size of the duct domain.

   vi. The overall results showed good correlation with the general findings from the literature.

  vii. For the individual mass fractions (carbon dioxide, water, oxygen, and methane fuel), increasing the length of the channel did not tangibly modify magnitudes and essentially similar distributions were computed with a smaller aspect ratio.

ANSYS FLUENT has shown impressive capabilities as a CFD simulation tool for turbulent combustion flows with flame propagation and pollutant production in this study. However, unsteady effects have been neglected. Furthermore, conjugate heat transfer has not been incorporated in the present simulations. The duct thermal properties may also have an impact on the combustion process, and this may be explored in future studies. Finally, fully 3-D configurations (cylindrical rocket chambers) with alternative turbulence models (LES, K-omega) and hydrogen enrichment [23] may also be simulated in future investigations.

REFERENCES

1. F. A. Williams, *Combustion Theory*, 2nd edn., Benjamin Cummings, California (1985).
2. A. C. Eckbreth, *Laser Diagnostics for Combustion Temperature and Species*, 2nd edn., Gordon and Breach, Amsterdam (1996).
3. A. F. Ghoniem, A. J. Chorin, and A. K. Oppenheim, Numerical modelling of turbulent flow in a combustion tunnel, *Philosophical Transactions of the Royal Society of London. Series A, Mathematical and Physical Sciences*, 304, 303–325 (1982).
4. X. Kang et al., A versatile numerical tool for simulating combustion features at small-scales, *Journal of Thermal Science*, 30, 343–361 (2021).
5. K. Yamamoto et al., Combustion simulation using the lattice Boltzmann method, *Japan Society of Mechanical Engineers- Series B*, 47(2), 403–409 (2004).
6. X. Kang, R. J. Gollan, P. A. Jacobs, and A. Veeraragavan, On the influence of modelling choices on combustion in narrow channels, *Computers and Fluids*, 144, 117–136 (2017).
7. V. Moreau, Progress in the self-similar turbulent flame premixed combustion model, *Applied Mathematical Modelling*, 34, 4074–4088 (2010).
8. M. E Usman Allauddin, Modelling of turbulent premixed combustion using LES and RANS methods, *PhD Thesis*, Department of Aerospace Engineering, Bundeswehr University, Munich, Germany (2017).
9. A. Gruber et al., Direct Numerical Simulation of hydrogen combustion at auto-ignitive conditions: Ignition, stability and turbulent reaction-front velocity, *Combustion and Flame*, 229, 111385 (2021).
10. M. Ayoobi and I. Schoegl, Numerical analysis of flame instabilities in narrow channels: Laminar premixed methane/air combustion, *International Journal of Spray and Combustion Dynamics*, 9(3), 155–171 (2017).
11. Z. J. He, C. T. Chang, and C. E. Follen, NOx emissions performance and correlation equations for a multipoint LDI injector, *53rd AIAA Aerospace Sciences Meeting 2015*, NASA Glenn Research Center, Cleveland, OH (2015).

12. D. R. Reddy, Overview of low emission combustion research at NASA Glenn, *ASME TURBO Expo 2016*, Seoul, S. Korea (2016).
13. R. R. Tacina, C. Wey, and K. J. Choi, Flame tube NOx emissions using a lean-direct-wall-injection combustor concept, *37th AIAA Joint Propulsion Conference and Exhibit*, Salt Lake City, Utah, July 8–11 (2001).
14. G. Leonard and J. Stegmaier, Development of an aeroderivative gas turbine dry low emissions combustion system, *ASME Journal of Engineering Gas Turbines and Power*, 116, 542–546 (1994).
15. T. Sattelmayer, W. Polifke, D. Winkler, and K. Dobbeling, NOx-abatement potential of lean-premixed gas turbine combustors, *ASME Journal Engineering for Gas Turbines and Power*, 120, 48–59 (1998).
16. C. J. Marek, T. D. Smith, and K. Kundu, Low emission hydrogen combustors for gas turbines using lean direct injection, *41st AIAA Joint Propulsion Conference and Exhibit*, Tuscon, Arizona, AIAA-2005-3776, July 10–13 (2005).
17. S. R. Turns, *An Introduction to Combustion: Concepts and Applications*, 3rd edn., McGraw-Hill Education; Mechanical Engineering Series, New York (2011).
18. K. Y. K. Kuo, *Principles of Combustion*, 2nd edn., John Wiley, Hoboken, NJ (2005).
19. G. P.Sutton and O. Biblarz, *Rocket Propulsion Elements*, 9th edn., John Wiley & Sons, Hoboken, NJ (2017).
20. *ANSYS FLUENT Theory Manual- Combustion Modelling*, ver 19.1, USA (2020).
21. *ANSYS FLUENT Theory Manual-Turbulence Models*, ver 19.1, USA (2020).
22. J. Hinze, *Turbulence*, MacGraw-Hill, New York (1975).
23. C. Lhuillier, P. Brequigny, F. Contino, and C. Mounaïm-Rousselle, Performance and emissions of an ammonia-fuelled SI engine with hydrogen enrichment. *14th International Conference on Engines & Vehicles*, Capri, Italy, September (2019).

10 Optimal Control Analysis on the Coupled Climate-Economy-Biosphere (CoCEB) Model

Godfrey Shoko and Sachin Shaw

10.1 INTRODUCTION

One of the most profound and urgent challenges of the 21st century is global warming, and it has effects on society, environment and the economy [1–3]. There have been many changes in the climate for many years most of which happened because of ice ages and post-glacial periods [4]. Starting with the industrial revolution of the 19th century, the earth has recorded a high increase of atmospheric carbon dioxide and temperature simultaneously. Reports from the Intergovernmental Panel on Climate Change (IPCC) [5–11] show that global warming is being caused by anthropogenic activities and carbon dioxide is the main contributor. The revolution meant a shift from human labour to machinery and started the era of combustion engines including the automobile, thereby resulting in the burning of excessive fossil fuels. Over the past century, human activities have released large amounts of carbon dioxide and other heat retaining greenhouse gases into the atmosphere resulting in the average global surface temperature to rise. The Paris Agreement as detailed in the IPCC (2018) report [12] is aiming to reduce the global surface temperature to below 2°C (preferably 1.5°C). To achieve this, effective mitigation policies achieved using technology and scientific research are needed to prevent carbon dioxide from further rising and are welcomed by the IPCC. Climate change represents one of the greatest environmental, social and economic threats facing planet Earth today, the global climate has been changing due to human activities and is projected to keep changing even more rapidly [1–3]. Our lives are connected to the earth and environment. The earth is a closed system full of intertwining cycles that require balance and we are dependent on it, without that balance, the impacts and consequences can be devastating. Increase in temperature both on land and sea, changes in precipitation, glacier and sea ice melt, ocean acidification, erosion, extreme weather events and coral bleaching can all come as a result of climate change and can have bad consequences on human health, economy, transportation (damages to bridges and roads), food and water security, infrastructure and the natural environment (changes in biological and physical that are irreversible) [4,13,14].

DOI: 10.1201/9781003473749-12

The Intergovernmental Panel on Climate Change (IPCC) is putting increasing emphasis on the use of time-dependent impact models that predict in time-dependent fashion important variables such as atmospheric concentration of greenhouse gases, temperature and precipitation [1,12]. Integrating these tools (greenhouse gas emissions, atmospheric processes, ecological impacts) into what is called integrated assessment models (IAM) will assist policymakers in the (IPCC) and elsewhere to assist the impact of a wide variety of emission strategies. To be useful to decision-makers, IAM must have a reasonable quick turnaround time and give results that are in good agreement with global circulation models (GCM). GCM although they are extremely useful as tools for scientific research, are too time consuming and thus too costly to be very useful for policy analyses. This means that IAM should not compete with GCM but rather be complementary to them and take advantage of specific results from them.

Low-order one-dimensional climate models can be found in McGuffie and Henderson-Sellers [15] and North et al. [16]. In particular, considering elementary thermodynamics, McGuffie and Henderson-Sellers [15] proposed low-order climate models to investigate the climate state as a function of the solar constant, whereas North et al. [16] investigated in detail global energy balance models using a general transport term. Bonetti and McIness [17] developed an energy balance partial differential equation from North et al. [16] in order to investigate the effects of solar radiation management (SRM) technologies for climate engineering. Ogutu [18] models a time evolution of the average surface temperature(AST) on Earth via an energy balance equation given by coupled system of ordinary differential equations with the goal of developing a reduced-complexity coupled climate-economy-biosphere (CoCEB) model to examine the interactions and feedbacks between the climate, economy, and biosphere systems. In this work, we extend the CoCEB model from Ogutu [18] to an optimal control problem.

Developed by the Russian mathematician L.S. Pontryagin (1908–1988), the optimal control theory is a branch of dynamic optimisation as well as a generalisation of the calculus of variations, its history and applications can be found in Refs. [19–23]. Being not just a research tool for mathematicians but also other disciplines such as engineering, industry and commerce shows the inherent beauty of control theory and its relevance to modern development playing a central role in systems and control engineering such as robotics and aeronautics and also in life sciences such as forest management, mathematical biology (modelling and controlling of infectious diseases) [24–26]. Significant progress has been done in coming up with various climate models and analysing them and also on optimal control theory of climate change models that can be found in Manoj et al. [27] and Nordhaus [28] who extended the dynamic integrated climate economy model (DICE) and the integrated assessment model (IAM) to optimal control problems respectively and some studies can be found in Sierra et al. [29] who worked on the congestion control of the global carbonate system. In this work we extend the CoCEB model from Ogutu et al. [1] to an optimal control problem with the aim of addressing the following: First, what are the best strategies to keep CO_2 emission levels to a minimum at a low cost and at the same time improving economic production and if the levels are already high how can they be steered down. Second, how can the resources be allocated across different control interventions in a cost-effective manner. The resulting model is an optimal control problem with six state variables and three control parameters which is then solved using the Pontryagin's maximum principle.

10.2 OPTIMAL CONTROL PROBLEM FOR COCEB MODEL

The coupled climate-economy-biosphere (CoCEB) model is a reduced complexity energy balance model that attempts to incorporate the climate-economy-biosphere interactions and feedbacks using the smallest number of variables and equations needed to capture the main mechanisms involved in the evolution of a coupled system. More details about the formulation and explanation of variables used can be found in Ogutu et al. [1,18]. We extend such model from Ogutu et al. [1] to an optimal control problem with the aim of reducing or controlling the amount of carbon dioxide concentration in the atmosphere at a minimum cost and at the same time maximising the production of an economy. Formally the model is an optimal control problem consisting of a dynamic system with six dimensional state vectors and three control variables. In this chapter, we outline the formulation of the optimal control problem for the CoCEB model which we then solve using Pontryagin's maximum principle (PMP). Some studies on the control theory of climate change models can be found in Manoj et al [27] and Nordhaus [28] who extended the Integrated Assessment Model (IAM) to an optimal control problem and Sierra et al. [29] who worked on the congestion control of the global carbonate system.

10.3 MATHEMATICAL FORMULATION

10.3.1 Schematic Diagram

In this formulation, we are going to consider the interaction between the economy and the biosphere. K and H are human and physical capital, respectively, and are found in the economy module of the model and are modelled using the Cobb-Douglas production function. B and E_y are the biomass and industrial emission, respectively, found in the biosphere, and lastly, C and T are carbon dioxide and temperature, respectively, found in the atmosphere. All these variables are interlinked and can be described by the equations below forming what is called the coupled climate-economy-biosphere (CoCEB) model as shown in Figure 10.1.

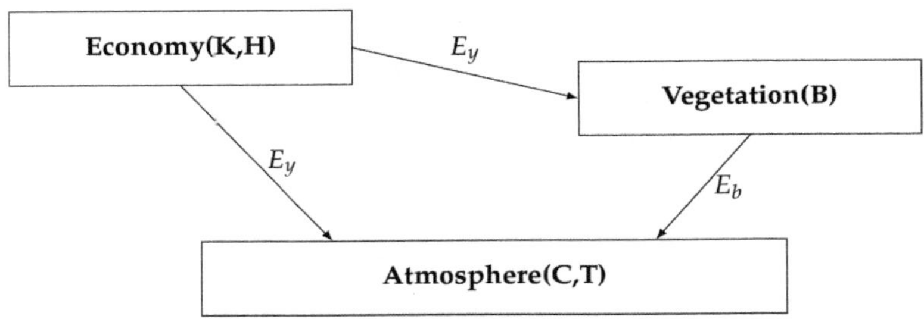

FIGURE 10.1 Schematic diagram for the CoCEB model.

10.3.2 ASSUMPTIONS

In this study, we made the following assumptions

- Climate change is mainly caused by anthropogenic activities.
- The earth emits heat like a blackboard surface.
- Industrial emission and biomass emission both influence atmospheric carbon dioxide concentration causing temperature rise and eventually global warming.
- Carbon dioxide emission and eventually temperature can be controlled and kept at a minimum.

10.3.3 GOVERNING EQUATIONS

$$\frac{dK}{dt} = A\left[1 - \tau(1 + \tau_b) - c(1 - \tau)\right]K^\alpha H^{1-\alpha} D\left(T - \hat{T}\right) - \left(\delta_k + n\right)K \tag{10.1}$$

$$\frac{dH}{dt} = A\left[1 - \tau(1 + \tau_b) - c(1 - \tau)\right]K^\alpha H^{1-\alpha} D\left(T - \hat{T}\right) - \left(\delta_H + n\right)H \tag{10.2}$$

$$C_h \frac{dT}{dt} = (1 - \alpha)Q - \varepsilon\sigma_T\tau_a T^4 + \beta_1\left(1 - \zeta\right)\ln\left(\frac{C}{\hat{C}}\right) \tag{10.3}$$

$$\frac{dC}{dt} = \beta_2\left(E_y + E_b\right) - \left(C - \hat{C}\right)\left[\mu_0 + \gamma_b B\right] \tag{10.4}$$

$$\frac{dB}{dt} = g_b B\left(1 - \frac{B}{\Lambda_b}\right) + \gamma_b B\left(C - \hat{C}\right) \tag{10.5}$$

$$\frac{dE_y}{dt} = \left[g_\sigma + g_y + n\right]E_y \tag{10.6}$$

The objective here is to minimise the amount of carbon dioxide emissions into the atmosphere and the cost of control, at the same time maximising production; we consider the maximisation principle of Pontryagin.

10.4 CONTROL INTERVENTIONS

These are methods or strategies employed to control the amount of Carbon dioxide emissions entering the atmosphere. Various methods can be used to control the emission and we shall consider a few below.

10.4.1 BIOLOGICAL CONTROL

The use of natural methods such as reforestation or afforestation and agricultural practices.

10.4.2 PHYSICAL CONTROL

This is the control by mechanical means such as the bio-energy carbon capture and storage, direct air and sea water capture.

10.4.3 CHEMICAL CONTROL

Examples include enhanced weathering where chemical carbonates are added to oceans thereby increasing the alkalinity and eventually carbon uptake.

10.5 OPTIMAL CONTROL PROBLEM

We present an optimal control equation using Pontryagin's maximum principle by incorporating into equations the system of differential equations, the control variables $u_i(t)$ where $i = 1, 2, 3$ and $0 \leq u_i(t) \leq 1$.

Because we want to minimise carbon dioxide concentration in the atmosphere at a minimum cost and at the same time maximising production of an economy we use Pontryagin's maximum principle and redefine our problem as follows:

$$\min_{0 \leq u_i \leq 1} J(u_i) = \int_0^t \left(Z_0 C(t) + \sum_{i=0}^3 \frac{Z_i u_i^2}{2} \right) dt$$

s.t

$$\frac{dK}{dt} = A\left[1 - \tau(1 + \tau b) - c(1 - \tau)\right] K^\alpha H^{1-\alpha} D\left(T - \hat{T}\right) - \left(\delta_K + n\right) K,$$

$$\frac{dH}{dt} = A\left[1 - \tau(1 + \tau b) - c(1 - \tau)\right] K^\alpha H^{1-\alpha} D\left(T - \hat{T}\right) - \left(\delta_H + n\right) H,$$

$$C_h \frac{dT}{dt} = (1 - \alpha)Q - \varepsilon \sigma_T \tau_a T^4 + \beta_1 (1 - \zeta)(1 - u_1) \ln\left(\frac{C}{\hat{C}} \right),$$

$$\frac{dC}{dt} = \beta_2 (1 - u_2) E_y + \beta_2 (1 - u_3) E_b - \left(C - \hat{C}\right)\left[\mu_0 + \gamma_b B\right],$$

$$\frac{dB}{dt} = g_b B \left(1 - \frac{B}{\Lambda_b} \right) + \gamma_b B(1 - u_1)\left(C - \hat{C}\right),$$

$$\frac{dE_y}{dt} = \left[g_\sigma + g_y + n \right](1 - u_2) E_y,$$

(10.7)

where Z_0 is the cost of pollution caused by carbon dioxide concentration in the atmosphere, $\dfrac{Z_1 u_1^2}{2}$ is the cost of effort to control carbon dioxide concentration by biological means, $\dfrac{Z_2 u_2^2}{2}$ is the cost of effort to control carbon dioxide concentration by mechanical means and $\dfrac{Z_3 u_1^3}{2}$ is the cost of effort to control carbon dioxide concentration by chemical means.

10.6 SOLUTION PROCEDURE

The main aim is to find the optimal control tuple $\left(u_1^*, u_2^*, u_3^*\right)$ such that:

$$\min_{0 \le u_i \le 1} J\left(u_1^*, u_2^*, u_3^*\right) = \left(Ju_1^*, u_2^*, u_3^*\right) \tag{10.8}$$

To find that optimal control tuple, we need to solve the necessary conditions for an optimal control problem by generating the Hamiltonian function. It is also imperative that we solve for the conditions that guarantees the existence and uniqueness of an optimal control solution.

10.6.1 EXISTENCE OF THE OPTIMAL CONTROL SOLUTION

To verify the existence of an optimal control solution, we need to prove the following:

1. The set of controls is non-empty.
2. The state system can be written as a linear function of the control variables with the coefficients dependent on time and control and state variables.
3. The integrand L from the objective functional is convex on the set of controls and $L(x, u, t) \ge \delta_1 |u_1, u_2, u_3|^{\beta} - \delta_2$, where $\delta_1 > 0$ and $\beta > 0$.

    In order to establish (1), we refer to the Picard-Lindelof theorem [33,34] which states that if the state equations are continuous then the solution exists and if the partial derivatives of the state equations are continuous also then the solution corresponding to every admissible control is unique.

It is easy to show that the state equations are continuous for example let;

$$\frac{dK}{dt} = f(t, K) = A[1 - \tau(1 + \tau_b) - c(1 - \tau)]K^{\alpha} H^{1-\alpha} D(T - \hat{T}) - (\delta_K + n)K \tag{10.9}$$

For $\left(t_0, K_0\right) \in \mathbb{R}, f\left(t_0, K_0\right)$ exists and hence $f(t, K)$ is continuous, the same can be done for all state equations and it is equally direct to show that the partial derivatives are continuous. Since the state equations and their partial derivatives are continuous, the solution exists and it is unique. Hence, the set of controls is not empty. Condition (2) is verified by observing the linear dependence of the state equations on controls

u_1, u_2, u_3. Now finally to verify condition (3), we use the fact that any constant, linear or quadratic function is convex. Therefore, $L(u,x,t)$ is convex on the set of controls. To prove the bound on L, we have:

$$Z_3 u_3^2 \leq Z_3 \text{ since } u_3 \in (0,1)$$

$$\Rightarrow \frac{Z_3 u_3^2}{2} \leq \frac{Z_3}{2},$$

$$\Rightarrow \frac{Z_3 u_3^2}{2} - \frac{Z_3}{2} \leq 0.$$

We can write:

$$L(u,x,t) = K(t) + H(t) + T(t) + C(t) + B(t) + E_y(t) + \frac{Z_1 u_1^2}{2} + \frac{Z_2 u_2^2}{2} + \frac{Z_3 u_3^2}{2}$$

Hence:

$$L(u,x,t) \geq \frac{Z_1 u_1^2}{2} + \frac{Z_2 u_2^2}{2} + \frac{Z_3 u_3^2}{2} - \frac{Z_3}{2}$$

$$\Rightarrow L(u,x,t) \geq \min\left(\frac{Z_1}{2}, \frac{Z_2}{2}, \frac{Z_3}{2}\right)\left(u_1^2 + u_2^2 + u_3^2\right) - \frac{Z_3}{2}$$

$$\Rightarrow L(u,x,t) \geq \min\left(\frac{Z_1}{2}, \frac{Z_2}{2}, \frac{Z_3}{2}\right)|u_1, u_2, u_3|^2 - \frac{Z_3}{2}$$

Therefore, $L(u,x,t) \geq \delta_1 |u_1, u_2, u_3|^\beta - \dfrac{Z_3}{2},$

where $\delta_1 = \min\left(\dfrac{Z_1}{2}, \dfrac{Z_2}{2}, \dfrac{Z_3}{2}\right)$, $\delta_2 = \dfrac{Z_3}{2}$ and $\beta = 2$.

10.6.2 The Necessary Conditions for an Optimal Control

The necessary conditions that an optimal control must satisfy come from Pontryagin's maximum principle (PMP). We want to minimise a Hamiltonian function with respect to u_1, u_2, u_3 together with the state equation and the ad-joint condition. Now we generate the Hamiltonian function and it is given by:

$$F\left(t,K,H,T,C,B,E_y,\lambda\right) = Z_0 C(t) + \frac{Z_1 u_1^2}{2} + \frac{Z_2 u_2^2}{2} + \frac{Z_3 u_3^2}{2} +$$

$$\lambda_1 \left(A\left[1 - \tau(1+\tau_b) - c(1-\tau)\right] K^\alpha H^{1-\alpha} D\left(T - \hat{T}\right) - (\delta_k + n)K \right) +$$

$$\lambda_2 \left(A\left[1 - \tau(1+\tau_b) - c(1-\tau)\right] K^\alpha H^{1-\alpha} D\left(T - \hat{T}\right) - (\delta_H + n)H \right) +$$

$$\lambda_3 \left((1-\alpha)Q - \varepsilon\sigma_T\tau_a T^4 + \beta_1(1-\zeta)(1-u_1)\ln\left(\frac{C}{\hat{C}}\right) \right) +$$

$$\lambda_4 \left(\beta_2(1-u_2)E_y + \beta_2(1-u_3)E_b - \left(C - \hat{C}\right)\left[\mu_0 + \gamma_b B\right] \right) +$$

$$\lambda_5 \left(g_b B\left(1 - \frac{B}{\Lambda_b}\right) + \gamma_b B(1-u_1)\left(C - \hat{C}\right) \right) +$$

$$\lambda_6 \left(\left[g_\sigma + g_y + n\right](1-u_2)E_y \right)$$

$$(10.10)$$

10.6.3 Optimality Conditions

We will consider the minimisation of the Hamiltonian with respect to the control variables. Since the cost function is convex, if the optimal control occurs in the interior region, we must have $\dfrac{\partial F}{\partial u_i} = 0$. After solving $,\dfrac{\partial F}{\partial u_i} = 0$ we get the following:

$$\begin{cases}
\dfrac{\partial F}{\partial u_1} = z_1 u_1 - \dfrac{\lambda_3\beta_1(1-\zeta)}{C_h}\ln\dfrac{C}{\hat{C}} - \lambda_5\gamma_b\left(C - \hat{C}\right), \\[2ex]
\Rightarrow u_1^*(t) = \dfrac{\lambda_3\beta_1(1-\zeta)\ln\dfrac{C}{\hat{C}} + \lambda_5\gamma_b\left(C - \hat{C}\right)}{C_h z_1}. \\[3ex]
\dfrac{\partial F}{\partial u_2} = z_2 u_1 - \lambda_4\beta_2 E_y + \lambda_6 E_y\left(g_\sigma + g_y + n\right), \\[2ex]
\Rightarrow u_2^*(t) = \dfrac{\lambda_4\beta_2 E_y - \lambda_6 E_y\left(g_\sigma + g_y + n\right)}{z_2}. \\[3ex]
\dfrac{\partial F}{\partial u_3} = z_3 u_3 - \lambda_4\beta_2 E_b, \\[2ex]
\Rightarrow u_3^*(t) = \dfrac{\lambda_4\beta_2 E_b}{z_3}.
\end{cases}$$

$$(10.11)$$

Now characterising the cost functions gives us the optimal control on the following bounded intervals:

$$u_1^*(t) = \max(0, \min(1, \tau_1)),$$

$$u_2^*(t) = \max(0, \min(1, \tau_2)),$$

$$u_3^*(t) = \max(0, \min(1, \tau_3)).$$

where

$$\begin{cases} \tau_1 = \dfrac{\lambda_3 \beta_1 (1-\zeta) \ln \dfrac{C}{\hat{C}} + \lambda_5 \gamma_b \left(C - \hat{C}\right)}{C_h z_1} \\[2em] \tau_2 = \dfrac{\lambda_4 \beta_2 E_y + \lambda_6 E_y \left(g_\sigma + g_y + n\right)}{z_2} \\[2em] \tau_3 = \dfrac{\lambda_4 \beta_2 E_b}{z_3} \end{cases} \tag{10.12}$$

10.6.4 The Ad-joint (Co-state) Equations

The second condition we are going to solve from Pontryagin's maximum principle is: $\dfrac{\partial F}{\partial x_i} = -\dfrac{\partial \lambda_i}{\partial t}$ at the optimal controls at each $i = 1, 2, 3 \ldots 6$. We need to calculate and solve the system:

$$\frac{\partial \lambda_1}{\partial t} = -\frac{\partial F}{\partial K} \qquad \frac{\partial \lambda_2}{\partial t} = -\frac{\partial F}{\partial H} \qquad \frac{\partial \lambda_3}{\partial t} = -\frac{\partial F}{\partial T}$$

$$\frac{\partial \lambda_4}{\partial t} = -\frac{\partial F}{\partial C} \qquad \frac{\partial \lambda_5}{\partial t} = -\frac{\partial F}{\partial B} \qquad \frac{\partial \lambda_6}{\partial t} = -\frac{\partial F}{\partial E_y}$$

together with the transversality conditions,

$\lambda_i \left(t_f\right) = 0$ for all $i = 1, 2, 3 \ldots 6$.

We now form the optimal system from the state system and ad-joint system by incorporating the control set together with the initial and transversality conditions.

The optimal system is given by:

$$
\begin{cases}
\dfrac{dK}{dt} = A\left[1-\tau(1+\tau_b)-c(1-\tau)\right]K^\alpha H^{1-\alpha}D\left(T-\hat{T}\right)-\left(\delta_k+n\right)K \\[2ex]
\dfrac{dH}{dt} = A\left[1-\tau(1+\tau_b)-c(1-\tau)\right]K_\alpha H^{1-\alpha}D\left(T-\hat{T}\right)-\left(\delta_H+n\right)H \\[2ex]
C_h\dfrac{dT}{dt} = (1-\alpha)Q-\varepsilon\sigma_T\tau T^4+\beta_1(1-\zeta)(1-u_1)\ln\left(\dfrac{C}{\hat{C}}\right) \\[2ex]
\dfrac{dC}{dt} = \beta_2(1-u_2)E_y+\beta_2(1-u_3)E_b-\left(C-\hat{C}\right)\left[\mu_0+\gamma b^B\right] \\[2ex]
\dfrac{dB}{dt} = g_bB\left(1-\dfrac{B}{\Lambda_b}\right)+\gamma_bB(1-u_1)\left(C-\hat{C}\right) \\[2ex]
\dfrac{dE_y}{dt} = \left[g_\sigma+g_y+n\right](1-u_2)E_y \\[2ex]
\dfrac{d\lambda_1}{dt} = -\lambda_1\left[\alpha AK^{\alpha-1}H^{1-\alpha}D\left(T-\hat{T}\right)\left(1-\tau(1+\tau_b)-c(1-\tau)-(\delta_k+n)\right)\right] \\[2ex]
\qquad -\lambda_2\left[(1-\alpha)AK^{\alpha-1}H^{1-\alpha}D\left(T-\hat{T}\right)\left(1-\tau(1+\tau_b)-c(1-\tau)\right)\right] \\[2ex]
\dfrac{d\lambda_2}{dt} = -\lambda_1\left[(1-\alpha)AK^\alpha H^{-\alpha}D\left(T-\hat{T}\right)\left(1-\tau(1+\tau_b)-c(1-\tau)\right)\right] \\[2ex]
\dfrac{d\lambda_3}{dt} = -\lambda_1\left(A\left[1-\tau(1+\tau_b)-c(1-\tau)\right]K^\alpha H^{1-\alpha}D\right) \\[2ex]
\qquad -\lambda_2\left(A\left[1-\tau(1+\tau_b)-c(1-\tau)\right]K^\alpha H^{1-\alpha}D\right)-\lambda_3\dfrac{4\varepsilon\sigma_T\tau T^3}{C_h} \\[2ex]
\dfrac{d\lambda_4}{dt} = -Z_0-\lambda_3\left(\dfrac{\beta_1(1-\zeta)(1-u_1)}{C_hC}\right)+\lambda_4\left((\mu_0+\gamma_bB)\right) \\[2ex]
\qquad -\lambda_5\left(\gamma_b(1-u_1)\right) \\[2ex]
\dfrac{d\lambda_5}{dt} = \lambda_4\gamma_b\left(C-\hat{C}\right)-\lambda_5\left(g_b\left(1-\dfrac{2B}{\Lambda b}\right)\right) \\[2ex]
\dfrac{d\lambda_6}{dt} = -\lambda_4\left(\beta_2(1-u_2)\right)-\lambda_6\left((g_\sigma+g_y+n)(1-u_2)\right)
\end{cases}
$$

$$(10.13)$$

10.7 RESULTS AND DISCUSSION

In this section, we shall carry out numerical simulations to demonstrate the output of optimal control values and the corresponding state variables. Table 10.1 shows the list of parameters and variables together with their values used in carrying out the simulations. The results are estimated using Runge Kutta fourth order method in MATLAB®. We solve the state equation system and the ad-joint (or co-state equation) system together with the optimality equation. Using t final to be 30 years, the system begins with the initial guess on control variables and uses the forward and backward sweep method until convergence is reached.

In Figure 10.2a and b, both human and physical capital will decrease over time and the introduction of controls makes it even decrease at a faster rate. In this case, human capital is the economic value in monetary terms (in million USD/year) of the skills and expertise possessed by human resources in an economy, these skills and

TABLE 10.1

Parameters and Symbols for the CoCEB Model

Parameter or Variable	Description	Value	Reference
n	Population growth rate	0.0157/year	[18,31]
A	Total productivity factor	2.9	[1,16]
c	Consumption share	0.8/year	[18,30]
δ_K	Depreciation rate of K	0.075/year	[18,32]
δ_H	Depreciation rate of H	0.072/year	[15,18]
α	Capital share	0.35	[18,30]
τ	Tax rate	0.2/year	[18,30]
τ_b	Abatement share	0.075	[16,18]
τ_a	Infra red transimissivity	0.6526	[17,18]
D	Damage factor	0.0067	[18, 30]
β_2	Emission taken up by oceans	0.49	[18, 31]
μ_0	Rate of CO_2 absorption by sea	0.05	[1, 31]
$\hat{C}$	Pre industrial CO_2 concentration	500 Gtc	[18, 30]
g_σ	Carbon intensity decline rate	0.03	[18, 30]
C_h	Specific heat capacity	16.7 W/(m² K)	[18, 31]
α_T	Planetary albedo	0.3	[18, 30]
ε	Emissivity	0.95	[18, 30]
σ_T	Steffan Boltzman constant	5.67×10^{-8} W/(m² K)	[18, 30]
Q	Solar flux	1,366 W/(m² K)	[18, 31]
ζ	Temperature rise absorbed by sea	0.23	[18, 30]
β_1	Feedback effect	3.3	[18, 31]
$\hat{T}$	Pre industrial temperature	287.17 Kelvin	[18, 30]
Λ_b	Biomass carrying capacity	900 GtC	[18, 30]
γ_b	Fertilisation parameter	0.0000053	[18, 30]
g_b	Biomass intrinsic growth rate	0.04	[18, 30]
E_b	Biomass emission	1.128	[18, 30]

knowledge decrease in value over time due to technological advancement if there is no other education and or training. The trajectories are showing that the introduction of control measures makes those skills deteriorate at a faster rate. The same applies for physical capital. From Figure 10.2c can be noted that there will be a significance rise in global surface temperature (in Kelvin) over time without controls, the introduction of controls makes a decrease in global surface temperature temperature over time.

In Figure 10.3b, carbon dioxide concentration (in GtC/year) will only continue to rise without controls, the introduction of controls makes a decrease to a minimum level that is not beyond the initial concentration. In Figure 10.3a, biomass emissions will continuously rise without controls, with controls it will again continue to rise but at a lower rate than that with no controls. In Figure 10.3c, the introduction of controls will keep industrial emission at a constant minimum but will continuously rise without controls although in real world it might be difficult to control.

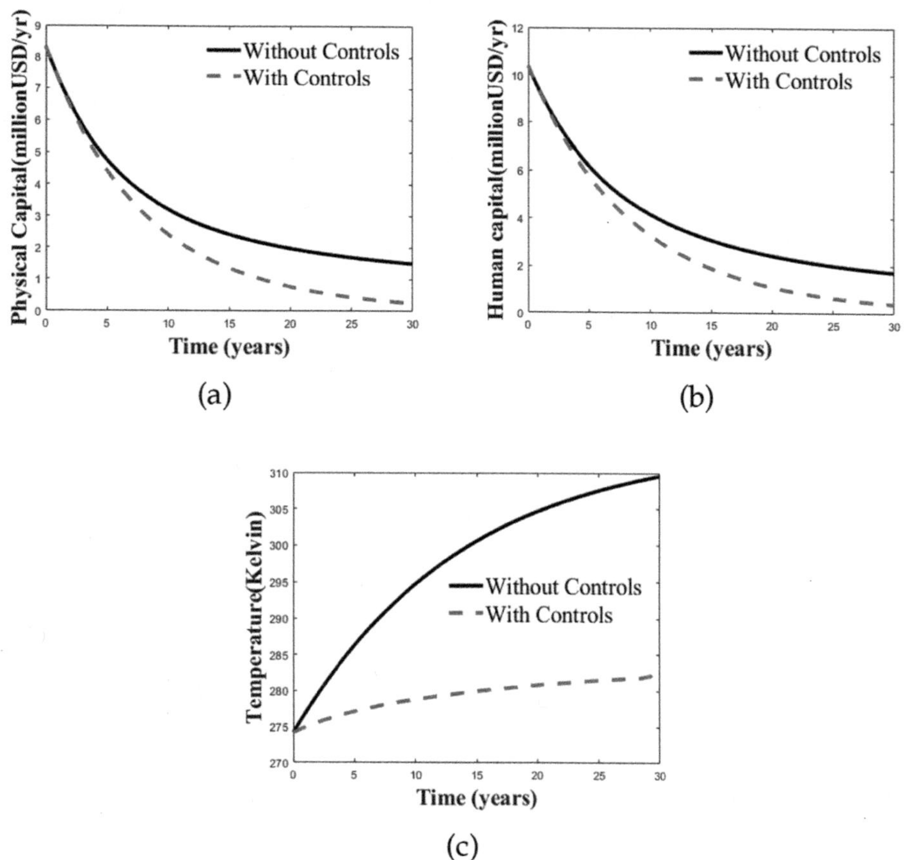

FIGURE 10.2   (a) Physical capital, (b) Human capital, (c) Temperature.

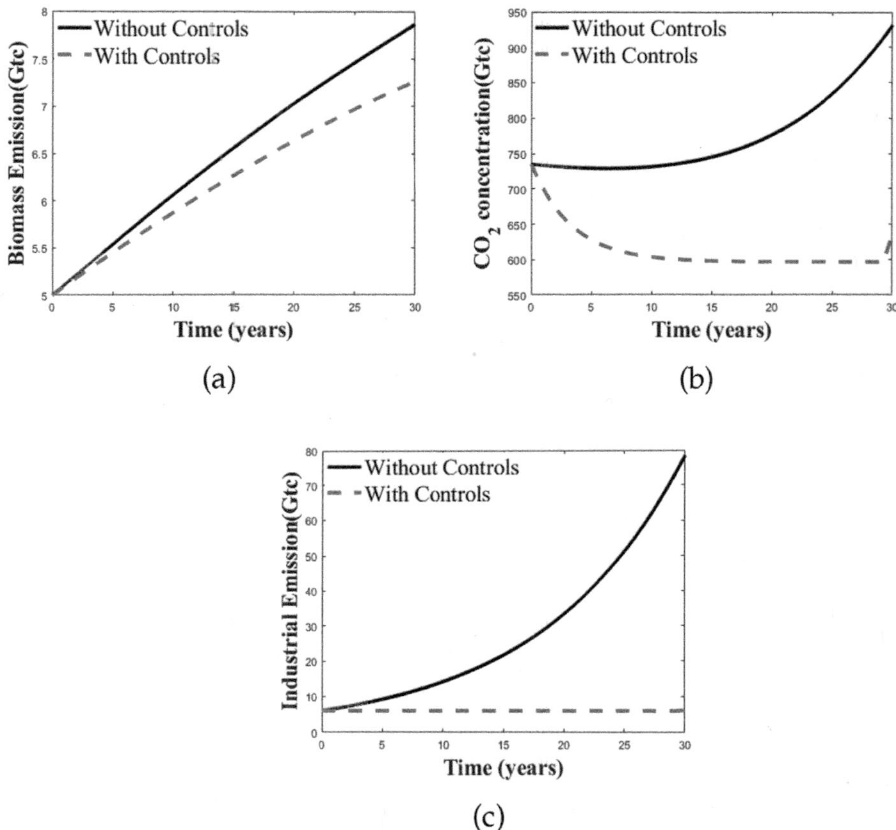

FIGURE 10.3 (a) Biomass, (b) CO_2 concentration, (c) Industrial emission.

The profile of controls in Figure 10.4 shows that for the optimal control to be reached, both the chemical and physical control measures have to be kept at their maximum for the first 27 years but the biological control (u_1) has no significant effect. This shows that as long as production is being carried out the control of carbon emissions is also an ongoing process and efforts should be aimed at methods such as mechanical and chemical means of carbon dioxide propagation. Biological efforts such as afforestation although helpful don't seem to remove large amounts of carbon dioxide from the atmosphere.

10.8 SENSITIVITY ANALYSIS OF THE MODEL

In this section, we carry out some sensitivity analysis of the model to ascertain the robustness of the results and clarify the degree to which they depend on key parameters such as τ_b. The value of this parameter is varied to gain insight in the extent to which particular model assumptions affect the results above. We first modify α_τ which is linked to τ_b by +50% and also by −50% to see how it affects model results.

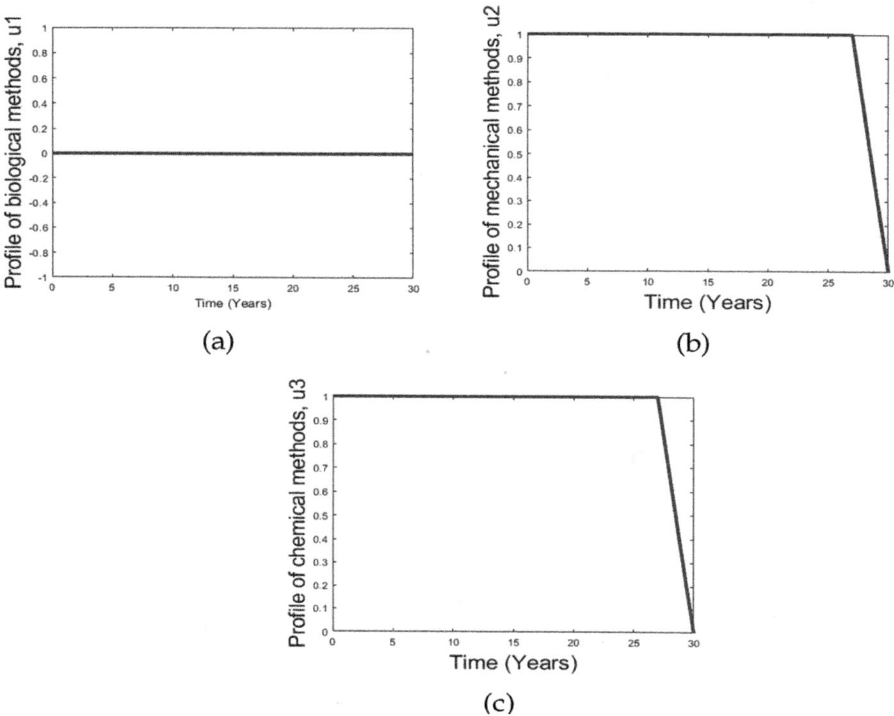

FIGURE 10.4   (a) Profile of u_1, (b) Profile of u_2, (c) Profile of u_3.

We take the lower value of $\alpha_\tau = 0.9$ and the upper value to be α_τ. The abatement parameter τ_b is also varied from $\tau_b = 0$ (Business as usual) to $\tau_b = 0.145$ (high abatement investment). We first run the simulations with $\alpha_{\tau_b} = 0.9$ and $\tau_b = 0.075$ (low abatement investment) to get the following results.

After running the simulations for low abatement investment, we observe from Figure 10.5a–c, the profile of the new optimal solution. The profile of the controls in Figure 10.5 shows that for the optimal control to be reached, both the chemical and physical control measures have to be kept at their maximum for the first 27 years but now a small investment in the biological control measures (u_1) has to be applied for the first 15 years, and the measures can be effective for the next 15 years without investment in biological controls. This shows that if we improve production, we also have to improve the investment in the control of carbon emissions, and efforts should be aimed at increasing the methods such as adding the biological methods to mechanical and chemical means of carbon dioxide propagation that were employed in the business as usual scenarios. Biological efforts such as afforestation can be used once the production of the economy starts to expand to cater for large amounts of carbon emissions in the atmosphere.

We also run simulations to check how the state variables evolve under the low investment in abatement activities. The evolution of the new trajectories is captured in Figures 10.6 and 10.7.

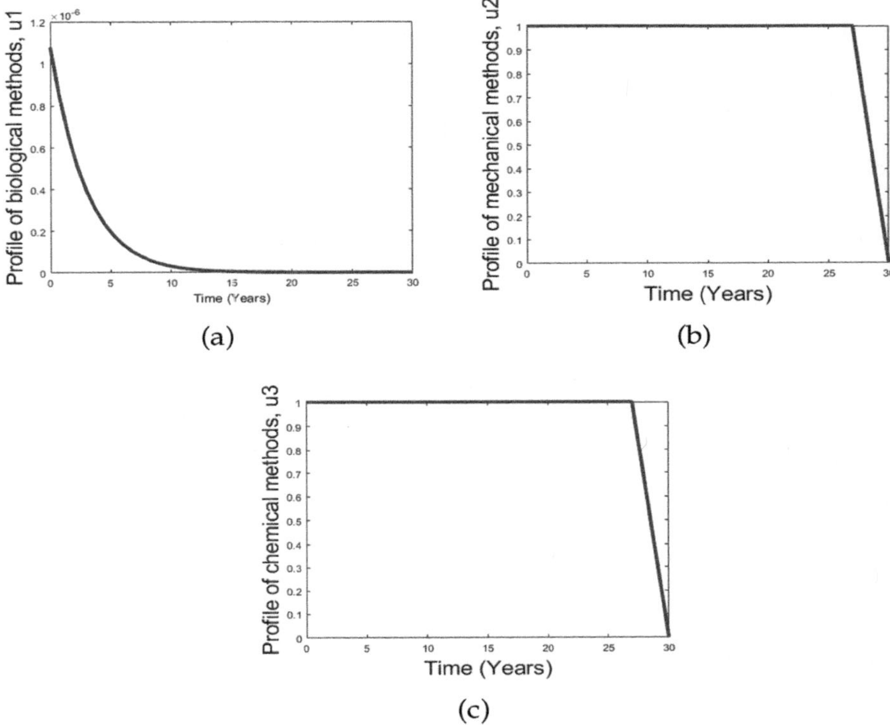

FIGURE 10.5 (a) Profile of u_1, (b) Profile of u_2, (c) Profile of u_3.

In Figure 10.6a and b, both human and physical capital will decrease over time and the introduction of controls under the low investment of abatement activities improves both the decline of both physical and human capital over time unlike in the business as usual scenario in Figure 10.2a and b. The trajectories are showing that the introduction of control measures makes those skills deteriorate at a slower rate. Figure 10.6c shows a significance rise of global surface temperature (in Kelvin) over time without controls even under the low investment scenarios, the introduction of controls makes a decrease in global surface temperature temperature over time, and in this scenario, the decrease is better compared to the business as usual scenario. In Figure 10.7a, carbon dioxide concentration (in GtC/year) will only continue to rise without controls, the introduction of controls makes a decrease to a minimum level which is not beyond the initial concentration like in the business-as-usual scenario. In Figure 10.7b, biomass emissions will continuously rise without controls, with controls it will again continue to rise but at a lower rate than the business-as-usual scenario. In Figure 10.7c, the introduction of controls will keep industrial emission at a constant minimum but will continuously rise without controls.

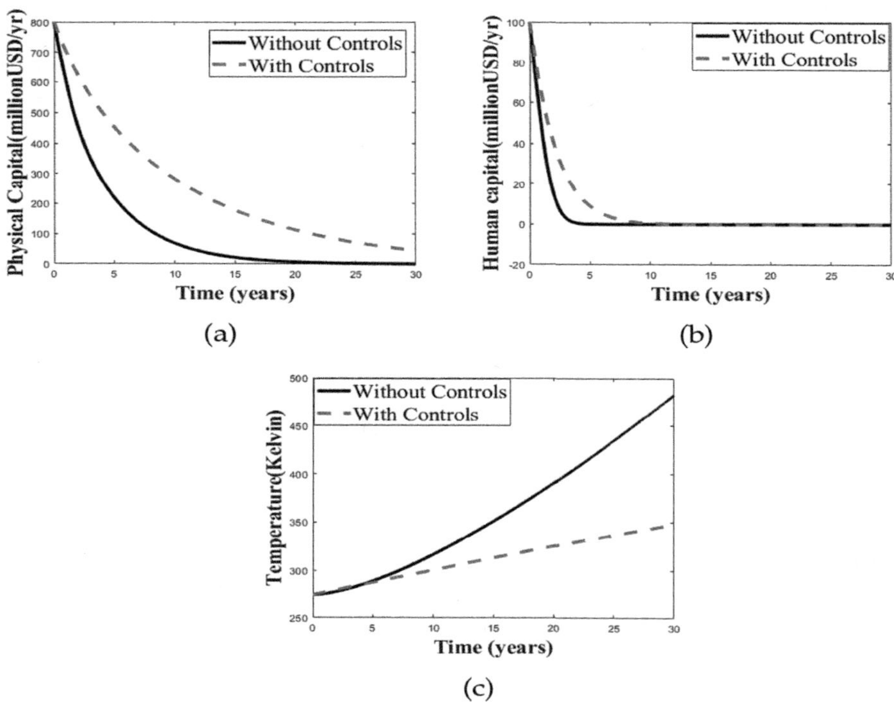

FIGURE 10.6   (a) Profile of u_1, (b) Profile of u_2, (c) Profile of u_3.

10.9   CONCLUSIONS

In this study, the CoCEB model has been extended to an optimal control problem and solved using Pontryagin's Maximum Principle. Numerical simulations were done in MATLAB for two scenarios, firstly the business as usual and secondly the scenario where there is investment in abatement activities; results show that:

- Both levels of CO_2 emissions and temperature will continue to rise due to human activities but can be kept to minimum levels or controlled using chemical and physical means when there is no investment in abatement activities. With a low investment in abatement activities, methods of controlling emissions have to include also biological methods of controlling emissions.
- Although the introduction of control measures to reduce CO_2 concentrations and eventually temperature is effective it increases the deterioration of physical and human capital under the business-as-usual scenario. With low investments in abatement activities, the introduction of controls improves economic growth.
- Physical and chemical control methods are more effective than biological control methods when it comes to controlling CO_2 emissions.

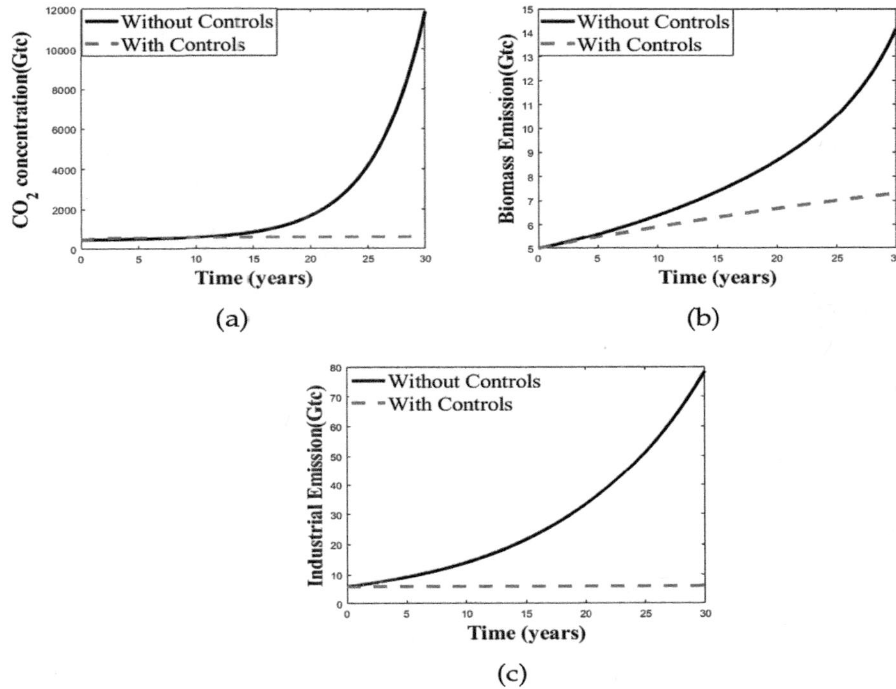

FIGURE 10.7   (a) Profile of u_1, (b) Profile of u_2, (c) Profile of u_3.

If control measures such as chemical and mechanical methods of controlling carbon dioxide emissions in the atmosphere are properly done, it is possible to keep carbon dioxide concentration levels to a minimum. To steer down further the carbon emission levels, there should be investments in abatement activities and more control measures.

REFERENCES

1. Ogutu, K.B.Z., D'Andrea, F., Ghil, M., and Nyandwi, C.: Coupled Climate-Economy Biosphere (CoCEB) model-Part 1: Abatement efficacy of low carbon technologies, *Earth System Dynamics Discussions* (2017) 6(1): 1–32.
2. Dong, W., Ren, F., Huang, J., and Guo, Y.: *The Atlas of Climate Change: Based on SEAPCMIP5: Super-Ensemble Projection and Attribution (SEAP) of Climate Change*, Springer (2013), Berlin, Heidelberg, Germany, 210 pp.
3. Chang, C.P., Ghil, M., Latif, M., and Wallace, J.M. (Eds.): *Climate Change: Multidecadal and Beyond*, WorldScientific Publ. Co./Imperial College Press (2015), Singapore, 388 pp.
4. IPCC (2007b) Climate Change 2007: Impacts, Adaptation and Vulnerability: Contribution of Working Group II to the Fourth Assessment Report of the IPCC, Parry, M.L., Canziani, O.F., Palutikof, J.P., van der Linden, P.J., and Hanson, C.E. (Eds.), Cambridge University Press, Geneva, Switzerland (2007), 976 pp.

5. IPCC: Climate Change 1995: The Science of Climate Change, Contribution of Working Group I to the Second Assessment Report of the IPCC, Cambridge University Press (1996a), Cambridge, 588 pp.

6. IPCC: Climate Change 1995: Economic and Social Dimensions of Climate Change, Contribution of Working Group III to the Second Assessment Report of the IPCC, Bruce, J.P., Lee, H., and Haites, E.F. (Eds.), Cambridge University Press (1996b), Cambridge, 438 pp.

7. IPCC: Climate Change 2001: The Scientific Basis, Contribution of Working Group I to the Third AssessmentReport of the IPCC, Houghton, J.T., Ding, Y., Griggs, D.J., Noguer, M., Van der Linden, P.J., Dai, X., Maskell, K., and Johnson, C.A. (Eds.), Cambridge University Press (2001), Cambridge, 881 pp.

8. IPCC: Climate Change 2007: The Physical Science Basis. Contribution of Working Group I to the Fourth Assessment Report of the IPCC, Solomon, S., Qin, D., Manning, M., Chen, Z., Marquis, M.,Averyt, K., Tignor, M.M.B., Miller Jr., H.L., and Chen, Z. (Eds.), Cambridge University Press (2007a), Cambridge, 1206 pp.

9. IPCC: Contribution of Working Groups I, II and III to the Fourth Assessment Report of the Intergovernmental Panel on Climate Change, Cambridge University Press (2007b), Cambridge.

10. IPCC: Summary for Policymakers, in: Climate Change 2007: Mitigation. Contribution of Working Group III to the Fourth Assessment Report of the IPCC, Metz, B., Davidson, O.R., Bosch, P.R., Dave, R., and Meyer, L.A. (Eds.), Cambridge University Press (2007c), Cambridge, United Kingdom and New York.

11. IPCC: Climate Change 2013: The Physical Science Basis, Contribution of Working Group I to the Fifth Assessment Report of the IPCC, Stocker, T.F., Qin, D., Plattner, G.-K., Tignor, M., Allen, S.K., Boschung, J., Nauels, A., Xia, Y., Bex, V., and Midgley, P.M. (Eds.), Cambridge University Press, Geneva, Switzerland (2013), 1535 pp.

12. Masson-Delmotte, V., Zhai, P., Pörtner, H.O., Roberts, D., Skea, J., Shukla, P.R., Pirani, A., Moufouma-Okia, W., Péan, C., Pidcock, R., Connors, S., Matthews, J.B.R., Chen, Y., Zhou, X., Gomis, M.I., Lonnoy, E., Maycock, T., Tignor, M., and Waterfield, T. (Eds.), IPCC, 2018: Summary for Policymakers. In: Global Warming of 1.5°C. An IPCC Special Report on the Impacts of Global Warming of 1.5°C above Pre-industrial Levels and Related Global Greenhouse Gas Emission Pathways, in the Context of Strengthening the Global Response to the Threat of Climate Change, Sustainable Development, and Efforts to Eradicate Poverty. In Press, Geneva, Switzerland (2018).

13. IPCC (2001b) Climate Change 2001: Synthesis report. A Contribution of Working Groups I,II, and III to the Third Assessment Report of the IPCC, Watson, R.T. (Ed.), Cambridge University Press, Geneva, Switzerland (2001).

14. Dell, M., Jones, B.F., and Olken, B.A.: What do we learn from the weather? The new climate economy literature. *Journal of Economic Literature* 52 (2014): 740–798.

15. McGuffie, K., Henderson-Sellers, A.: *A Climate Modelling Primer*, 3rd edn. Wiley (2005), Chichester, Ch 6.

16. North, G.R., Cahalan, R.F., and Coakley, J.A.: Energy balance climate models, *Reviews of Geophysics* 19 (1981): 91–121.

17. Bonetti, F. and McInnes, C.: A continuous latitudinal energy balance model to explore nonuniform climate engineering strategies, *Climate Dynamics* 52 (2019): 5739–5757.

18. Ogutu, B.K.Z.N.: Energy balance mathematical model on climate change. Continental interfaces, environment, *Doctoral dissertation*, Universite Pierre et Marie Curie - Paris VI (2015).

19. Athans, M. and Falb, P.L.: *Optimal Control*, Mcgraw-Hill (1996), New York.

20. Boltyyanskii, V.G.: *Mathematical Methods of Optimal Control*, Holt, Rinehart and Winston (1971), New York.

21. Fattorini, H. O.: *Infinite Dimensional Optimization and Control Theory*, Cambridge University Press (1999), London.

22. Glowinski, R., *Numerical Methods for Nonlinear Variational Problems*, 2nd edn, Springer-Verlag Publications (1984), New York.

23. Pinch, E.R.: *Optimal Control and the Calculus of Variations*, Oxford University Press Inc (1993), New York.

24. Pontryagin, L., Boltyyanskii, L.S., Gamkrelidze, R.V., and Mishchenko, E.F.: *The Mathematical Theory of Optimal Processes*, Pergamon Press (1964), Oxford.

25. Stodola, P. and Mazal, J.: Optimal location andmotion of autonomous unmanned ground vehicles, *Wseas Transactions on Signal Processing* 6(2) (2010): 68–77.

26. Vinter, R.B.: *Optimal Control*, Birkhauser (2000), Boston, MA.

27. Manoj, A., Prakash, L., Helmut, M., and Willi, S.: Optimal Control of a Global Model of Climate Change with Adaptation and Mitigation, 2018 International Monetary Fund WP/18/270 (2018).

28. Nordhaus. W.D.: Rolling the DICE– an optimal transition path for controlling greenhouse gases, *Resource and Energy Economics* 15 (1993): 27–50.

29. Sierra, C., Metzler, H., Müller, M., and Kaiser, E.: Closed-loop and congestion control of the global carbon-climate system, *Climatic Change* 165 (2021): 15.

30. Nordhaus, W.D.: Integrated Economic and Climate Modeling. In: *Handbook of Computable General Equilibrium Modeling*, Dixon, P.B. and Jorgenson, D.W. (Eds.), Elsevier B.V (2013a), 1069–1131.

31. Nordhaus W.D.: *RICE-2010 Model*, Yale University (2010), New Haven, CT.

32. Nordhaus, W.D.: Rolling the DICE– an optimal transition path for controlling greenhouse gases. *Resource and Energy Economics* 15 (1993): 27–50.

33. Coddington, E.A. and Levinson, N.: *Theory of Ordinary Differential Equations*, McGraw Hill Co Inc. (1955), New York.

34. Grass, D., Caulkins, J.P., Feichtinger, G., Tragler, G., and Behrens, D.A.: *Optimal Control of Nonlinear Processes, with Applications in Drugs, Corruption, and Terror*, Springer (2008), Berlin/Heidelberg, Germany.

11 Enhancement of Double-Pipe Heat Exchangers with Al_2O_3 Nanofluid

Numerical Investigations with Two Configurations

Iman Bashtani and Javad Abolfazli Esfahani

11.1 INTRODUCTION

Heat exchangers, which are extensively employed in industries like air conditioning, chemical processes, and heat recovery, serve as crucial equipment for transferring thermal energy between fluids. Considering widespread applications, the enhancement of their heat transfer efficiency is important. To enhance heat exchanger efficiency, various approaches are investigated, such as forming turbulence through turbulators, increasing the heat transfer surface area via surface corrugation, and employing nanofluids.

A lot of research has been investigated to improve heat transmission by nanofluids. Pak and Cho [1] conducted experimental investigations to assess the influence of Al_2O_3 and TiO_2 nanoparticles on turbulent flow within a constant heat flux tube. The findings revealed that Al_2O_3 exhibited more effective heat transfer compared to TiO_2. Subsequently, they established a correlation for the Nusselt number based on their observations. Bianco et al. explored the numerical study of Al_2O_3 nanoparticles with concentrations of 1, 4, and 6 percent in a pipe under laminar flow [2] and turbulent flow with constant temperature [3] and constant heat flux [4]. Bahmani et al. [5] numerically evaluated the turbulent alumina-water nanofluid passing through a double-pipe heat exchanger. The findings revealed that increasing nanofluid concentration and Re number resulted in an augmentation of the Nusselt number by about 33%. Milani Shirvan et al. [6] numerically studied laminar alumina-water nanofluid flow passing through a double-pipe heat exchanger. Their findings revealed an increase of 57.7% in the Nusselt number when employing a nanofluid with a concentration of 0.03 within the Re number range of 50–150.

Fin turbulators are one of the methods of augmentation of thermal devices, and a part of this chapter is dedicated to evaluating a gear disc fin turbulator in a heat exchanger. To express the importance of fin turbulators, Azeez mohammed Hussein et al. [7] conducted a review of different approaches aimed at enhancing heat exchangers.

DOI: 10.1201/9781003473749-13

Their findings showed that fins were considered a feasible and widely accepted method. Aissa et al. [8] provided the following explanations for the improvements in heat transfer with the turbulator: employing a turbulator and reducing the diameter of the flow passage results in the acceleration of flow speed, leading to amplified heat transmission. Additionally, the turbulator transforms the fluid flow from uniform to spiral, leading to an increase in the contact between the flow and the wall, consequently increasing heat transfer. Furthermore, the presence of turbulators generates secondary flows that disrupt the boundary layer and promote the enhancement of heat transfer between the wall and the fluid. Yadav and Sahu [9–11] conducted experiments to study a water-to-air double-pipe heat exchanger with helical surface disc turbulators with three diameter ratios and helix angles. Hot water and cold air flowed through the inner and outer tubes. They concluded that the reduction of diameter ratios and the increase in helix angles resulted in higher Nusselt numbers and friction factors. Additionally, Yadav, Paulraj, and Sahu [12] conducted experimental and numerical studies to explore the influence of diameter and pitch ratios in a double-pipe heat exchanger with plain surface disc turbulators. The numerical analysis investigated the flow within the shell of the heat exchanger, employing a fixed temperature boundary condition for the inner tube. Bashtani et al. [13] investigated a water-to-water double-pipe heat exchanger with Dolphin dorsal fin turbulators by applying the hybrid CFD-ANN method. The results indicated that heat transfer increased as a result of the flow encountering the turbulator and breaking the boundary layer. The Nu and ε ratios, compared to the conventional heat exchanger, were reported as 2.48 and 1.63 for heat exchangers with turbulators. Tafarroj et al. [14] employed an artificial neural network to predict the performance of various cuts on twisted tape. The results showed that the neural network reduced the calculation costs.

To show the effects of nanofluid inside the corrugated tube, Rabienataj Darzi et al. [15] experimentally investigated the effects of SiO$_2$-water nanofluid in a helically corrugated tube. They reported that the heat transfer improved significantly with a little pressure drop penalty. Also, they experimentally [16] and numerically [17] investigated the effect of Al$_2$O$_3$-water nanofluid in the helically corrugated tube. They reported that adding 2% and 4% nanoparticles by volume to water improved heat transfer by 21%–58% [17]. Qi et al. [18] numerically and experimentally examined the effect of TiO$_2$ nanofluid on heat transfer and flow resistance in a corrugated tube. The results indicated that adding nanoparticles and corrugating increased the heat transfer by 53.95% in comparison with distilled water in a tube. Ekrani et al. [19] numerically investigated the influence of SiO$_2$-water nanofluid flow in a tube with a delta winglet turbulator. They found that the delta winglet improved heat transfer by destroying the boundary layer. Adibi et al. [20] numerically studied the influence of Al$_2$O$_3$-water nanofluid flow in a tube with perforated anchors. They concluded that the nanofluid increased the Nu number by 21.36%. Naphon and Wiriyasart [21] experimentally investigated the effect of the flow and magnetic field on heat transfer in a spirally corrugated tube in the presence of titanium dioxide nanofluids. It was reported that concurrently employing multiple heat transfer enhancement methods resulted in a greater increase in heat transfer than utilizing a single method. Bartwal et al. [22] studied the quadratic convective flow of tangent hyperbolic fluid over a flat stretched sheet with variable thickness using the Legendre wavelet collocation technique (LWCT). The results revealed a decrease in fluid temperature as the relaxation parameter was increased. Gupta et al. [23] numerically studied the unsteady magnetic hybrid nanofluid flow over a stretching surface with suction and viscous dissipation by LWCT. The findings indicated that the heat transfer rate of the nanofluid increased with increasing temperature-dependent viscosity. Upreti et al. [24] investigated the effect of gold-blood nanofluid over the stagnation point flow of Casson nanofluid over a stretching sheet. They observed that the magnetic field and the porosity parameter contributed to an enhancement in heat transfer. In another study, Upreti et al. [25] tested the effect of different shapes of gold nanoparticles, including cylindrical, platelet, and blade shapes. They determined that blade-shaped nanoparticles significantly affect temperature increase owing to their heightened thermal properties. Upreti et al. [26] examined nanofluid flow's entropy generation and heat transfer characteristics passing over a Riga plate. Pandey et al. [27] investigated the influence of magnetic nanofluid flow over a cone with volumetric heat generation on heat and mass transfer. They concluded that the magnetic field improved the temperature profiles. Bartwal et al. [28] numerically solved the flow of tangent hyperbolic fluid through a stretching disk. They examined the effects of various external forces, including magnetic field, ohmic heating, heating due to porous media, and viscous heating. The results indicated that increasing the magnetic parameter values resulted in an increase in radial and tangential velocities, as well as a decrease in axial velocity.

To show the influence of corrugating shape, Jin et al. numerically investigated the effects of corrugating on flow resistance [29] and heat transfer [30] in a six-start spirally wavy tube. They reported that the values of $\dfrac{f_e}{f_s}$ and $\dfrac{\mathrm{Nu}_e}{\mathrm{Nu}_s}$ were 1.7–5.1 and 1.05–1.33, respectively, compared to a circular tube. Balla [31] studied heat transfer in a six-wave spiral wave tube numerically. The results indicated that the corrugated tube's heat transfer and flow resistance were 2.4–3.7 and 1.7–2.3 times compared to a circular tube. Zhang et al. [32] carried out flow and heat transfer in a corrugated tube considering uniform and non-uniform waves. Wang et al. [33] numerically investigated the effect of transverse and helically corrugated tubes on heat transfer in turbulent flow. They reported that the maximum increase in Nusselt number was 1.77. Sun and Zeng [34] studied numerically and experimentally the effects of heat transfer and pressure drop in turbulent flow for smooth and corrugated tubes. Results indicated that the heat transfer in the corrugated tube increased by 50%. Bashtani and Esfahani [35] numerically investigated the effects of corrugating the inner tube of a water-to-water double-pipe heat exchanger. The problem was solved in 3D and 2D, assuming an axisymmetric condition. The results indicated that two- and three-dimensional solutions were coincident. The findings indicated that corrugating increased the Nusselt number and effectiveness by 1.75 and 1.73 compared to the simple heat exchanger.

Motivated by the literature mentioned above and industrial applications, the aim of this chapter is to investigate the effects of the turbulent flow of alumina-water nanofluid through enhanced double-pipe heat exchangers with two different configurations. The first configuration is about improving a double-pipe heat exchanger with the combination of the nanofluid and gear disc fin turbulators. The velocity and temperature fields, along with the local Nusselt number in the heat exchanger with gear turbulators, are interpreted to reveal the turbulators' effect in improving the heat exchanger's efficiency. Also, a detailed explanation is provided regarding the influence of the turbulator on the flow's physics and the reasons behind the observed enhancement in heat transfer. The heat exchanger is solved by conjugating heat transfer and evaluated using the ε-NTU method. The second configuration is about enhancing a double-pipe heat exchanger by combining the nanofluid and sinusoidal corrugated inner tube. The heat exchanger is solved by conjugated heat transfer considering both the hot and cold sides of the heat exchanger. A detailed investigation is carried out to examine the effects of nanofluid volume concentration, wave amplitude, and Reynolds number on important parameters of the heat exchanger, including the overall heat transfer coefficient, effectiveness, and the number of transfer units. Finally, the heat exchanger is evaluated with the ε-NTU method in detail.

11.2 PROBLEM DEFINITION

The influence of turbulent Al$_2$O$_3$-water nanofluid flow is examined in improved double-pipe heat exchangers, involving one configuration with gear disc turbulators and another configuration with corrugation of the inner tube.

11.2.1 ADDING AL$_2$O$_3$-WATER NANOFLUID TO HEAT EXCHANGER WITH GEAR DISC TURBULATORS

The first configuration is about evaluating the effects of alumina nanoparticles addition to the water in a water-to-water parallel flow double-pipe heat exchanger featuring gear disc turbulators under turbulent flow condition. The nanofluid with concentrations of 1%, 4%, and 6% is added to hot flow inside the tube under conditions of $3{,}000 \leq \mathrm{Re} \leq 13{,}000$ and $T_i = 350$ K. Simultaneously, cold pure water flows inside the shell under conditions of $\mathrm{Re} = 500$ and $T_i = 285$ K. Figure 11.1 shows the geometrical shape and dimensions of the heat exchanger. Also, geometrical parameters are defined in Table 11.1.

11.2.2 ADDING AL$_2$O$_3$-WATER NANOFLUID TO HEAT EXCHANGER WITH WAVY INNER TUBE

The second configuration deals with assessing the influence of adding Al$_2$O$_3$ to water in a water-to-water sinusoidal corrugated counterflow double-tube heat exchanger, assuming three different wave amplitudes at turbulent flow. The nanoparticles with concentrations of 1%, 4%, and 6% are added to the hot fluid and streams in the tube with the Re number ranging from 3,000 to 13,000. The pure water, as the cold fluid, passes through the shell in the Re number of 5,000. The schematic view of the heat exchanger, along with the geometric dimensions, is shown in Figure 11.2. The wave equation for the corrugated tube is derived from the Equation (11.1). In this equation, $\mathscr{T}$ represents the wavelength, set to 20 mm, and A represents the wave amplitude, taking values of 1, 2, and 3 mm.

$$\text{Wave equation} = A\sin\left(\frac{2\pi}{\tau}X\right) \tag{11.1}$$

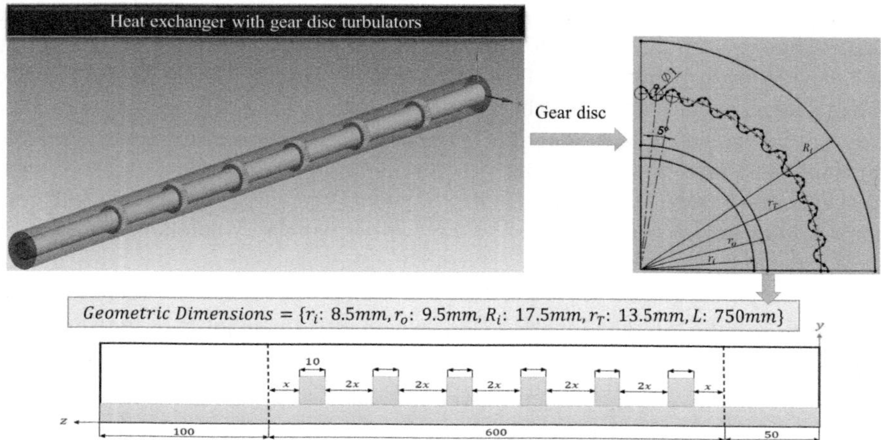

FIGURE 11.1 Geometrical shape and dimensions of the heat exchanger with gear turbulators.

TABLE 11.1
Geometrical Parameters of the Heat Exchanger with Gear Turbulators

Parameter	Symbol	Size
Tube inner radius (mm)	r_i	8.5
Tube outer radius (mm)	r_o	9.5
Shell radius (mm)	R_i	17.5
Turbulator radius (mm)	r_T	13.5
Heat exchanger length (mm)	L	750

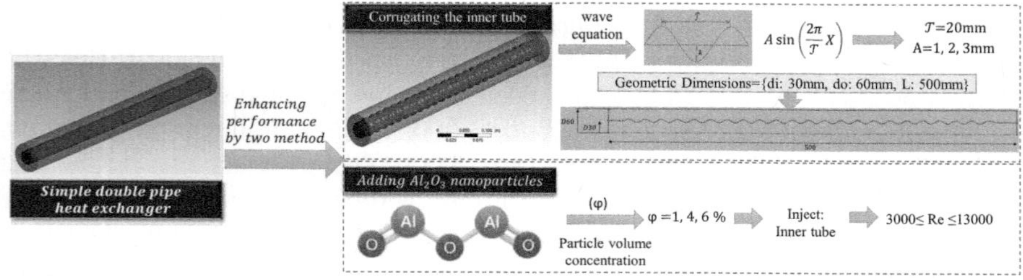

FIGURE 11.2 Geometrical shape and dimensions of the heat exchanger with corrugated tube.

11.3 MATHEMATICAL MODEL

11.3.1 Governing Equations

The governing equations for steady-state conservation of mass, momentum, and energy are expressed as follows [36].

Continuity equation:

$$\nabla \cdot V = 0 \tag{11.2}$$

Momentum equation:

$$\rho \frac{D\bar{V}}{Dt} = \rho g - \nabla \bar{p} + \nabla \cdot \tau \tag{11.3}$$

$$\tau_{ij} = \mu \left(\frac{\partial \bar{u}_i}{\partial x_j} + \frac{\partial \bar{u}_j}{\partial x_i} \right) - \rho \overline{\acute{u}_i \acute{u}_j} \tag{11.4}$$

Energy equation:

$$\rho C_p \frac{D\bar{T}}{Dt} = -\frac{\partial q_i}{\partial x_i} + \bar{\Phi} \tag{11.5}$$

$$\bar{\Phi} = \frac{\mu}{2} \overline{\left(\frac{\partial \bar{u}_i}{\partial x_j} + \frac{\partial \acute{u}_i}{\partial x_j} + \frac{\partial \bar{u}_j}{\partial x_i} + \frac{\partial \acute{u}_j}{\partial x_i} \right)^2} \tag{11.6}$$

$$q_i = -k \frac{\partial \bar{T}}{\partial x_i} + \rho C_p \overline{\acute{u}_i T'} \tag{11.7}$$

The Wilcox k-ω model exhibits the capacity to simulate near-wall regions (inner boundary layer) more effectively than the k-ε model, enabling higher accurate simulation of flows experiencing a reverse pressure gradient. Nonetheless, the ω equation demonstrates a significant dependency on the value of ω in the outer boundary layer region. Conversely, the ε equation in the k-ε model demonstrates less dependence on external flow conditions. Menter [37] introduced the Shear Stress Transport model (SST k-ω), incorporating the advantages of both the k-ω and k-ε models through the definition of weight functions. The function was devised to activate the k-ω model in the near-wall region and switch to the k-ε model outside the boundary layer. The SST k-ω model is a two-equation model encompassing equations for turbulence kinetic energy (11.8) and specific dissipation rate (11.9), which are as follows:

Turbulence kinetic energy:

$$U_j \frac{\partial \rho k}{\partial x_j} = \tau_{ij} \frac{\partial u_i}{\partial x_j} - \beta^* k \rho \omega + \frac{\partial}{\partial x_j} \left[\left(\mu + \sigma_k \mu_t \right) \frac{\partial k}{\partial x_j} \right] \tag{11.8}$$

Specific dissipation rate:

$$U_j \frac{\partial \rho \omega}{\partial x_j} = \frac{\gamma}{V_t} \tau_{ij} \frac{\partial u_i}{\partial x_j} - \beta \rho \omega^2 + \frac{\partial}{\partial x_j}\left[(\mu + \sigma_\omega \mu_t)\frac{\partial \omega}{\partial x_j}\right] + 2(1-F_1)\rho \sigma_{\omega 2}\frac{1}{\omega}\frac{\partial k}{\partial x_j}\frac{\partial \omega}{\partial x_j}$$

$$(11.9)$$

The constants associated with the inner boundary layer are outlined below:

$$\sigma_{k1} = 0.85, \sigma_{\omega 1} = 0.5, \beta_1 = 0.0750, \beta^* = 0.09, \kappa = 0.41,$$

$$\gamma_1 = \frac{\beta_1}{\beta^*} - \frac{\sigma_{\omega 1}\cdot \kappa^2}{\sqrt{\beta^*}}, a_1 = 0.31$$

$$(11.10)$$

The constants associated with the outer boundary layer are outlined below:

$$\sigma_{k2} = 1, \sigma_{\omega 2} = 0.856, \beta_2 = 0.0828, \beta^* = 0.09, \kappa = 0.41,$$

$$\gamma_2 = \frac{\beta_2}{\beta^*} - \frac{\sigma_{\omega 2}\cdot \kappa^2}{\sqrt{\beta^*}}, a_1 = 0.31$$

$$(11.11)$$

The constants are consolidated for the internal model $\varnothing_1$ and the outer model $\varnothing_2$, resulting in the derivation of a new constant $\varnothing$:

$$\varnothing = F_1\varnothing_1 + (1-F_1)\varnothing_2$$

$$(11.12)$$

Where F_1 is equivalent to:

$$F_1 = \tanh\left(\arg_1^4\right)$$

$$(11.13)$$

$$\arg_1 = \min\left[\max\left(\frac{\sqrt{k}}{0.09\omega y}; \frac{500v}{y^2\omega}\right); \frac{4\rho\sigma_{\omega 2}k}{CD_{k\omega}y^2}\right]$$

$$(11.14)$$

$$CD_{k\omega} = \max\left(2\rho\sigma_{\omega 2}\frac{1}{\omega}\frac{\partial k}{\partial x_j}\frac{\partial \omega}{\partial x_j}, 10^{-20}\right)$$

$$(11.15)$$

The eddy viscosity is characterized as:

$$V_t = \frac{a_1 k}{\max\left(a_1\omega, \varnothing F_2\right)}$$

$$(11.16)$$

$$F_2 = \tanh\left(\arg_2^2\right)$$

$$\text{arg}_2 = \max\left[\frac{2\sqrt{k}}{0.09\omega y}; \frac{500v}{y^2\omega}\right] \tag{11.17}$$

11.3.2 Computational Model and Boundary Conditions

The solution is regarded as the steady condition, and the problem is solved numerically by conjugated heat transfer, in which both cold and hot streams are solved. The SIMPLE algorithm is carried out for coupling pressure-velocity terms, and the second-order upwind is used to discretize the governing equations. The inlet and outlet boundary conditions are velocity inlet and pressure outlet. The adiabatic boundary condition is used for the shell, and the condition of the coupled wall is used for conjugated heat transfer. Thermophysical properties are presented in Table 11.2 for hot and cold fluid flow at 350 K and 285 K, respectively. The *SST K-ω* turbulence model is used to capture the turbulent fluid flow characteristics. The boundary layer mesh configuration is done with the y^+ value below one.

11.3.3 Required Equations

The properties of the nanofluid are computed using Equations (11.18) and (11.19) [4]. Equations (11.18) and (11.19) are formulated according to the conventional concept of mixture. Equation (11.20) [39–41] is obtained through fitting experimental data [39], and Equation (11.21) [39–41] is formulated based on a classical model [39]. The volumetric fraction values used in this study are inspired by references found in the literature, such as [4]. These equations are selected considering the acceptable accuracy between the experimental correlation of Pak and Cho and the numerical simulation of nanofluid, which is reported in the validation section:

$$\rho_{nf} = (1-\varphi)\rho_{bf} + \varphi\rho_p \tag{11.18}$$

$$Cp_{nf} = (1-\varphi)Cp_{bf} + \varphi Cp_p \tag{11.19}$$

$$\mu_r = \frac{\mu_{nf}}{\mu_{bf}} = 123\varphi^2 + 7.3\varphi + 1 \tag{11.20}$$

TABLE 11.2

Thermophysical Properties [38]

Thermophysical Properties	Hot Stream – 350 K	Cold Stream– 285 K
$\rho\left(\text{kg/m}^3\right)$	973.46	999.58
$C_p\left(\text{kJ/kg}\cdot\text{K}\right)$	4.195	4.189
$K\left(\text{W/m}\cdot\text{K}\right)$	0.666	0.59
$\mu\left(\text{N}\cdot\text{s/m}^2\right)$	0.000365	0.001225

$$k_r = \frac{k_{nf}}{k_{bf}} = 4.97\varphi^2 + 2.72\varphi + 1 \tag{11.21}$$

Reynolds number (Re) of the tube and shell sides of the heat exchanger are defined according to Equations (11.22) and (11.23), respectively [42]:

$$\text{Re}_{\text{Tube Side}} = \frac{\rho u_m d_i}{\mu} \tag{11.22}$$

$$\text{Re}_{\text{Shell Side}} = \frac{\rho u_m (d_o - d_i)}{\mu} \tag{11.23}$$

Where d_i and d_o are the diameter of the inner and outer tubes, and u_m is the mean velocity.

Friction factor coefficient (f) is defined as follows [42]:

$$f = \frac{\Delta P}{\dfrac{L}{D_h} * \dfrac{1}{2}\rho u_m^2} \tag{11.24}$$

Nusselt number (Nu) is defined as follows [42]:

$$\text{Nu} = \frac{\bar{h} * D_h}{K_f} \tag{11.25}$$

$$\bar{h} = \frac{q''}{T_{\text{surface}} - T_{\text{bulk}}} \tag{11.26}$$

Where ΔP, L, and D_h are the pressure drop, length, and hydraulic diameter of the heat exchanger. Also, h- and k_f are the mean heat transfer coefficient and fluid thermal conductivity.

Logarithmic mean temperature difference (LMTD) is defined as follows [42]:

$$\text{LMTD} = \frac{\Delta T_1 - \Delta T_2}{\text{Ln}\left[\dfrac{\Delta T_1}{\Delta T_2}\right]} \tag{11.27}$$

Where $\Delta T_1 = T_{h,i} - T_{c,i}$ and $\Delta T_2 = T_{h,o} - T_{c,o}$ are for parallel flow heat exchanger, and $\Delta T_1 = T_{h,i} - T_{c,o}$ and $\Delta T_2 = T_{h,o} - T_{c,i}$ are for counterflow heat exchanger.

The overall heat transfer coefficient (U) is defined as follows [42]:

$$U = \frac{q}{\text{LMTD} * \bar{A}} \tag{11.28}$$

Effectiveness [42] is the ratio of the actual heat transfer (q) to the maximum possible heat transfer (q_{max}) in a heat exchanger, and it is calculated as follows:

$$\varepsilon = \frac{q}{q_{\text{max}}} \tag{11.29}$$

$$q_{\max} = C_{\min}\left(T_{h,i} - T_{c,i}\right) \qquad (11.30)$$

$$C_h = \dot{m}_h c_{p,h} \qquad (11.31)$$

$$C_c = \dot{m}_c c_{p,c} \qquad (11.32)$$

$$C_{\min} = \min\left(C_h, C_c\right) \qquad (11.33)$$

Number of transfer units (NTU) is defined as follows [42]:

$$\mathrm{NTU} = \frac{U * \overline{A}}{C_{\min}} \qquad (11.34)$$

11.3.4 Mesh View and Grid Independence Test

Figure 11.3 shows the grid of the heat exchangers with gear disc turbulators and the sinusoidal corrugated tube. As can be seen, the appropriate boundary layer grid is designed according to the turbulence model and the desired value of y^+. The grid independence study (Table 11.3) shows that the number of suitable cells for the heat exchanger with gear turbulators and the heat exchanger with sinusoidal corrugated tube is equal to $2*10^6$ and $12*10^4$, respectively.

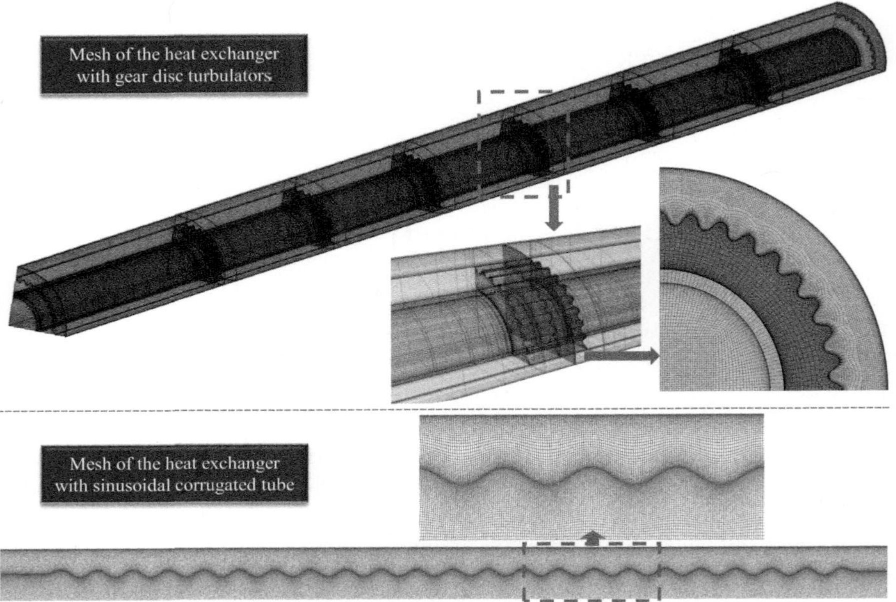

Figure 11.3   The mesh of the heat exchanger.

TABLE 11.3
Grid Independence Study

Type	Grid	ΔP_h (kPa)	ΔP_c (kPa)	Nu	$E - \Delta P_h$ (%)	$E - \Delta P_c$ (%)	E–Nu (%)
Gear turbulators HE	$1*10^6$	0.16856	0.01829	78.2912	1.57	3.18	1.49
	$2*10^6$	0.17125	0.01889	79.4754	0.88	1.25	0.68
	$4*10^6$	0.17277	0.01913	80.0195			
Sinusoidal corrugated HE	$6*10^4$	0.03304	0.03161	100.6570	0.93	0.28	1.06
	$12*10^4$	0.03335	0.03170	99.6057	0.09	0.44	0.93
	$25*10^4$	0.03332	0.03184	98.6833			

11.4 VALIDATION

The Nu number and friction factor coefficient are numerically calculated and compared with experimental correlations of (11.35), (11.36), and (11.37) in a tube for fully developed turbulent flow. The maximum error of comparing numerical values with experimental correlations of Blasius, Petukhov, and Gnielinski is about 3.5%, 2.9%, and 4.9%, respectively (Figure 11.4):

$$f_{\text{Blasius}} = 0.316 \, \text{Re}^{-\frac{1}{4}} \quad \text{Re}_d \leq 2*10^4 \tag{11.35}$$

$$f_{\text{Petukhov}} = \left[0.79 \, \text{Ln} \, \text{Re} -1.64\right]^{-2} \quad 3,000 \leq \text{Re}_d \leq 5*10^6 \tag{11.36}$$

$$\text{Nu}_{\text{Gnielinski}} = \frac{\frac{f}{8}*\left[\text{Re}_d - 1,000\right]*\text{pr}}{1 + 12.7\left(\frac{f}{8}\right)^{\frac{1}{2}}*\left(\text{Pr}^{\frac{2}{3}} - 1\right)} \quad \begin{array}{c} 3,000 < \text{Re}_d < 5*10^6 \\ \\ 0.5 < \text{Pr} < 2*10^3 \end{array} \tag{11.37}$$

The turbulent Al_2O_3-water nanofluid flow passing through a tube is numerically solved for concentrations of 1%, 4%, and 6% in the Re number ranging from 10^4 to $4*104$. The numerical data are compared with the experimental correlation of Pak and Cho [1], in which the maximum error of the Nu number is about 5.7% (Figure 11.5).

Additional validation is done to reveal the precision of the numerical procedure in passing the turbulent flow over turbulators. So, the experimental results of Ruengpayungsak et al. [43] are numerically solved for turbulent flow passing over seven ring turbulators in a tube at Re number ranging from $10*10^3$ to $18*10^3$. The maximum error for the friction factor compared with the experimental results is about 6.2% (Figure 11.6). The accuracy of the numerical solution is acceptable, considering the 5% uncertainty of the experimental results.

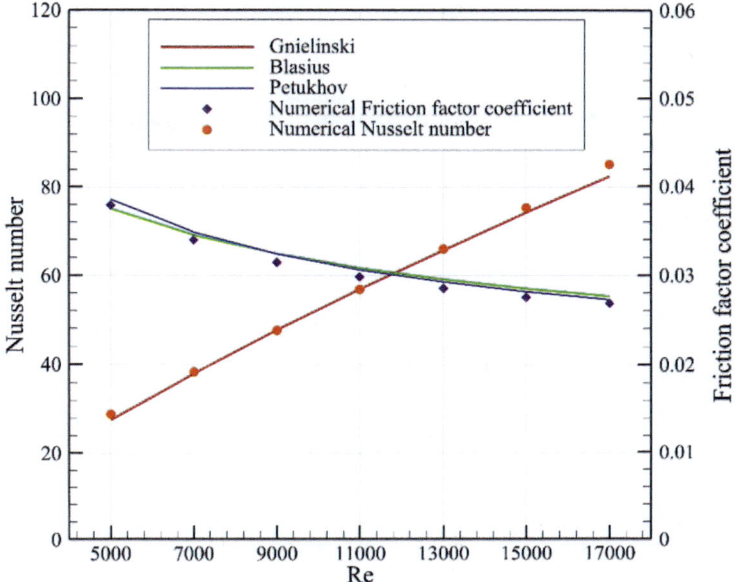

FIGURE 11.4 Validation of Nusselt number and friction factor coefficient.

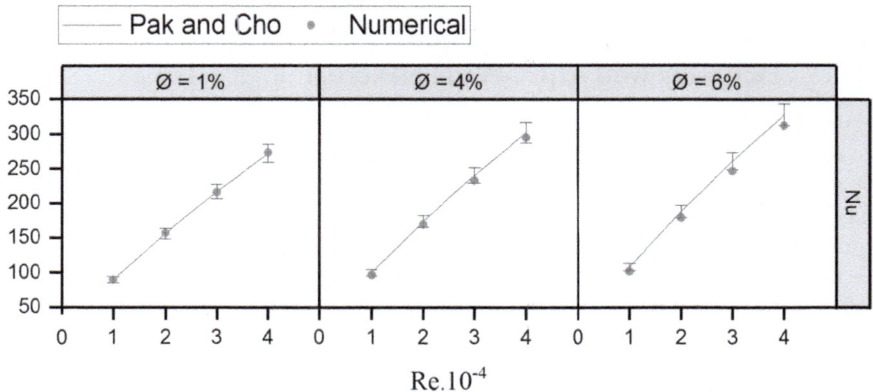

FIGURE 11.5 Validation of the numerical solution of Al_2O_3-water nanofluid.

11.5 RESULTS AND DISCUSSION

This section investigates the effect of the nanofluid flow of different concentrations and Reynolds numbers for two improved configurations, including a heat exchanger with gear turbulators and a heat exchanger with sinusoidal wavy tube.

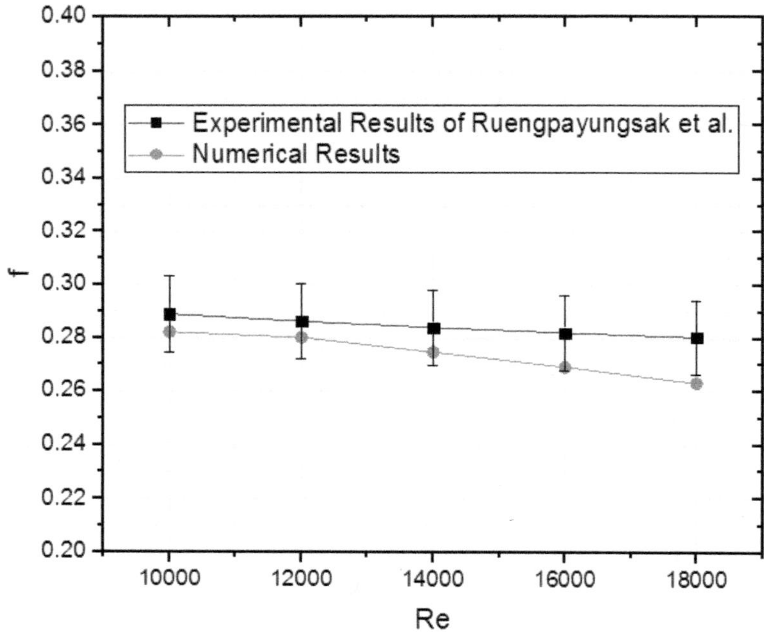

FIGURE 11.6   Validation of the numerical solution of fluid flow over ring turbulators.

11.5.1   Assessment of Gear Turbulators in a Heat Exchanger with Al_2O_3-Water Nanofluid

Figure 11.7 illustrates the temperature and velocity contours of a double-pipe heat exchanger featuring six-gear turbulators. The nanoparticles with a concentration of 6% are added to the hot fluid and streams in the tube with the Re number 13,000, and pure water as the cold fluid passes through the shell with the Re number 500. Figure 11.7a shows the temperature distribution in sections from $Z = 0.1\,m$ to $Z = 0.7\,m$, and Figure 11.7b provides an enlarged view of the temperature distribution at $Z = 0.6\,m$. Also, temperature distribution along the heat exchanger is presented in Figure 11.7c. As revealed in the temperature contours, when the fluid flows through the heat exchanger, the cold fluid is impacted by the hot fluid, resulting in an increase in its temperature.

Figure 11.7d illustrates velocity distributions in sections from $Z = 0.1\,m$ to $Z = 0.7\,m$, and Figure 11.7e provides an enlarged view of the velocity distribution in $Z = 0.6\,m$. Also, velocity distribution along the heat exchanger is presented in Figure 11.7f. A part of Figure 11.7f is selected, and velocity contour and streamlines are shown in the enlarged view in Figure 11.7g. The streamlines and velocity distribution (Figure 11.7g) show that the collision of the fluid flow with the turbulator leads to breaking the boundary layer and forming the vortex flow behind the turbulator. The impinging flow on the gear turbulators can recover heat from them and transfer to the mainstream by turning its direction with the turbulators. In summary, turbulators

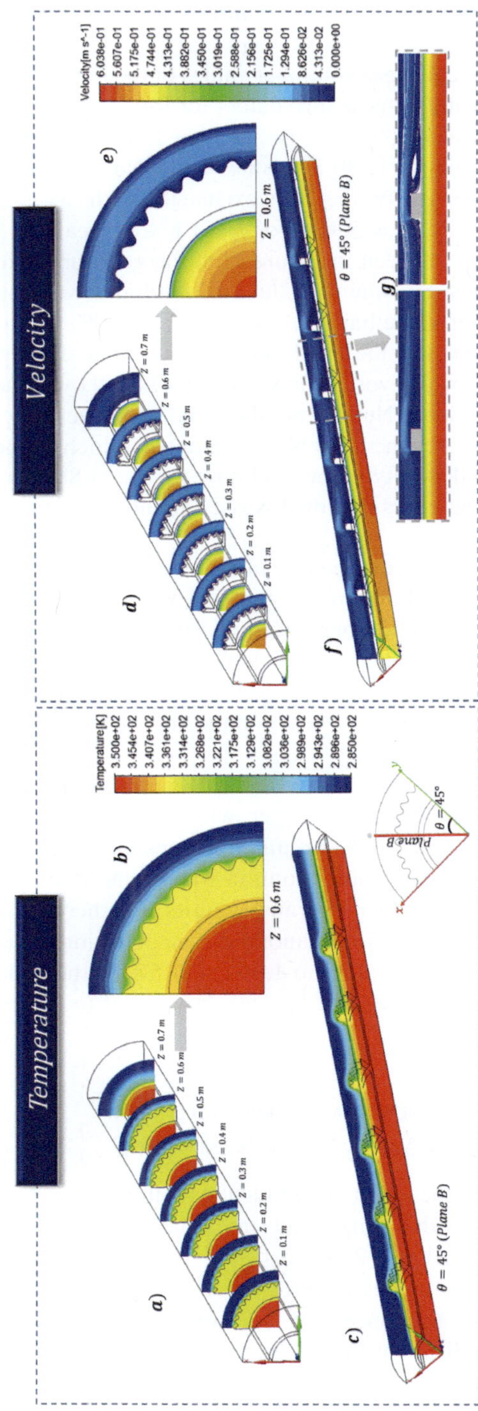

FIGURE 11.7 Temperature and velocity contours of the heat exchanger with six-gear turbulators.

improve heat transfer (Figure 11.7c) by breaking the boundary layer (Figure 11.7g) and increasing the heat transfer surface.

Figure 11.8a shows local Nusselt number values of the inner side versus the length of the heat exchanger with six-gear turbulators for $Re_{tube} = 13,000$ and $Re_{shell} = 500$ for volume concentrations of 0%, 1%, 4%, and 6% to reveal the influence of nanofluid on heat transfer. Figure 11.8b illustrates local Nusselt number values of the inner side versus the length of the heat exchanger with six-gear turbulators for $3,000 \leq Re_{tube} \leq 13,000$, $Re_{shell} = 500$, and $\varphi = 1\%$ to reveal the influence of Re number on heat transfer. As seen in Figure 11.8, the gear turbulator has effectively affected the flow physics. So, when the fluid flow collides with the gear turbulator, the local Nusselt number value has increased by about 70% due to the breaking of the boundary layer. However, the local Nusselt number is slightly reduced behind the turbulator because of the vortex flow. Figure 11.8a shows that the addition of nanofluid increases the local Nusselt number, and also, the increase in the nanofluid concentration increases the local Nusselt number, which is due to the improvement of the thermal conductivity of the fluid flow. Figure 11.8b shows that increasing the Reynolds number increases the local Nusselt number due to the increase in the kinetic energy of the flow.

Figure 11.9 shows the overall heat transfer coefficient (U) values of the heat exchanger with six-gear turbulators for hot flow Reynolds numbers in the range of 3,000–13,000 and nanofluid concentrations in the range of 0%–6%. It can be seen that the overall heat transfer coefficient increases with increasing Re number and nanofluid concentration. The overall heat transfer coefficient experiences an approximately 58% increase when comparing $Re = 13,000$ and $\varphi = 6\%$ to $Re = 3,000$ with pure water.

Figure 11.10 illustrates the effectiveness and number of transfer units of the parallel flow double-pipe heat exchanger with six-gear turbulators versus the Reynolds number of the inner tube for nanofluid concentrations of 0%, 1%, 4%, 6% with $Re_{shell} = 500$. As evident, using alumina-water nanofluid leads to an increase in effectiveness and the number of transfer units. Furthermore, these improvements are intensified by increasing nanofluid concentration. The results indicate that the ε and NTU ratios are equal to 1.66 and 1.57 for the heat exchanger with gear turbulators at $Re = 13,000$ and $\varphi = 6\%$ compared to the conventional heat exchanger at $Re = 3,000$.

11.5.2   ASSESSMENT OF SINUSOIDAL CORRUGATED HEAT EXCHANGER WITH AL_2O_3-WATER NANOFLUID

This section investigates the effect of turbulent nanofluid flow in a counterflow double-pipe heat exchanger with a simple and sinusoidal corrugated tube with three different wave amplitudes of $A = 1, 2, 3\,mm$.

Figure 11.11 shows the values of the overall heat transfer coefficient according to the wave amplitude in the range of 0–3 versus the concentration of nanofluid in the range of 0%–6% in different Reynolds numbers of the inner tube in the range of 3,000–13,000. With a cursory glance, it can be figured out that increasing the wave

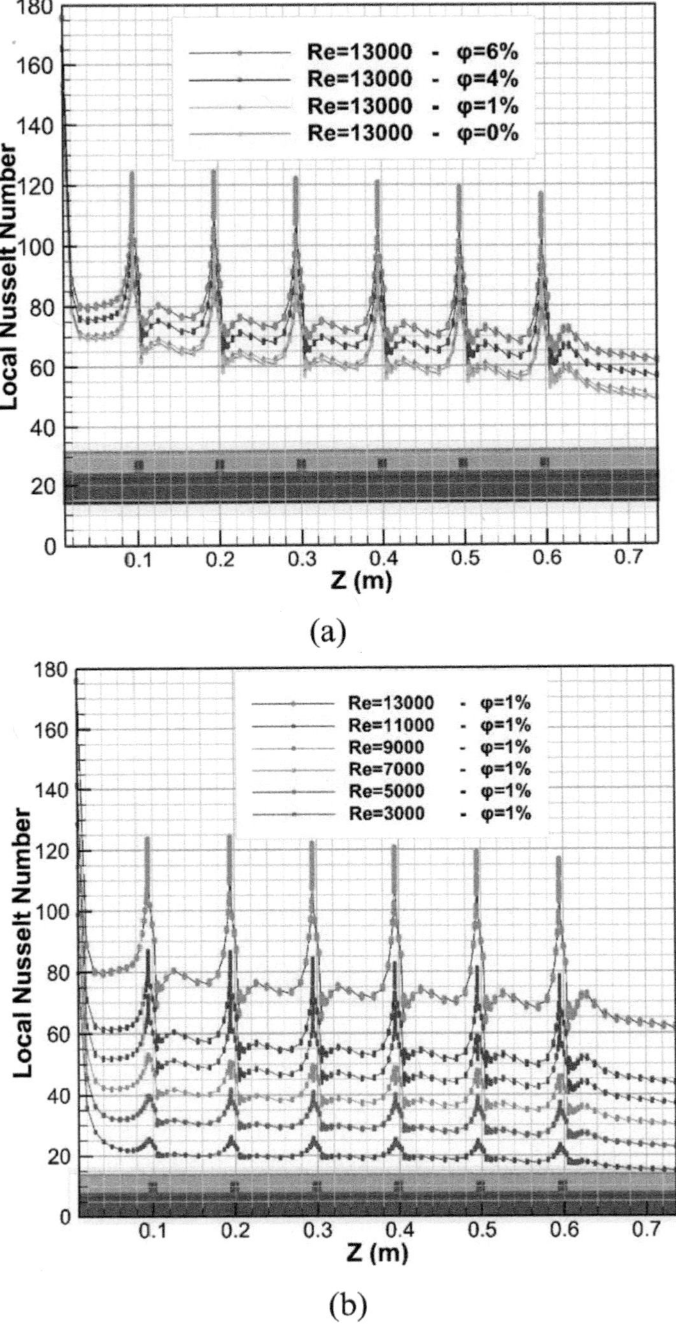

FIGURE 11.8 Local Nusselt number versus the length of the heat exchanger for (a) different nanofluid concentrations and (b) different Re numbers of the inner flow.

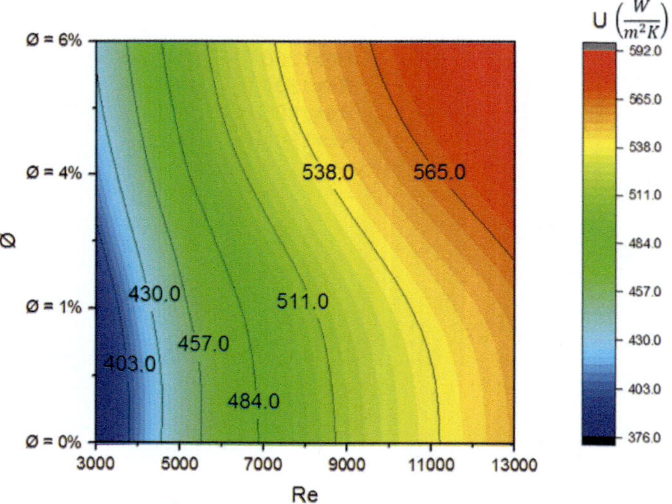

FIGURE 11.9 Overall heat transfer coefficient of the heat exchanger with gear turbulators.

amplitude and nanofluid concentration increases the overall heat transfer coefficient. Also, increasing the Reynolds number increases the overall heat transfer coefficient. The maximum value of the overall heat transfer coefficient is for a corrugated heat exchanger with the amplitude of $A = 3$ and $\varphi = 6\%$, which is equal to $1,125\dfrac{W}{m^2 k}$. The results show that corrugating the inner tube and adding nanoparticles enhance the overall heat transfer coefficient of the heat exchanger, in which the values of $\dfrac{U_{c,\,\varphi=6\%}}{U_{s,\,\varphi=0\%}}$ for $A = 1$ are 1.61–1.69, for $A = 2$ are 1.73–1.87, and for $A = 3$ are 1.76–1.89.

The first and second rows of Figure 11.12 show the values of effectiveness (ε) and number of transfer units (NTU) versus the Reynolds number of the internal flow of the heat exchanger. The first to third columns of this figure display the ε and NTU values for wavelengths of 1, 2, and 3 mm for the simple heat exchanger with water and the corrugated heat exchanger with different concentrations of 0%, 1%, 4%, and 6%. The last column shows the effect of wave amplitudes on ε and NTU at a constant nanofluid concentration of 4%.

As can be seen in the first row of Figure 11.12, the corrugation of the tube increases the effectiveness of the heat exchanger in each wave amplitude in comparison with the simple one with pure water. The maximum value of $\dfrac{\varepsilon_{c,\varphi=0\%}}{\varepsilon_{s,\varphi=0\%}}$ for $A = 1$, 2, 3 is equal to 1.40, 1.58, and 1.72 at the same Re number condition, respectively. Additionally, adding nanoparticles increases the effectiveness in the corrugated heat exchanger, in which the maximum value of $\dfrac{\varepsilon_{c,\varphi=x\%}}{\varepsilon_{c,\varphi=0\%}}$ for $x = 1, 4, 6$ is equal to 1.04, 1.20, and 1.33 for wave amplitude of 2 mm and at the same Re number condition,

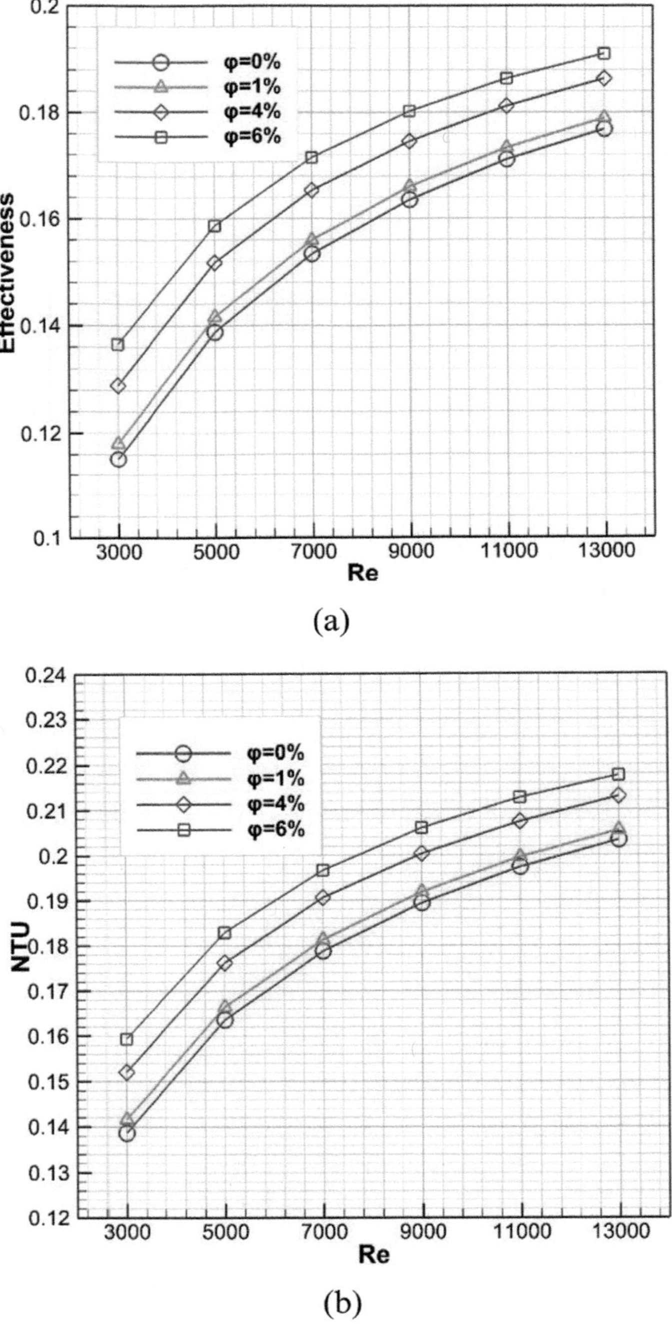

FIGURE 11.10 The values of (a) effectiveness and (b) number of transfer units for the heat exchanger with six gear turbulators versus the Re number of the inner tube.

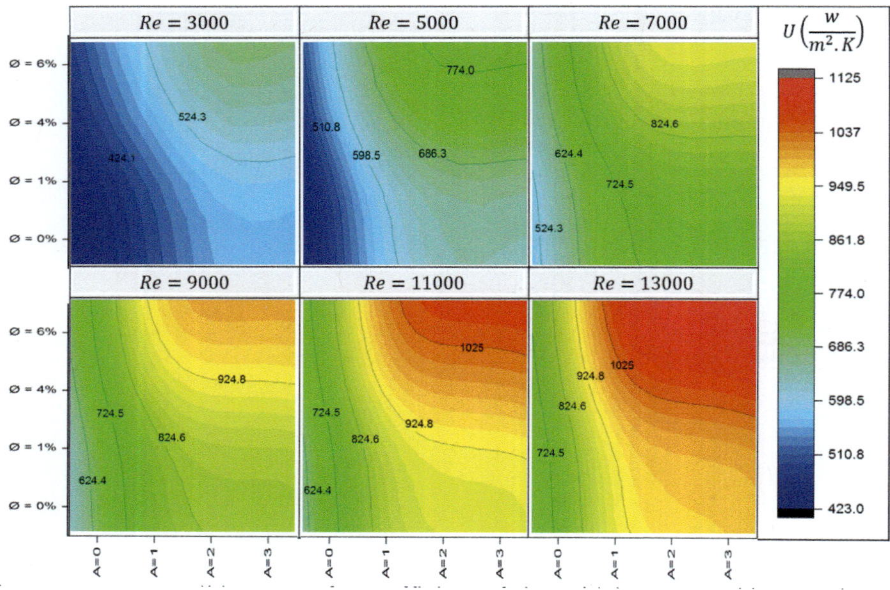

FIGURE 11.11   Overall heat transfer coefficient of sinusoidal corrugated heat exchanger.

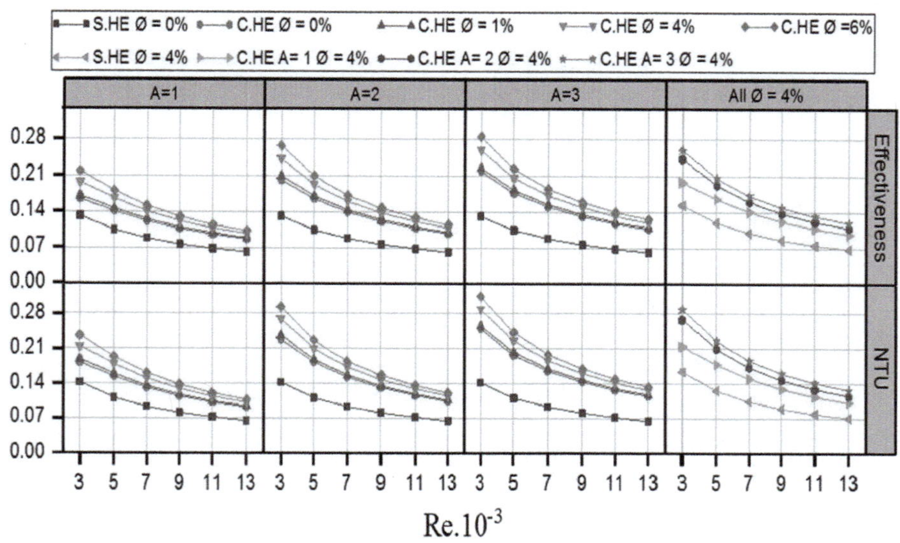

FIGURE 11.12   Effectiveness and number of transfer units of the corrugated heat exchanger.

respectively. In total, the results show that corrugating the inner tube and adding nanoparticles enhance the effectiveness of the heat exchanger, in which the values of $\dfrac{\varepsilon_{c,\,\varphi=6\%}}{\varepsilon_{s,\,\varphi=0\%}}$ for $A = 1$ are 1.66–1.73, for $A = 2$ are 1.86–2.03, and for $A = 3$ are 2.04–2.16 in the same Re number condition.

The second row of Figure 11.12 illustrates the enhancement of the number of transfer units by corrugating the heat exchanger and adding nanoparticles. So, increasing the wave amplitudes and nanofluid concentrations increases NTU. The results show that the values of $\dfrac{\mathrm{NTU}_{c,\,\varphi=6\%}}{\mathrm{NTU}_{s,\,\varphi=0\%}}$ for $A = 1$ are 1.65–1.74, for $A = 2$ are 1.88–2.06, and for $A = 3$ are 2.07–2.21. The last column shows the effect of wave amplitudes on NTU at a constant nanofluid concentration of 4%, which reveals that increasing wave amplitude increases the NTU.

11.6 CONCLUSIONS

The purpose of this chapter was to evaluate the effects of turbulent alumina-water nanofluid flow passing through enhanced double-pipe heat exchangers with two configurations. The first configuration was about the nanofluid flow passing through a parallel flow double-pipe heat exchanger with six-gear disc turbulators located in the shell side. The influence of the gear turbulators on flow physics was investigated by local Nusselt number, as well as velocity and temperature fields. Then, the heat exchanger was examined by ε-NTU method. The second configuration was about the nanofluid flow passing through a sinusoidal corrugated counterflow double-pipe heat exchanger, assuming three different wave amplitudes. The analysis of the heat exchanger involved the investigation of the effects of wavelengths and nanofluid concentrations using the ε-NTU method. The results revealed that:

- The gear disc turbulator efficiently enhanced heat transfer. The impingement of the fluid flow with the turbulator surface resulted in the destruction of the boundary layer, leading to an approximately 70 percent increase in the local Nusselt number.
- Using the gear disc turbulators and nanofluid increased the overall heat transfer coefficient by about 58 percent by increasing the heat transfer surface, breaking the boundary layer via turbulators, and improving the thermal conductivity of the fluid flow via nanofluid.
- Increasing the wave amplitude and nanofluid concentration increased U, ε, and NTU. The value of $\left[\dfrac{\varepsilon_{c,\varphi=0\%}}{\varepsilon_{s,\varphi=0\%}}\right]_{A=1,2,3}$ was equal to 1.40, 1.58, and 1.72, and the value of $\left[\dfrac{\varepsilon_{c,\varphi=x\%}}{\varepsilon_{c,\varphi=0\%}}\right]_{x=1,4,6}^{A=2}$ was equal to 1.04, 1.20, and 1.33.
- The values of $\dfrac{\varepsilon_{c,\,\varphi=6\%}}{\varepsilon_{s,\,\varphi=0}}$ for $A = 1$ were 1.66–1.73, for $A = 2$ were 1.86–2.03, and for $A = 3$ were 2.04–2.16 in the same Re number for sinusoidal corrugated double-pipe heat exchanger.

REFERENCES

1. Pak BC, Cho YI. Hydrodynamic and heat transfer study of dispersed fluids with submicron metallic oxide particles. *Exp Heat Transf* 1998;11:151–70. https://doi.org/10.1080/08916159808946559.
2. Bianco V, Chiacchio F, Manca O, Nardini S. Numerical investigation of nanofluids forced convection in circular tubes. *Appl Therm Eng* 2009;29:3632–42. https://doi.org/10.1016/j.applthermaleng.2009.06.019.
3. Bianco V, Manca O, Nardini S. Numerical simulation of water/ Al2O3 nanofluid turbulent convection. *Adv Mech Eng* 2010;2010. https://doi.org/10.1155/2010/976254.
4. Bianco V, Manca O, Nardini S. Numerical investigation on nanofluids turbulent convection heat transfer inside a circular tube. *Int J Therm Sci* 2011;50:341–9. https://doi.org/10.1016/j.ijthermalsci.2010.03.008.
5. Bahmani MH, Sheikhzadeh G, Zarringhalam M, Akbari OA, Alrashed AAAA, Shabani GAS, et al. Investigation of turbulent heat transfer and nanofluid flow in a double pipe heat exchanger. *Adv Powder Technol* 2018;29:273–82. https://doi.org/10.1016/j.apt.2017.11.013.
6. Milani Shirvan K, Mamourian M, Mirzakhanlari S, Ellahi R. Numerical investigation of heat exchanger effectiveness in a double pipe heat exchanger filled with nanofluid: A sensitivity analysis by response surface methodology. *Powder Technol* 2017;313:99–111. https://doi.org/10.1016/j.powtec.2017.02.065.
7. Azeez mohammed Hussein H, Zulkifli R, Faizal Bin Wan Mahmood WM, Ajeel RK. Structure parameters and designs and their impact on performance of different heat exchangers: A review. *Renew Sustain Energy Rev* 2022;154:111842. https://doi.org/10.1016/j.rser.2021.111842.
8. Aissa A, Qasem NAA, Mourad A, Laidoudi H, Younis O, Guedri K, et al. A review of the enhancement of solar thermal collectors using nanofluids and turbulators. *Appl Therm Eng* 2023;220:119663. https://doi.org/10.1016/j.applthermaleng.2022.119663.
9. Yadav S, Sahu SK. Heat transfer augmentation in double pipe water to air counter flow heat exchanger with helical surface disc turbulators. *Chem Eng Process-Process Intensif* 2019;135:120–32. Elsevier B.V. https://doi.org/10.1016/j.cep.2018.11.018.
10. Yadav S, Sahu SK. Effect of helical surface disc turbulators on heat transfer and friction factor characteristics in the annuli of a double-pipe heat exchanger. *Chem Eng Technol* 2019;42:1205–13. https://doi.org/10.1002/ceat.201800251.
11. Yadav S, Sahu SK. Effect of helical surface disc turbulator on thermal and friction characteristics of double pipe water to air heat exchanger. *Exp Heat Transf* 2020;6152. https://doi.org/10.1080/08916152.2020.1714819.
12. Yadav S, Paulraj MP, Sahu SK. Thermal performance of air to water heat exchanger with plain surface disc turbulators: Experimental and numerical study. *Heat Mass Transf/Waerme- Und Stoffuebertragung* 2020;56:1777–91. https://doi.org/10.1007/s00231-020-02824-x.
13. Bashtani I, Esfahani JA, Kim KC. Hybrid CFD-ANN approach for evaluation of bio-inspired dolphins dorsal fin turbulators of heat exchanger in turbulent flow. *Appl Therm Eng* 2023;219:119422. https://doi.org/10.1016/j.applthermaleng.2022.119422.
14. Tafarroj MM, Zarabian Ghaeini G, Esfahani JA, Kim KC. Multi-purpose prediction of the various edge cut twisted tape insert characteristics: Multilayer perceptron network modeling. *J Therm Anal Calorim* 2021;145:2005–20. https://doi.org/10.1007/s10973-021-10904-1.
15. Darzi AAR, Farhadi M, Sedighi K, Shafaghat R, Zabihi K. Experimental investigation of turbulent heat transfer and flow characteristics of SiO 2/water nanofluid within helically corrugated tubes. *Int Commun Heat Mass Transf* 2012;39:1425–34. https://doi.org/10.1016/j.icheatmasstransfer.2012.07.027.

16. Rabienataj Darzi AA, Farhadi M, Sedighi K. Experimental investigation of convective heat transfer and friction factor of Al2O3/water nanofluid in helically corrugated tube. *Exp Therm Fluid Sci* 2014;57:188–99. https://doi.org/10.1016/j.expthermflusci.2014.04.024.

17. Darzi AAR, Farhadi M, Sedighi K, Aallahyari S, Delavar MA. Turbulent heat transfer of Al2O3-water nanofluid inside helically corrugated tubes: Numerical study. *Int Commun Heat Mass Transf* 2013;41:68–75. https://doi.org/10.1016/j.icheatmasstransfer.2012.11.006.

18. Qi C, Wan YL, Li CY, Han DT, Rao ZH. Experimental and numerical research on the flow and heat transfer characteristics of TiO2-water nanofluids in a corrugated tube. *Int J Heat Mass Transf* 2017;115:1072–84. https://doi.org/10.1016/j.ijheatmasstransfer.2017.08.098.

19. Ekrani SM, Ganjehzadeh S, Esfahani JA. Multi-objective optimization of a tubular heat exchanger enhanced with delta winglet vortex generator and nanofluid using a hybrid CFD-SVR method. *Int J Therm Sci* 2023;186. https://doi.org/10.1016/j.ijthermalsci.2023.108141.

20. Adibi O, Rashidi S, Abolfazli Esfahani J. Effects of perforated anchors on heat transfer intensification of turbulence nanofluid flow in a pipe. *J Therm Anal Calorim* 2020;141:2047–59. https://doi.org/10.1007/s10973-020-09705-9.

21. Naphon P, Wiriyasart S. Pulsating flow and magnetic field effects on the convective heat transfer of TiO2-water nanofluids in helically corrugated tube. *Int J Heat Mass Transf* 2018;125:1054–60. https://doi.org/10.1016/j.ijheatmasstransfer.2018.05.015.

22. Bartwal P, Upreti H, Pandey AK, Joshi N, Joshi BP. Application of modified Fourier's law in a fuzzy environment to explore the tangent hyperbolic fluid flow over a non-flat stretched sheet using the LWCM approach. *Int Commun Heat Mass Transf* 2024;153. https://doi.org/10.1016/j.icheatmasstransfer.2024.107332.

23. Gupta T, Pandey AK, Kumar M. Numerical study for temperature-dependent viscosity based unsteady flow of GP-MoS2/C2H6O2-H2O over a porous stretching sheet. *Numer Heat Transf A Appl* 2023. https://doi.org/10.1080/10407782.2023.2195689.

24. Upreti H, Bartwal P, Pandey AK, Makinde OD. Heat transfer assessment for Au-blood nanofluid flow in Darcy-Forchheimer porous medium using induced magnetic field and Cattaneo-Christov model. *Numer Heat Transf B Fund* 2023;84:415–31. https://doi.org/10.1080/10407790.2023.2209281.

25. Upreti H, Mishra SR, Pandey AK, Bartwal P. Shape factor analysis in stagnation point flow of Casson nanofluid over a stretching/shrinking sheet using Cattaneo-Christov model. *Numer Heat Transf B Fund* 2023. https://doi.org/10.1080/10407790.2023.2265555.

26. Upreti H, Pandey AK, Joshi N, Makinde OD. Thermodynamics and heat transfer analysis of magnetized casson hybrid nanofluid flow via a Riga plate with thermal radiation. *J Comput Biophys Chem* 2023;22:321–34. https://doi.org/10.1142/S2737416523400070.

27. Pandey AK, Upreti H, Uddin Z. Magnetic SWCNT-Ag/H2O nanofluid flow over cone with volumetric heat generation. *Int J Mod Phys B* 2023;37. https://doi.org/10.1142/S0217979223502533.

28. Bartwal P, Upreti H, Mishra SR, Pandey AK. Exploring the features of Von-Karman flow of tangent hyperbolic fluid over a radially stretching disk subject to heating due to porous media and viscous heating. *Mod Phys Lett B* 2023. https://doi.org/10.1142/S0217984924501355.

29. Jin ZJ, Liu BZ, Chen FQ, Gao ZX, Gao XF, Qian JY. CFD analysis on flow resistance characteristics of six-start spirally corrugated tube. *Int J Heat Mass Transf* 2016;103:1198–207. https://doi.org/10.1016/j.ijheatmasstransfer.2016.08.070.

30. Jin ZJ, Chen FQ, Gao ZX, Gao XF, Qian JY. Effects of pitch and corrugation depth on heat transfer characteristics in six-start spirally corrugated tube. *Int J Heat Mass Transf* 2017;108:1011–25. https://doi.org/10.1016/j.ijheatmasstransfer.2016.12.091.

31. Balla HH. Enhancement of heat transfer in six-start spirally corrugated tubes. *Case Stud Therm Eng* 2017;9:79–89. https://doi.org/10.1016/j.csite.2017.01.001.
32. Zhang D, Tao H, Xu Y, Sun Z. Numerical investigation on flow and heat transfer characteristics of corrugated tubes with non-uniform corrugation in turbulent flow. *Chin J Chem Eng* 2018;26:437–44. https://doi.org/10.1016/j.cjche.2017.03.038.
33. Wang W, Zhang Y, Li B, Li Y. Numerical investigation of tube-side fully developed turbulent flow and heat transfer in outward corrugated tubes. *Int J Heat Mass Transf* 2018;116:115–26. https://doi.org/10.1016/j.ijheatmasstransfer.2017.09.003.
34. Sun M, Zeng M. Investigation on turbulent flow and heat transfer characteristics and technical economy of corrugated tube. *Appl Therm Eng* 2018;129:1–11. https://doi.org/10.1016/j.applthermaleng.2017.09.136.
35. Bashtani I, Esfahani JA. ε-NTU analysis of turbulent flow in a corrugated double pipe heat exchanger: A numerical investigation. *Appl Therm Eng* 2019;159. https://doi.org/10.1016/j.applthermaleng.2019.113886.
36. White FM, Majdalani J. *Viscous Fluid Flow*. New York: McGraw-Hill; 2006.
37. Menter FR. Two-equation eddy-viscosity turbulence models for engineering applications. *AIAA J* 1994;32:1598–605. https://doi.org/10.2514/3.12149.
38. Incropera FP, DeWitt DP. *Fundamentals of Heat and Mass Transfer*. New York: John Wiley & Sons, Inc.; 1996. https://doi.org/10.1016/j.applthermaleng.2011.03.022.
39. Maïga SEB, Nguyen CT, Galanis N, Roy G. Heat transfer behaviours of nanofluids in a uniformly heated tube. *Superlattices Microstruct* 2004;35:543–57. https://doi.org/10.1016/j.spmi.2003.09.012.
40. El Bécaye Maïga S, Palm SJ, Nguyen CT, Roy G, Galanis N. Heat transfer enhancement by using nanofluids in forced convection flows. *Int J Heat Fluid Flow* 2005;26:530–46. https://doi.org/10.1016/j.ijheatfluidflow.2005.02.004.
41. Maïga SEB, Nguyen CT, Galanis N, Roy G, Maré T, Coqueux M. Heat transfer enhancement in turbulent tube flow using Al2O 3 nanoparticle suspension. *Int J Numer Methods Heat Fluid Flow* 2006;16:275–92. https://doi.org/10.1108/09615530610649717.
42. Kakac S, Liu H, Pramuanjaroenkij A. *Heat Exchangers: Selection, Rating, and Thermal Design*. Boca Raton, FL: CRC Press; 2020.
43. Ruengpayungsak K, Wongcharee K, Thianpong C, Pimsarn M, Chuwattanakul V, Eiamsa-ard S. Heat transfer evaluation of turbulent flows through gear-ring elements. *Appl Therm Eng* 2017;123:991–1005. https://doi.org/10.1016/j.applthermaleng.2017.05.108.

Part III

Sustainable Future Approaches

12 Applications of Ionic Liquid-Based Electrolytes in Batteries

Bibi Shaguftah Khatoon, Himanshu Arora,
Ritika Gera, Md Abrar Siddiquee,
Manoj Kumar Banjare, Kamalakanta Behera,
and Kamal Nayan Sharma

12.1 INTRODUCTIONS

12.1.1 IONIC LIQUIDS

Ionic liquids (ILs) are molten ionic salts that consist of cations and anions; cations are generally found in organic form, while anions are found in both organic and inorganic forms and have a melting point below 100°C [1,2]. ILs are entirely composed of ions, and they have appealing and distinctive qualities such as little vapor pressure, high ionic conductivities, thermal stability (about 300°C), solubility in water, non-flammability, and hydrophobicity. ILs have a large number of fascinating characteristics [3,4]. Thus, during the past several years, ILs have drawn a lot of attention in a variety of academic fields. The strong ion pair interaction, which is absent in ordinary salts, is one of the other significant characteristics linked to ILs.

12.1.2 HISTORY OF IONIC LIQUIDS

The discovery of ethyl ammonium nitrate ([EtNH$_3$][NO$_3$], melting point approx. 13°C–14°C, which was produced by a combination of ethylamine with strong nitric acid, marked the beginning of the field of ionic liquids by Paul Walden in 1914 [5]. Ionic liquids containing halide or tetra halogen aluminate anions and imidazolium and pyridinium cations were developed as potential battery electrolytes in the 1970s and 1980s [6]. This initial IL was not widely utilized due to its air sensitivity and moisture resistance, which made handling difficult. This changed in 1992, when Wilkes and Zaworotko reported the first set of imidazolium-based ILs that were both air- and moisture-stable [7]. After that, ILs have been extensively utilized in various fields including biotechnology and materials chemistry due to a variety of advantageous characteristics. Tunability is one of the most unique properties of IL which occurs from changes in the solvent's properties, like conductivity, viscosity, and miscibility in water [8]. They increased in popularity due to their simplicity in the synthesis and the fact that they were mostly made of accessible, affordable, and

DOI: 10.1201/9781003473749-15

frequently biocompatible materials. In the late 1990s and early 2000s, ILs were used in biomedical applications to increase the thermostability of enzymes and model proteins, as well as enzymatic catalytic performance. In addition, ILs were utilized in the 2000s for the production of anti-cancer drugs and as antibacterial agents [9,10].

12.1.3 STRUCTURE AND PROPERTIES OF IONIC LIQUIDS

Structurally, IL consists of complete ions. It is a combination of organic cation and organic or inorganic anion. The commonly studied structures having bulky asymmetrically substituted N-alkyl cation of variable heteroatom are associated with a charge-diffuse anion in the majority of systems that have been examined. Numerous kinds of cations have been studied so far, including tetraalkylammonium, N-alkyl pyridinium, N-methyl-N-alkyl pyrrolidinium, and 1-alkyl-3-methylimidazolium. Numerous inorganic or organic counter ions including $[(CF_3SO_2)2N]$, $[SCN]$, $[CF_3SO_3]$, $[BF_4]^-$, $[(C_2F_5)_3PF_3]$, $[PF_6]^-$, $[(C_2F_5SO_2)_2N]$, $[NiCl_3]$, $[CF_3CO_2]$, and $[(CN)_2N]$ have been utilized as anionic species. A charge-diffuse anion and a developing more alkyl-substituted cation can be used to create a more lipophilic ionic liquid. Nowadays, a broad spectrum of semi-organic salts having low melting properties or salt mixtures with a broad liquid spectrum are referred to as ILs, or we can say that an ionic liquid is a broad category of organic salts or salt mixtures having a vast liquid range and a low melting point.

ILs are distinguished from other salts, such as sodium chloride, because of their tightly packed structure and robust ionic connections, which only cause them to melt at extremely high temperatures (>800°C). The bulky and asymmetrical structure of the constituent ions, which can be combined in many ways to adjust their thermophysical properties, is thought to be the cause of an IL's lower melting point [11,12]. Figure 12.1 displays some typical cations and anions of ILs.

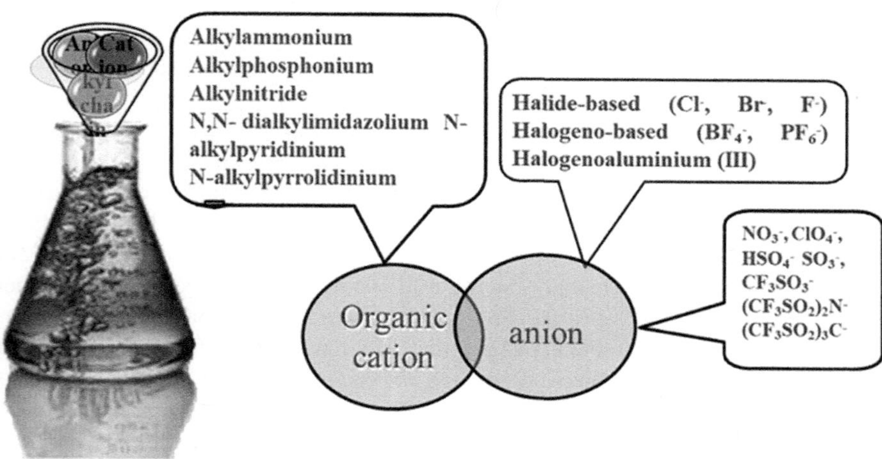

FIGURE 12.1 Some common ionic liquids of different cations and anions.

ILs carry exceptional features such as better chemical and thermal stability, seemingly low vapor pressure, and higher solvation ability with organic or inorganic compounds [10–12]. The physicochemical properties of ILs could be well modified by choosing various amalgamations of ions like cations, anions, and their substituents to accomplish a specific need. Non-flammability, non-volatility, and better activity as solvents than conventional organic solvents make ILs environment-friendly and desirable for a spectrum of biological, chemical, and industrial uses; therefore, it is also referred as "green solvents" [1]. The self-organization property of ILs to form micellar aggregates or micellar systems can dissolve several compounds/drugs that are either poorly soluble or insoluble in common organic solvents. Furthermore, it assists in the transport of these compounds/drugs to the target site with enhanced bioavailability by lowering drug degradation. ILs have unique solvent properties that allow for customized and energy-efficient bioprocessing, making them promising for biocatalysis applications. ILs are gaining popularity as solvents for biocatalysis, extraction, and electrochemistry due to their high adjustable properties [4–6].

12.1.4 APPLICATIONS OF IONIC LIQUIDS

Environmentally friendly ILs have exceptional properties and are used in a wide range of fields, including the pharmaceutical industry, drug delivery in biomedical science, chemical synthesis, extractions, analytical separation, catalysis, chromatography as a stationary phase, separation techniques, synthesis of nanoparticles as templates, membrane preparation, and the creation of high conductivity materials, and several energy-related applications due to their distinctive set of characteristics [13–18].

ILs have the potential to replace organic solvents as effective reaction media [19,20]. Focus started to include several other sectors as well a decade after they were initially produced. ILs were used as formulation excipients and controlled release systems for small compounds with weak water solubility [21,22]. The essential molecular characteristics of ILs make them distinct and very effective substances for pharmaceutical applications. For a biomolecule to function at its best in a variety of situations, such as biocatalysis, drug transport, therapies, and biosensing, the solvent environment in which it is found is essential. Therefore, ILs have gained popularity among researchers due to several benefits as substitute solvent systems for biomolecules [23]. The fundamental chemical characteristics that make ILs distinct and useful substances have played a significant role in their expansion into various sectors across the years.

Since the first-known IL-based method was commercialized in 1996, its application has increased significantly. This section aims to describe many processes that have been used on a commercial or pilot basis. ILs are excellent prospects for several energy-related applications due to their distinctive set of characteristics. It is now feasible to design suitable electrolytes for dye-sensitized solar cells, batteries, super-capacitors, actuators, and thermo electrochemical cells because of cation-anion combinations that exhibit low volatility, good thermal and electrochemical stability, and ionic conductivity. Protic IL family members have been employed for generating some of the most effective water oxidation catalysts for hydrogen production by water splitting, and some of them have also been used to co-catalyze a rare and very energy-efficient water oxidation process. The ability of protic ionic liquid electrolytes to conduct protons well makes it possible for fuel cells to operate at optimal temperature. Due to their low vapor pressure, tunable functionality, and additional applications outside of electrochemistry, these liquids are ideal for use as carbon dioxide absorbent materials for post-combustion capture of carbon dioxide. Moreover, a crucial feature in electrochemical applications is their innate ionic conductivity, which may make them suitable as solvents and electrolytes.

ILs are influencing many different electrochemical techniques, including fuel cell technology, water splitting, solar cell dye-sensitization, actuators, developed batteries, and thermo-cells. They are also influencing non-electrochemical fields like the capture of carbon and novel uses for thermal power storage. Lithium ion batteries and other high-energy electrochemical systems are ideal applications for both anion and cation forms of ILs due to their exceptional stability in electrolysis. It was then shown that cyclic pyrrolidinium and piperidinium [NTf$_2$] ILs, which are generally less viscous and a bit more cathodically stable than the aliphatic cations, provided the high-efficiency lithium cycling [24]. Even while ILs for lithium batteries have been the subject of most research, it's probable that their application as a solvent in the development of inorganic electrode components will be their most remarkable and significant contribution [24]. Some of the major energy application areas of ILs are depicted in Figure 12.2.

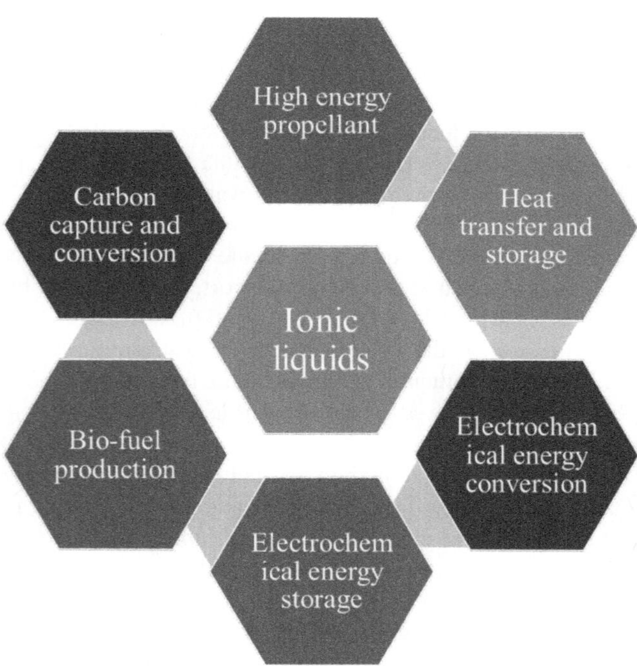

FIGURE 12.2 Some application areas of ILs.

12.2   IONIC LIQUIDS IN ENERGY STORAGE DEVICES

Energy storage plays a vital role in our society. There are numerous approaches to energy storage, such as batteries, that play a significant part in everyday life due to their high energy density, environmental sustainability, and affordability. However, significant performance and safety concerns remain unresolved. This chapter provides an overview of ILs and the electrode materials used in Li-ion battery systems, including representative elements of group-1 or alkali metals, p-block elements, and inner-transition metals, as well as their applications and attributes. The use of ILs as alloys, transition metal oxides, and GO, i.e., graphene oxide, is discussed. This chapter covers each component's advantages and disadvantages as well as the derived compounds and blends of ILs for use in next-generation battery systems. This can open doors for sustainable, affordable, flexible, and environmentally friendly energy storage technologies. Alkali metals and p-block elements form strong covalent bonds with highly electronegative atoms like O and F. The outermost electrons of p orbitals form both σ and π-bonds, resulting in several inorganic polymers. B, C, and N can create nanomaterials in one, two, and three dimensions, such as fullerene, graphene, and carbon nanotubes. The nanotechnology revolution has led to the rapid growth of this group of nanomaterials. Oxygen is a powerful oxidant that is ubiquitous in nature, and it is required for the synthesis of transition metal oxides (TMOs) that can be employed as significant material in electrode application in Lithium-ion batteries. Further, it is used to build super-capacitors, sensors, and other innovative energy technologies [25]. ILs have been discovered to use as phase-change materials for the transfer of thermal energy and preservation, and electrolytes for storage of electrochemical energy and conversion. Additionally, it is employed in highly energetic propellants for aerospace use, bio-fuel extraction, and treatment of biomass and can be used as catalysts for CO_2 collection and conversion. ILs have shown tremendous potential for various energy applications, such as fuel cells, batteries, solar cells, and super-capacitors [25].

12.2.1   IONIC LIQUIDS APPLICATION IN SOLAR CELLS

Solar cells are rapidly advancing as a viable energy-harvesting medium. Within a short period of time, it has attained power conversion efficiency equivalent to other established technologies on the market. Unfortunately, solar cells exhibit instability in real-time operating settings, posing a barrier to commercialization [26]. Various strategies and engineering have been used to address this issue, and ILs have recently attracted the interest of researchers. ILs offer unique and diverse features, including high ionic conductivity, thermal stability, and electrochemical stability, making them appropriate for use in solar cells. These features make ILs appropriate for application in perovskite solar cells, and ILs were employed as a solvent or additive in perovskite precursor. ILs-assisted development of perovskite films leads to enhanced shape and crystallinity, and it can prevent power conversion efficiency from increasing proportionally. The residual IL in the film hinders charge dissociation and carrier movement. When ILs are utilized as a solvent or additive in a precursor solution, their polar nature allows for homogenous and uniform perovskite film production.

Additionally, the interaction of IL ions with perovskite precursors slowed down crystal formation. The slower reaction rate resulted in improved perovskite crystal and film quality, with fewer flaws, pinholes, and surface roughness.

12.2.2 IONIC LIQUIDS IN MODERN BATTERY TECHNOLOGY

The fascinating properties of ILs, including high thermochemical stability and high electrical conductivity of various anion/cation types, make them ideal for use in highly energetic electrochemical gadgets like lithium-ion batteries. The major advantage of electrolytes is their low volatility and flammability, which can improve safety and stability.

12.2.2.1 ILs in Lithium-Ion Batteries

Goodenough and Park's [27] latest viewpoint essay provides a comprehensive overview of lithium-ion rechargeable batteries and their evolution. A comprehensive overview of Li-ion rechargeable batteries and their evolution has been represented by Bashir et al. [26]. Some ILs like piperidinium- and cyclic pyrrolidinium–based ILs are low in viscosity and somewhat highly catholically stable as compared to aliphatic cations, which are found to be very efficient for lithium cycling. Several studies have reported examining how IL cations and anions interact with various lithium battery anodes and cathodes [26]. Addition of IL indicates that this smaller can facilitate faster transport rates. This anion may play a role in generating a surface layer on graphitic carbon electrodes, as it eliminates the requirement for chemicals to prevent irreversible IL cation contact. Small phosphonium cations have been used in several applications, including high-voltage cathodes. This particular class of ILs has recently shown dependable cycling in cells with high-capacity and high-voltage electrode materials like polyaniline, $LiV-PO_4F$, and sulfur silicon. The destiny and speciation of dissolved lithium ions in ILs electrolytes for lithium batteries are crucial for ion transport and electrode surface interactions. Designing cation and anion functionality, as well as using additives and diluents, can affect electrolyte characteristics. Protective surface layers, also known as the solid electrolyte interphase (SEI), are essential for charge transfer, electrochemical stability, and ion transport While much research on ILs for Li batteries has focused on their usage as electrolytes, it's possible that their most significant contribution will be as solvent for the production of electrode materials. Ion thermal synthetic methods have revolutionized the synthesis of inorganic materials, opening new possibilities. Lower temperature processes and the capacity of ILs to stabilize metastable phases open new cost-effective energy materials that were before unavailable. The need for high-demand energy storage material in energy device installations will drive the development of IL electrolytes. Safety, robustness, long life, and capacity to operate at high temperatures will take precedence over high energy density requirements. ILs have the potential to address current technology's inadequacies due to their inherent stability. This presents an ideal opportunity. The feasibility of IL electrolytes has been shown by recent attempts to build prototype energy sources for this kind of application, especially in setups with photovoltaic cells where their thermal stability is essential and rate limits are not very significant.

For the most popular ILs, Lane et al. carried out a thorough investigation of elimination processes at negative electrodes. Li ion transport and electrode interactions are influenced by several factors including cation and anion combination, variation, stability, breakdown products, and interfacial properties. However, how the addition of Li ions results in a rise of the cathodic reduction limit is yet unresolved.

The significance of the chemical decomposition of ILs ingredients to generate cation-anion interfacial structure, as well as speciation to remove reactive compounds, remains unclear. The demand for large energy storage in renewable energy installations will drive the development of IL electrolytes. The inherent stability of ILs presents a unique opportunity to address current technology's inadequacies. Recent efforts to construct batteries and their application have focused on the viability of ionic liquid electrolytes, particularly in integration in solar panels where heat and electrical stability of ILs are crucial, and velocity constraints are not significant.

Although these ILs have been used in numerous Li-cell arrangements having different capacities of power and energy, since then no industrial energy storage devices with IL as electrolytes have evolved so far. The high cost of ILs compared to traditional carbonate solvents, especially for fluorinated amide anions, may be a contributing factor. If the expense of IL as electrolytes at the moment prevents the development of commercially available devices, certain measures to reduce costs or boost value are needed. The primary approach is to continue researching and producing ecofriendly cations and anions, as proven by Li cells that contain pyrrolidinium dicyanamide IL. The intrinsic stability of ILs opens the possibility of developing batteries that can withstand high charge-discharge cycles, even in extreme temperatures. Devices with stable intercalation electrode materials ($Li_4Ti_5O_{12}$ and $LiFePO_4$) have shown high cycle stability, as evidenced by prototypes. The lower capability rate of IL-based design batteries is mostly due to the higher viscosity of the electrolyte, especially with the addition of more Li salts, which leads to stronger ion interaction. Recently, the bis(fluorosulfonyl)imide anion family of ILs has made significant progress in addressing this issue. Li-ion concentrations can support high charge-discharge rates, but this comes at a cost of lower stability. Further research is needed to address concerns concerning the thermal stability and safety of these anion-based ILs [27]. Recent studies have explored blending ILs with carbonate-based electrolytes to address cost and rate limitations, especially at low temperatures. The incorporation of ILs increases the electrolyte's ionic strength, affecting ion dynamics, interfacial stability, and flammability [28,29]. Research is exploring high-capacity electrode materials like Lithium-air and Lithium-sulfur rechargeable batteries to overcome constraints in current Li-ion systems. $Li-O_2$ batteries face challenges in the decomposition of electrolyte at the cathode surface and require highly efficient performance. Anode lithium cycling requires a large amount of work in the formulation of the electrolyte [29]. Mizuno et al. evaluated electrolyte stability by calculating Mulliken charges for individual atoms. By employing an ether-functionalized cation (N, N-diethyl-N-methyl-N-(2-methoxyethyl)), a piperidinium [NTf$_2$] IL shows enhanced efficiency and reduced cathodic polarization as compared to a carbonate electrolyte, demonstrating its stability toward the O_2 radical anion [30].

Allen et al. found a significant correlation between the ionic charge density of the IL cation and the products of the oxygen reduction reaction. The Lewis acidity of the cation in the electrolyte provided an explanation for this, showing that the word "SO" IL cations may stabilize the anion of superoxide in the presence of strong acids[31]. Safronov et al. investigated the kinetics of electrochemical lithium intercalation and deintercalation at various currents in composite materials made of lithium iron phosphate and lithium titanate with fine carbon particles. The study found that lithium intercalation and deintercalation in electrode materials are characterized by an overvoltage lithium iron phosphate cell. Another, Vijayaraghavan et al. studied the relationship between ionic liquids accelerated rate calorimetric (ARC) at elevated temperatures and examine the exothermic behaviour and intrinsic safety of ionic liquids for battery and solar cell applications. The ARC results indicate that pyrrolidinium-based ionic liquids have more exothermic activity than imidazolium ionic liquids, which do not exhibit any exothermic activity[32,33].

12.2.2.2 ILs in Sodium–Ion Batteries

Researchers are exploring alternative anodic metals like sodium (Na), magnesium (Mg), and zinc (Zn) due to the sustainability and cost of Li supplies. These metals are environmentally and eco-friendly and safer than Li. Demand for lithium will rise due to the growing use of lithium-ion batteries and electric automobiles for static energy storage applications including load balancing and variable renewable electricity generation. While they might not have the same energy density as lithium, other battery technologies like Na-ion, Zn-air, and Mg-air might be appropriate for specific uses. Electrolytes of IL are appropriate for these kinds of devices due to their less volatility, greater electrochemical window, and strong thermal stability, and more advantages highlighted in Figure 12.3 which are similar to those utilized in lithium devices.

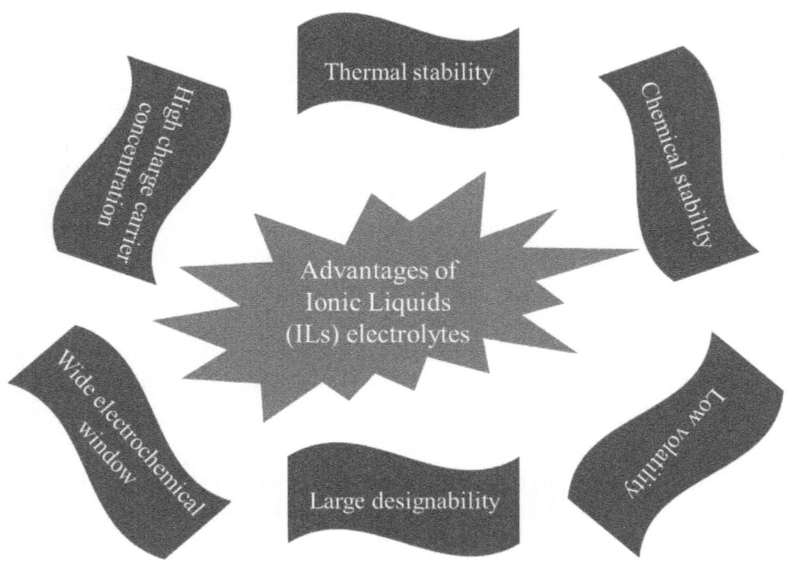

FIGURE 12.3 Advantages of IL electrolytes in electrochemical energy storage.

There have been few articles on IL electrolytes in developing applications. Buchner et al. studied the dielectric characteristics, conductivity, and viscosity of sodium tetrafluoroborate ($NaBF_4$) in 1-butyl-3-methyl imidazolium ILs. They found that the Na-salt had minimal impact on the IL's transport properties unlike Li. Egashira et al. found that sodium tetrafluoroborate ($NaBF_4$) is not well mixable in tetraalkylammonium-based ILs. They investigated a ternary mixture using a methyl ether-capped polyether as a coordinating agent [34]. Nohira et al. [42] discovered intermediate-temperature IL systems using inorganic IL systems and binary combinations. These ILs had a large electrochemical window, reaching up to 5.2 V at a temperature of 363 K. The Na-ion-based devices performed effectively at temperatures ranging from 333 to 393 K. Emerging research suggests that ILs may provide similar benefits to lithium-based electrolytes for Na-based battery systems. However, it's essential to highlight that the development of Na in ILs may differ. Further studies are going on to fully grasp the idea of this phenomenon [35,36].

12.2.2.3 ILs in Magnesium-Ion Battery

Recent research suggests that IL mixtures can be used as electrolytes for Mg anodes in both primary and reversible Mg devices [37–43]. More research is needed to determine the potentially reversible nature of Mg electrochemistry, which is at present unknown in the literature. Mg may be easily reduced from the electrolyte to a metal substrate, but its subsequent dissolution is less obvious. Wang et al. revealed that the combining 1:4 ratio of N-methyl-N-propylpiperidinium [NTf_2] and [C_4mim] [BF_4] ILs with 0.3M $MgNTf_2$ resulted in the dissolution and addition of Mg electrode reversibly [38]. Water plays a vital role in the reduction of oxygen in experimental air electrodes. An aqueous with 8 mol% water in the presence of IL demonstrated the highest performance. It was suggested that this resulted in a gel-like interphase that enhanced Mg ion transit while also enhancing the electrolyte's fluidity. The reduction of oxygen requires a proton source, which can be obtained from the IL cation. This results in a $1e^-$ reduction of O^{2-}. Water and other protic additions can enhance the efficiency of the $1e^-$, $2e^-$, and $4e^-$ reduction processes, resulting in reduced overpotential and higher current densities. The goal of current research is to create IL mixes that can change the solubility and surface properties of Mg^{2+} ions during air discharge.

In order to successfully cycle Li anodes in Mg-air devices, the interphase layer is necessary while designing the phase transition on the Mg electrode could allow for reversibility. This requires forming and retaining compounds in the interphase that can be easily decreased. A rechargeable magnesium-air device would be a substantial advancement for the storage of energy devices [39]. According to recent investigations, quaternary phosphonium chloride-based electrolytes can sustain Mg discharge at current densities as high as $1\,mA/cm^2$ and maintain a steady cell voltage of 1.5–1.6 V for up to three weeks in a Mg-air primary device [40]. Recent research indicates that IL electrolytes can reversibly deposit and dissolve Zn, making them suitable for secondary zinc batteries. Currently, Zn-air primary cells are commonly employed using electrolytes [41]. These appliances, although exhibiting a fairly low cell voltage (1.2 V), offer a substantial amount of energy density (500 Wh/kg) and are built from readily available low-cost components. A reversible Zn-air cell may provide an appropriate replacement for Li-ion battery technology. Using Zn[dca]$_2$ instead of ZnCl$_2$ in 1-Ethyl-3-methylimidazolium dicyanamide ([C$_2$mim][dca]) has been demonstrated to significantly impact the electrochemistry of Zn^{2+}. This results in greater currents and positive potentials for deposition [41]. The difficulty with supercapacitors is to boost energy without sacrificing power. To enhance operating voltage, use electrolytes based on ILs instead of aqueous or molecular non-aqueous electrolytes. ILs offer safety benefits and can be tailored to client requirements. Capacitance is another way to improve energy efficiency. The emergence of heavily porous carbon materials has limited the potential for improving EDLC by developing a more ion-exposed surface area.

12.2.2.4 Advantages and Disadvantages of Using Ionic Liquids Compared to Other Electrolytes

Advantages: Ionic liquids offer remarkable qualities for application in electrochemistry: they possess a large potential stability window, excellent electrical conductivity, and may substitute water in operations where water is not a suitable solvent. As a result, the number of studies has expanded dramatically in recent years [44–47]. When electrocatalytic applications are used, the shape of the interphase affects the electrode's electrocatalytic reaction. Therefore, these materials are used as electroactive materials and enhance the internal resistance of battery. In battery electrolyte applications, an ideal cation-anion combination should produce an electrolyte with several great qualities, including but not limited to strong ionic conductivity and thermal and electrochemical stability. Another important property is that they lack transferrable hydrogen ions, allowing them to survive wide operational voltage windows and being appropriate for sodium-ion battery applications. ILs are a versatile solvent and electrolyte, making it a promising alternative for renewable energy storage devices such as lithium-ion batteries and supercapacitors. It can be used as a designer solvent to replace combustible organic carbonate and enhance the green credentials and performance of energy storage devices. Additionally, it serves as a co-solvent in electrolyte preparation, preventing the solvent effect in films [46,47].

Disadvantages: One of the major disadvantages of utilizing ILs in electrochemistry is their relatively high viscosity, which prevents the migration of electroactive species toward the electrode surface and hence reduces reaction rates. This has prompted research into low-viscosity ILs, which are often based on the imidazole cation ([Im$^+$]) and/or the bis(trifluoromethyl)sulfonyl imide anion ([NTf$_2^-$]). It should be emphasized that the interphase between ionic liquid and electrodes is more complex than that of an aqueous medium. Ionic liquids contain a higher electrostatic charge and interaction between ions, complicating the use of simple double-layer models. Inflammable organic solvent electrolytes can cause fires or explosions during battery operations due to hazardous circumstances or short circuits. Operating safety concerns need to be solved before widespread implementation for practical applications. To improve operating safety, ILs have been combined with an organic carbonate-based electrolyte, which has improved the thermal stability of lithium cells [46–50].

12.3 CONCLUSIONS AND FUTURE PROSPECTS

The next generation of energy systems will be heavily dependent on electrochemical devices for storing energy. Enhanced electrochemical storage of energy devices relies primarily on innovative electrode material. ILs add new attributes and maximize the finest chemical properties of elements in order to convert them to a usable plane. In conclusion, we found that ILs play a significant role in addressing current energy challenges. There is still more work to be done, and we encourage researchers to prioritize what is truly beneficial in each area. IL-based electrolytes have demonstrated commercial potential due to their advanced laboratory performance in terms of economics and sustainability. Furthermore, ILs can be combined with other materials to form suitable hybrid electrolytes for better performance. For this purpose, ILs of suitable cation and anion combinations should be synthesized and their properties should be analyzed thoroughly before any applications related to energy storage and conversion devices. The advantages of using ILs are their capacity to dissolve a wide range of electrolytes and solvents which boosts their energy efficiency. Better ILs can be developed and created to minimize environmental impact. ILs can be used not only as electrolytes but also as solvents if required.

REFERENCES

1. T. Welton, Ionic liquids: a brief history, *Biophysical Reviews*, 2018, Vol. 10, Issue 3, Pages 691–706.
2. L. Miao, Z. Song, D. Zhu, L. Li, L. Gan and M. Liu, Ionic liquids for supercapacitive energy storage: A mini-review, *Energy & Fuels*, 2021, Vol. 35, Issue 10, Pages 8443–8455.
3. Y. Pei, Y. Zhang, J. Ma, M. Fan, S. Zhang and J. Wang, Ionic liquids for advanced materials, *Materials Today Nano*, 2022, Vol. 17, Pages 100159.
4. H. K. Stassen, R. Ludwig, A. Wulf and J. Dupont, Imidazolium salt ion pairs in solution, *Chemistry-A European Journal*, 2015, Vol. 21, Issue 23, Pages 8324–8335.
5. Z. Lei, B. Chen, Y.-M. Koo and D. R. MacFarlane, Introduction: Ionic liquids, *Chemical Reviews*, 2017, Vol. 117, Pages 6633–6635.
6. P. Hapiot and C. Lagrost, Electrochemical reactivity in room-temperature ionic liquids, *Chemical Reviews*, 2008, Vol. 108, Issue 7, Pages 2238–2264.
7. J. S. Wilkes, A short history of ionic liquids-from molten salts to neoteric solvents, *Green Chemistry*, 2002, Vol. 4, Issue 2, Pages 73–80.
8. C. A. Angell, Y. Ansari and Z. Zhao, Ionic liquids: Past, present and future, *Faraday Discussions*, 2012, Vol. 154, Pages 9–27.
9. M. Freemantle, *An Introduction to Ionic Liquids*, Royal Society of Chemistry, 2010, https://doi.org/10.1039/9781839168604.
10. M. B. Shiflett and A. M. Scurto, *Ionic liquids: Current State and Future Directions*, **DOI:** 10.1021/bk-2017-1250.ch0012017, 1–13.
11. M. A. Siddiquee, J. Saraswat, M. ud din Parray, P. Singh, S. Bargujar and R. Patel, Spectroscopic and DFT study of imidazolium based ionic liquids with broad spectrum antibacterial drug levofloxacin, *Spectrochimica Acta Part A: Molecular and Biomolecular Spectroscopy*, 2023, Vol. 285, Pages 121803.
12. M. Vazquez, M. Liu, Z. Zhang, A. Chandresh, A. B. Kanj and W. Wenzel, Structural and dynamic insights into the conduction of lithium-ionic-liquid mixtures in nanoporous metal-organic frameworks as solid-state electrolytes, *ACS Applied Materials & Interfaces*, 2021, Vol. 13, Issue 18, Pages 21166–21174.

13. G. Damilano, D. Kalebić, K. Binnemans and W. Dehaen, One-pot synthesis of symmetric imidazolium ionic liquids N, N-disubstituted with long alkyl chains, *RSC Advances*, 2020, Vol. 10, Issue 36, Pages 21071–21081.

14. A. N. Masri, A. M. Mi and J.M. Leveque, A review on di cationic ionic liquids: Classification and application, *Industrial Engineering and Management*, 2016, Vol. 5, Pages 197–204.

15. G. A. Baker, S. N. Baker, S. Pandey and F. V. Bright, An analytical view of ionic liquids, *Analyst*, 2005, Vol. 130, Issue 6, Pages 800–808.

16. N. Adawiyah, M. Moniruzzaman, S. Hawatulaila and M. Goto, Ionic liquids as a potential tool for drug delivery systems, *MedChemComm*, 2016, Vol. 7, Issue 10, Pages 1881–1897.

17. A. J. Greer, J. Jacquemin and C. Hardacre, Industrial applications of ionic liquids, *Molecules*, 2020, Vol. 25, Issue 21, Pages 5207.

18. A. M. Curreri, S. Mitragotri and E. E. Tanner, Recent advances in ionic liquids in biomedicine, *Advanced Science*, 2021, Vol. 8, Issue 17, Pages 2004819.

19. N. Nasirpour, M. Mohammadpourfard and S. Z. Heris, Ionic liquids: Promising compounds for sustainable chemical processes and applications, *Chemical Engineering Research and Design*, 2020, Vol. 160, Pages 264–300.

20. E. E. Tanner, A. M. Curreri, J. P. Balkaran, N. C. Selig-Wober, A. B. Yang and C. Kendig, Design principles of ionic liquids for transdermal drug delivery, *Advanced Materials*, 2019, Vol. 31, Issue 27, Pages 1901103.

21. T. M. Dhameliya, P. R. Nagar, K. A. Bhakhar, H. R. Jivani, B. J. Shah, K. M. Patel, et al., Recent advancements in applications of ionic liquids in synthetic construction of heterocyclic scaffolds: A spotlight, *Journal of Molecular Liquids*, 2022, Vol. 348, Pages 118329.

22. W. Huang, X. Wu, J. Qi, Q. Zhu, W. Wu and Y. Lu, Ionic liquids: Green and tailor-made solvents in drug delivery, *Drug Discovery Today*, 2020, Vol. 25, Issue 5, Pages 901–908.

23. M. Sivapragasam, M. Moniruzzaman and M. Goto, Recent advances in exploiting ionic liquids for biomolecules: Solubility, stability and applications, *Biotechnology Journal*, 2016, Vol. 11, Issue 8, Pages 1000–1013.

24. D. Z. Troter, Z. B. Todorović, D. R. Đokić-Stojanović, O. S. Stamenkovic and V. B. Veljko Vic, Application of ionic liquids and deep eutectic solvents in biodiesel production, *Renewable and Sustainable Energy Reviews*, 2016, Vol. 61, Pages 473–500.

25. D. R. MacFarlane, N. Tachikawa, M. Forsyth, J. M. Pringle, P. C. Howlett and G. D. Elliott, Energy applications of ionic liquids, *Energy & Environmental Science*, 2014, Vol. 7, Issue 1, Pages 232–250.

26. T. Bashir, S. A. Ismail, Y. Song, R. M. Irfan, S. Yang, S. Zhou, A review of the energy storage aspects of chemical elements for lithium-ion based batteries, *Energy Materials*, 2021, Vol. 1, Pages 100019.

27. J. B. Goodenough and K. S. Park, The Li-ion rechargeable battery: A perspective, *Journal of the American Chemical Society*, 2013, Vol. 135, Issue 4, Pages 1167–1176.

28. S. Seki, Y. Kobayashi, H. Miyashiro, Y. Ohno, Y. Mita and A. Usami, Reversibility of lithium secondary batteries using a room-temperature ionic liquid mixture and lithium metal, *Electrochemical and Solid-State Letters*, 2005, Vol. 8, Issue 11, Pages A577.

29. N. Recham, J. N. Chotard, J. C. Jumas, L. Laffont, M. Armand and J.-M. Tarascon, Ionothermal synthesis of Li-based fluorophosphates electrodes, *Chemistry of Materials*, 2010, Vol. 22, Issue 3, Pages 1142–1148.

30. D. R. MacFarlane, N. Tachikawa, M. Forsyth, J. M. Pringle, P. C. Howlett and G. D. Elliott, Energy applications of ionic liquids, *Energy & Environmental Science*, 2014, Vol. 7, Issue 1, Pages 232–250.

31. G. T. Kim, S. Jeong, M. Z. Xue, A. Balducci, M. Winter and S. Passerini, Development of ionic liquid-based lithium battery prototypes, *Journal of Power Sources*, 2012, Vol. 199, Pages 239–246.
32. D. Safronov, S. Novikova, A. Skundin and A. Yaroslavtsev, Lithium intercalation and deintercalation processes in Li4Ti5O12 and LiFePO4, *Inorganic Materials*, 2012, Vol. 48, Pages 57–61.
33. R. Vijayaraghavan, M. Surianarayanan, V. Armel, D. R. Macfarlane and V. Sridhar, Exothermic and thermal runaway behaviour of some ionic liquids at elevated temperatures, *Chemical Communications*, 2009, Issue 41, Pages 6297–6299.
34. Y. Wang, K. Zaghib, A. Guerfi, F. F. Bazito, R. M. Torresi and J. Dahn, Accelerating rate calorimetry studies of the reactions between ionic liquids and charged lithium-ion battery electrode materials, *Electrochimica Acta*, 2007, Vol. 52, Issue 22, Pages 6346–6352.
35. F. Mizuno, S. Nakanishi, A. Shirasawa, K. Takechi, T. Shiga, H. Nishikoori, et al. Design of non-aqueous liquid electrolytes for rechargeable Li-O2 batteries, *Electrochemistry*, 2011, Vol. 79, Issue 11, Pages 876–881.
36. C. J. Allen, S. Mukerjee, E. J. Plichta, M. A. Hendrickson and K. Abraham, Oxygen electrode rechargeability in an ionic liquid for the Li-air battery, *The Journal of Physical Chemistry Letters*, 2011, Vol. 2, Issue 19, Pages 2420–2424.
37. L. L. Zhang, Y. X. Zhou, T. Li, D. Ma and X.-L. Yang, Multi-heteroatom doped carbon coated Na3V2(PO4)3 derived from ionic liquids, *Dalton Transactions*, 2018, Vol. 47, Issue 12, Pages 4259–4266.
38. X. Hu, Z. Li and J. Chen, Flexible Li-CO2 batteries with liquid-free electrolyte, *Angewandte Chemie*, 2017, Vol. 129, Issue 21, Pages 5879–5883.
39. V. A. Nikitina, A. Nazet, T. Sonnleitner and R. Buchner, Properties of sodium tetrafluoroborate solutions in 1-butyl-3-methylimidazolium tetrafluoroborate ionic liquid, *Journal of Chemical & Engineering Data*, 2012, Vol. 57, Issue 11, Pages 3019–3025.
40. M. Egashira, T. Asai, N. Yoshimoto and M. Morita, Ionic conductivity of ternary electrolyte containing sodium salt and ionic liquid, *Electrochimica Acta*, 2011, Vol. 58, Pages 95–98.
41. T. Yamamoto, T. Nohira, R. Hagiwara, A. Fukunaga, S. Sakai and K. Nitta, Charge-discharge behavior of tin negative electrode for a sodium secondary battery using intermediate temperature ionic liquid sodium bis (fluorosulfonyl) amide-potassium bis (fluorosulfonyl) amide, *Journal of Power Sources*, 2012, Vol. 217, Pages 479–484.
42. T. Nohira, T. Ishibashi and R. Hagiwara, Properties of an intermediate temperature ionic liquid NaTFSA-CsTFSA and charge-discharge properties of NaCrO2 positive electrode at 423 K for a sodium secondary battery, *Journal of Power Sources*, 2012, Vol. 205, Pages 506–509.
43. G. Vardar, A. E. Sleightholme, J. Naruse, H. Hiramatsu, D. J. Siegel and C. W. Monroe, Electrochemistry of magnesium electrolytes in ionic liquids for secondary batteries, *ACS Applied Materials & Interfaces*, 2014 Vol. 6, Issue 20, Pages 18033–18039.
44. T. Khoo, P. C. Howlett, M. Tsagouria, D. R. MacFarlane and M. Forsyth, The potential for ionic liquid electrolytes to stabilize the magnesium interface for magnesium/air batteries, *Electrochimica Acta*, 2011, Vol. 58, Pages 583–588.
45. P. Wang, Y. NuLi, J. Yang and Z. Feng, Mixed ionic liquids as electrolyte for reversible deposition and dissolution of magnesium, *Surface and Coatings Technology*, 2006, Vol. 201, Issue 6, Pages 3783–3787.
46. E. Jónsson, Ionic liquids as electrolytes for energy storage applications-A modelling perspective, *Energy Storage Materials*, 2020, Vol. 25, Pages 827–835.

47. K. Sirengo, A. Babu, B. Brennan and S. C. Pillai, Ionic liquid electrolytes for sodium-ion batteries to control thermal runaway, *Journal of Energy Chemistry*, 2023, Vol. 81, Pages 321–338.
48. K. Karuppasamy, J. Theerthagiri, D. Vikraman, C.-J. Yim, S. Hussain and R. Sharma, Ionic liquid-based electrolytes for energy storage devices: A brief review on their limits and applications, *Polymers*, 2020, Vol. 12, Issue 4, Pages 918.
49. M. Forsyth, L. Porcarelli, X. Wang, N. Goujon and D. Mecerreyes, Innovative electrolytes based on ionic liquids and polymers for next-generation solid-state batteries, *Accounts of Chemical Research*, 2019, Vol. 52, Issue 3, Pages 686–694.
50. R.S. Kühnel, N. Böckenfeld, S. Passerini, M. Winter and A. Balducci, Mixtures of ionic liquid and organic carbonate as electrolyte with improved safety and performance for rechargeable lithium batteries, *Electrochimica Acta*, 2011, Vol. 56, Issue 11, Pages 4092–4099.

13 TiO₂ Nanotube Synthesis, Properties, and Applications

Partha Roy and Hemant Joshi

13.1 INTRODUCTION

Among the metals and metal oxides, TiO_2 is the most comprehensively researched material in recent times because it is readily available, non-toxic, and environmentally friendly. Despite this, being ceramic TiO_2 is very hard and remains inert in contact with a wide range of acids (except HF) and bases. Consequently, TiO_2 has emerged as a potential candidate in various industrial applications as corrosion, erosion, and fouling-resistant materials [1,2]. Nevertheless, because TiO_2 has the highest refractive index of any known material, it is ideally suited for engineering opto-electronic device [3]. Moreover, these unique properties have enabled TiO_2 pigment to be widely used in paint and plastic [4]. Its non-toxic, inexpensive, biocompatibility, and highly reactive nature make TiO_2 an attractive candidate for use in environmental remediation. In order to develop cost-effective solar cells with high efficiency, several studies are being focused in the past few years. Incidentally, dye-sensitized TiO_2 solar cell storage has surfaced as a viable option to meet the demand [5–7]. Nevertheless, TiO_2 can be further utilized in energy storage, electrochemical devices, and sensing applications [8–10].

13.2 WHY TIO₂ NANOTUBES?

Evidentially, tube-shaped nanomaterials possess unique properties over other morphologies which make them attractive to a wide range of applications. However, the large surface area and broadened band gap of 0D TiO_2 nanoparticles have enabled them to be used in hydrogen production, photocatalysis, adsorbents, solar cells, and sensors in the past. Despite this, they had to endure inevitable drawbacks such as rapid recombination of electrons and holes, a hetero-high cost of recycling, and slow charge carrier transfer. Contextually, the synthesis of 2D TiO_2 nanostructured materials can overcome some of these disadvantages, but it is not a consumer-friendly process. Conversely, 1D nanostructures have been extensively studied because of their distinct advantages. The diameter of 1D nanostructures typically ranges from 1 to 100 nm, and the aspect ratio is generally high. Hence, 1D TiO_2 nanostructured materials exhibited all the typical TiO_2 nanoparticle characteristics and included a large surface area that facilitates the transfer of carriers along an axis. In addition, the relative ease of their production adds to their advantages. Furthermore, these materials are ideal building blocks for constructing various surface heterostructures because of

DOI: 10.1201/9781003473749-16

their large surface area and chemical firmness, making them ideal for photocatalysis, water splitting, and nanotechnology applications [11,12]. Due to several critical issues, however, 1D TiO_2 nanostructured materials have been limited in their applications across many fields. Owing to the wide band gap, TiO_2 (anatase: 3.2 eV, rutile: 3.0 eV) makes up only 3%–5% of the total solar spectrum. Moreover, photo-generated electron-hole pairs often recombine rapidly, resulting in decreased photocatalysis efficiency. Historically, the very first example of water spiriting using TiO_2 electrodes by Fujishima and Honda in 1972 was a significant milestone [13]. In the 80s–90s, the outcome of various experiments has revealed that solar energy can be potentially harvested by using TNT [14–16]. Over the last few decades, the potential application of TNT has been widened toward various sophisticated device fabrication, self-cleaning properties, sensor, bio-medical coating, and smart surface coatings [2,17–27].

13.3 CRYSTAL STRUCTURES

Mainly, TiO_2 exists in three different polymorphous: rutile (tetragonal; a = 0.46 nm and c = 0.27 nm), anatase (tetragonal; a = 0.54 nm and c = 0.95 nm) and brookite (orthorhombic; a = 0.915 nm, b = 0.54 nm, and c = 0.51 nm) [28] as shown in Figure 13.1; additionally, synthetic layered phase namely TiO_2(B), as well as some high-pressure polymorphous phase also exist. Argumentatively, rutile is deemed to be the thermodynamically most durable bulk phase; whereas, at the nanoscale (≤20 nm), anatase is interpreted as a stable one. Notably, nanoscale materials are significantly influenced by surface energy and stress. An uncoordinated titanium cation site has a higher surface energy (e.g., a fourfold-coordinated site has a higher surface energy than a fivefold-coordinated site). The anatase crystal has lower surface energy than that of the rutile, but the stress energy of the rutile is lower than that of the nano-anatase. There is a compensational effect between the two energies, which explains why there is some uncertainty about which phase is dictating nanoscale materials.

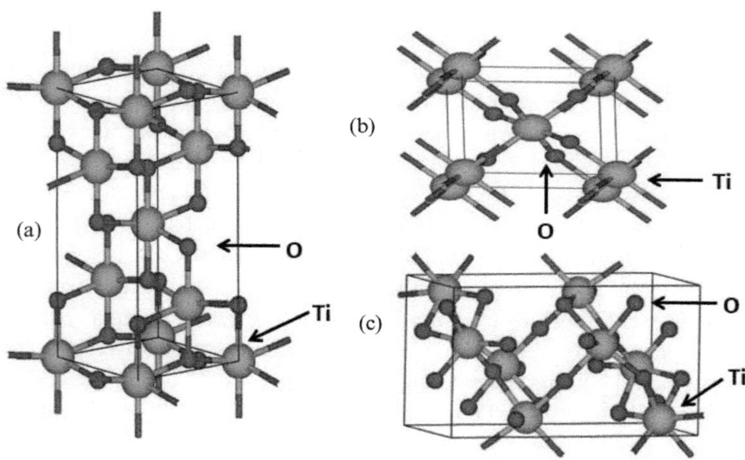

FIGURE 13.1   Schematic representation of unit cells of (a) anatase, (b) rutile, and (c) brookite TiO_2. (The red spheres represent the O atoms and blue spheres Ti atoms.)

In the nanoscale range, the anatase phase has been of immense interest because of its key feature in the injection and transportation of electrons.

Brookite, among these three phases, is the rarest form of this mineral and not so easy to synthesize. Table 13.1 lists a few of the crucial properties of the three prevalent titania polymorphous.

Aside from stoichiometric TiO_2, various suboxide phases, formulated by Ti_nO_{2n-1} ($4 \leq n \leq 10$) (*e.g.*, Ti_4O_7, Ti_5O_9, Ti_6O_{11}, Ti_7O_{13}, Ti_8O_{15}, and Ti_9O_{17}) exist and they are referred to as "Magneli" phases. In such phases, titania octahedra are arranged in two-dimensional chains, as in Ti_4O_7, for instance, which is composed of each unit cell that has three TiO_2 octahedra and one TiO octahedron with oxygen vacancies. This kind of phases have been obtained under reductive high temperature in the H_2 atmosphere. Remarkably, these suboxide phases demonstrate high conductivity (almost like graphite). Nonetheless, many of the properties of TiO_2 are strongly influenced by the presence or absence of bulk or surface structural defects, especially Ti^{3+} or O_2 vacancies. During vacuum or inert-gas annealing, there is a possibility that TiO_2's surface may generate unsaturated titanium cations, such as Ti^{3+}, Ti^{2+}, and Ti^+. Mostly, Ti^{3+} states are formed (especially sputtering and e-beam treatment), which is the main contributor to electronic and optical properties and surface reactivity as well. Under the exposure of UV radiation, adsorbed oxygen can turn into O^{2-}, which impedes the formation of Ti^{3+} states. It was noticed that Ti^{3+} may diffuse into the bulk from the surface due to the long time annealing.

13.4 PROCESSING TECHNIQUES OF TNT

Vast research efforts have been put forward to manufacture 1D TiO_2 nanotubes to manipulate its exceptional characteristics in many application fields in the last few decades. Herein, we are discussing a few well-known methods of their synthesis, like the hydrothermal method, the solvothermal method, the electrochemical anodization method and the template-assisted method.

13.4.1 HYDROTHERMAL METHOD [30]

The hydrothermal method is one of the most popular ones to synthesize 1D TiO_2 nanotubes as it's a modest, cost-effective process and the yield of the conversion of nanotubular shape is almost 100%. Characteristically, amorphous TiO_2 powder or other precursors (*e.g.*, titanium iso-propoxide, titanium butoxide, $TiCl_4$ etc.) are used to make a sol by gradually adding a highly concentrated aqueous solution of NaOH.

TABLE 13.1

Different TiO₂ Polymorphs and Some Physical Properties [29]

Crystallinity	Crystal System	Density (g/cm³)	Optical Band-Gap (eV)	Refractive Index
Rutile	Tetragonal	4.13–4.26	3.0	2.72
Brookite	Orthorhombic	3.99–4.11	3.11	2.63
Anatase	Tetragonal	3.79–3.84	3.19	2.52

Afterward, the mixture was treated at high temperature and pressure in a Teflon-line stainless steel autoclave. Finally, the nanotubular shape is attained by thoroughly washing the resultant materials with 0.1 M HCl solution or deionized (DI) water till pH ~ 7.0. This washing procedure (washing times, acid concentration) plays a crucial role in defining the structural (crystal lattice), morphological, physical, and chemical properties of as synthesized TNT products. Figure 13.2 illustrates TNT formed by hydrothermal treatment. It is important to note that most often, TiO_2 is formed during the drying or calcination of as-synthesized titanate. It was observed that titanate nanotubes transform into TiO_2 nanotubes in the following sequence while the calcination was done at 150°C to 500°C.

$$H_2Ti_3O_7 \cdot xH_2O \rightarrow H_2Ti_3O_7 \rightarrow H_2Ti_6O_{13} \rightarrow TiO_2(B) \rightarrow TiO_2(anatase)$$

Essentially, the morphology of TiO_2 nanostructures can be altered by the influence of pH of the reaction mixture, temperature of the reaction and the intrinsic nature of the solvents.

13.4.1.1 Mechanism of Formation of Nano-Tubular Structure during Hydrothermal Process

Apparently, during the acid treatment at the time of washing, electrostatic repulsion between the same charged ions is removed, producing lamellar sheets from which nanotubes are formed. Elaborately, during the treatment of TiO_2 by NaOH, Ti-O-Ti

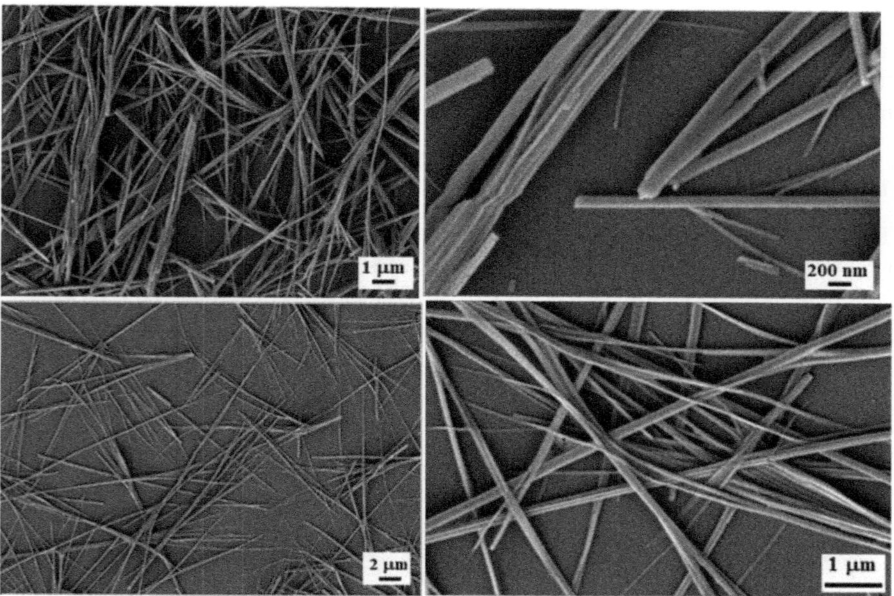

FIGURE 13.2   FESEM images of TNT at different resolution.

bonds break into intermediates where Ti-O-Na and Ti-OH are formed, and these intermediates transform into sheets because of the rearrangement of Na$^+$ and H$^+$ ions. Owing to the ion exchange of Na$^+$ with H$^+$, the surface charge variation has taken place, resulting in the scrolling of the sheet into nanotubes. According to the experimental findings, the overall morphology of the as-prepared nanotubes is not considerably changed after washing. Therefore, hydrothermal treatment appears to play a key role in the formation of TNTs, whereas acid washing produces protonated titanates by ion exchange of Na$^+$ and H$^+$.

13.4.2 SOLVOTHERMAL METHOD [31]

Unlike hydrothermal, where water was used as a medium, organic solvents (*e.g.*, ethanol, ethylene glycol, n-hexane, etc.) are employed as a medium in this process. Advantageously, using organic solvents with higher boiling points gives better control over agglomeration, crystallization, size distribution, and nano-sized products than hydrothermal methods. The solvothermal method is heavily utilized for synthesizing various metal oxides *i.e.* Fe$_3$O$_4$, ZnO, ZeO$_2$ etc. At the same time, titanium nano-tubular structures can be fabricated using the same method. Likewise, the hydrothermal process also requires high temperature and pressure treatment. Moreover, solvents having different physical and chemical properties have a great impact on the solubility, reactivity, diffusion behavior, and crystallization behavior of TNTs. Compared to the hydrothermal method, choosing the correct solvents for the solvothermal process is the key factor, which actually limits its wide application. Noteworthy, in solvothermal process, TNTs are composed of bilayers or multilayers due to the scrolling of adjoined multilayers of TiO$_2$ nanosheets.

13.4.3 ELECTROCHEMICAL ANODIZATION METHOD [32]

Self-organized oxide nanotubes can be made by using the anodization method which is widely used for the anodization of various metals such 1D TiO$_2$ nanotube arrays (NTA). By contrast to hydrothermal nanotubes, the size controllable NTA are highly ordered and oriented perpendicularly to their substrate surface. This unique feature of highly ordered structure of NTA makes it ideal for employed in photoanodic materials (harvesting large spectrum of visible range light) that can be used for long term because they are stable, non-toxic, environmentally friendly, cheap, reusable and easily synthesized. Electrochemical anodization process has been carried out using two electrodes dipped in electrolytic solution and applied potential at constant voltage between 1–30 V in an aqueous electrolyte or 5–150 V in a non-aqueous electrolyte containing fluoride ions (0.1–1 wt%). In this process, the cathode must be conductive and can be the same metal, platinum, or graphite with similar dimensions as the anode. Finally, the crystallized TiO$_2$ nanotubes are accomplished by treating the anodised Ti plate at 300°C–500°C for ~6 h in presence of oxygen.

13.4.3.1 Stages in TiO$_2$ Nanotubes Growth in Anodization Process

The formation of TiO$_2$ nanotubes arrays is obtained in acidic electrolytes (usually, various acids are used to adjust the acidity) having optimum concertation of fluoride or chloride ions. However, in the absence of fluoride ions, a compact oxide layer grows on TiO$_2$ surface, and this growth of the oxide layer can be observed by monitoring the current–time behaviour (Figure 13.3a).

Customarily, the formation of oxide layer on anode follows the following steps [32]:

$$M \rightarrow M^{z+} + ze^- \tag{13.1}$$

$$M + \frac{z}{2} H_2O \rightarrow MO_{\frac{z}{2}} + zH^+ + ze^- \tag{13.2}$$

$$M^{z+} + ZH_2O \rightarrow M(OH)_z + zH^+ \tag{13.3}$$

$$M(OH)_z \rightarrow MO_{\frac{z}{2}} + \frac{z}{2} H_2O \tag{13.4}$$

At the same time, hydrogen evolution takes place at the counter electrode due to reduction of hydrogen ions (step - v).

$$ZH_2O + ze^- \rightarrow \frac{z}{2}H_2 + zOH^- \tag{13.5}$$

Step – (i) suggests the formation of Ti^{4+}, reaction with O^{2-}. This step is named as "Field Assist Oxidation"

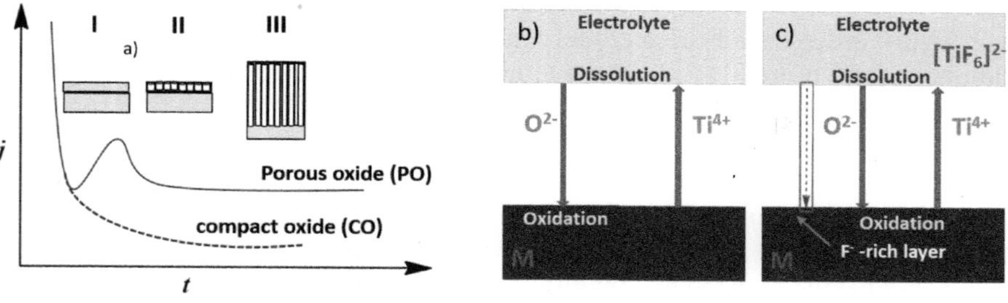

FIGURE 13.3 (a) Plot of current (j) vs. time (t) in the absence (broken line) and presence (solid line) of F$^-$ ions. Either oxide (F$^-$ ions free) or porous/tubular metal oxide formation (with F$^-$ ions) structures by distinct structural stages (I–III). (b and c) Schematic drawing depicting field aided transport of mobile ions through the oxide layers in the absence and presence of fluoride ions.

Steps – (ii), (iii) & (iv) indicate deprotonation due to the field-aided (Field Assist Oxidation).

Afterward, field-assist migration and dissolution take place (Figure 13.3b). Nonetheless, fluoride ions present in the solution are very small and compete with O^{2-} migration through the TiO$_2$ bottom oxide (Figure 13.3c). The growth of nano-tubular structures can be accomplished by a trade-off between the dissolution of the electrolyte and growth oxide layer. Besides, parameters like electrolyte, voltage, temperature, pH, water and fluoride content can regulate the growth of nanotube. Contextually, TNT can have smooth tube wall in organic electrolyte. This is assumed to happen due to the low content of water in the electrolyte. On the other side, V-shaped morphology is predominantly detected in case of aqueous and some nonaqueous (*e.g.*, ethylene glycol, glycerol) electrolytes. It is important to note that anodization voltage has great influence on tube diameter length. At lower applied voltages, due to less electric field dissolution, titania nanotubes with smaller diameters form. In the absence of nanotubular structures, the oxide layer thickens when the voltage is too low. On the contrary, too high voltage results in sponge-like structures or, in some cases, discrete, hollow and tube-like nanostructures rather than nano-tubular structures. Noteworthy, the influence of applied voltage on the diameter of nanotube is trivial in non-aqueous electrolytes due to low conductivity. On the other hand, the aqueous electrolytes affect the nanotube diameter in the range of anodization voltage usually between 10 and 25 V. It's worth to mention that electrolyte temperature can effectively alter the wall thickness, rate of oxide growth and length of the tube. The increase in electrolyte temperature increases the pore size and wall thickness due to the faster movement of fluoride ions which leads to a faster etching rate. Another key factor in the formation of titania nanotubes is, the fluoride ions concentration present in electrolyte. A compact oxide layer forms if there are no fluoride ions in the electrolyte. Eventually, nanotube length increases with fluoride ion concentration. Similarly, anodization duration has a huge impact on determining the length on nanotubes. However, this effect depends on the nature of electrolytes. For instance, after about an hour the nanotube length becomes independent of anodization duration in aqueous electrolytes, and it was found that the nanotube lengths were limited up to 500 nm in HF containing water-based electrolyte. However, much longer nanotubes can be achieved with same anodization voltage in longer duration in organic electrolyte. Interestingly, longer period of anodization may lead to collapse the nanotube structures because of the over chemical dissolution (Figure 13.4).

13.4.4 Template-Assisted Method [33]

Template method is a very facile technique in which nanostructures are created based on the morphology defined templates. Templates such as anodic aluminium membranes (AAM), ZnO, silica etc. are usually used because they are easy to remove via chemical etching or combustion, leaving the resultants with predetermined porosities. Depending on the growth of the material, the templet can be divided into two positive and negative templet synthesis. In positive template synthesis, the outer surface coating of the materials is grown, while in negative template synthesis, the materials deposited inside the template interspace are grown. Typically, in this process, titanium

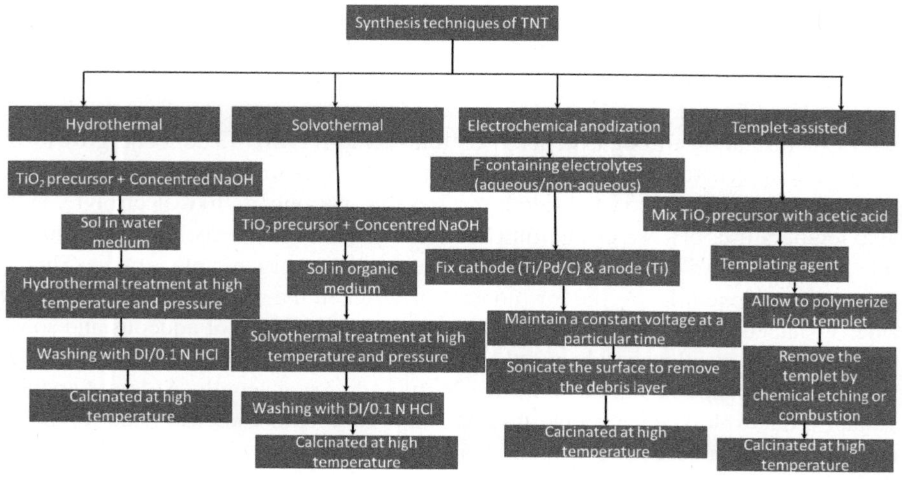

FIGURE 13.4 Flowchart of synthetic methods.

isopropoxide or tetrabutyl titanate is mixed with acetic acid in presence of templating agent to prepare TiO_2 sol-gel. Afterward, TiO_2 is allowed to polymerize in the templet and in the final step, encoded morphology is obtained by selectively excluding the templet agent and then calcinating it. The advantages of the template method are the uniformity and manageable scale of nanotube growth and size of the applied template. However, this process has several unavoidable disadvantages. As can be seen, as-synthesized nanotubes need to be stripped from the templets which may compromise the structure. Moreover, in most cases used templet cannot be reinstated into further processing, creating waste and increasing operational costs. Practically, it requires tedious steps, including prefabrication of the template and post-removal of template, which are time-consuming and labour-intensive. Each of the fabrication techniques has its unique advantages and functional features. These are summarized in the following table (Table 13.2) [34].

13.5 APPLICATIONS OF TIO₂ NANOTUBES

The low cost of production, chemical inertness, and biocompatibility of TiO_2 nanotubes make them an attractive material for energy conversion and storage, as well as biological applications and many others. Herein, some of these applications are briefly discussed.

13.5.1 ENERGY CONVERSION APPLICATIONS

13.5.1.1 Solar Cell Application

Preferably, silicon is used as a base material of most commercially available solar cells. The crystallinity nature (mono- or multi-crystalline) of the silicon governs the solar energy to electricity conversion efficiency. Nonetheless, numerous researchers are being engaged to achieve maximum conversion efficiency with profitable

TABLE 13.2

Comparison Table for Various Methods for Preparation of TNT

Synthesis Method	Reaction Parameters	Advantages	Disadvantages
Hydrothermal	i. Aqueous solvents. ii. High temperature and pressure.	i. Nanotube production rate is high. ii. Easy to improve the features of titanium nanotubes.	i. Harsh reaction condition. ii. Long reaction time. iii. Non-uniform size distribution.
Solvothermal	i. Organic solvents. ii. High temperature and pressure.	i. Better control in size distribution and crystal phase. ii. Varieties of organic solvents can be chosen.	i. Harsh reaction condition. ii. Long reaction time
Electrochemical anodization	i. 5–50 V and 0.2–10 h under ambient conditions. ii. Preferably, F$^-$-based electrolytes and organic electrolytes,	i. Nanotubes are ordered and aligned. ii. Manageable dimension of nanotubes by varying the voltage, electrolyte, pH and anodizing time.	i. Limited large scale production.
Templet assisted	i. AAO, ZnO etc. as sacrificial template under specific conditions.	i. Manageable scale of nanotube by applied template. ii. Uniform distribution.	i. Tedious steps involved in the process. ii. Distortion or contamination of tubes may occur during process.

materials. In this context, TiO$_2$ has emerged as a promising candidate to utilize in dye-sensitized solar cell (DSSC). The transformation of solar energy to electricity with 80% quantum proficiency was reported while ruthenium bypyridyl was sensitized on TiO$_2$ nanoparticles [35]. Principally, in DSSC, a dye absorbs light in the visible region, thereby exciting electrons from the HOMO to LUMO level, and then injecting the excited electron into TiO$_2$'s conduction band; it implies LUMO of the dye must be energetically higher than the TiO$_2$ conduction band (schematic representation in Figure 13.5).

Meanwhile, oxidised dye is brought to life through the redox reaction catalysed by I$^-$/I$_3^-$ electrolytes. Moreover, several thermodynamic considerations must be taken into account, and these are listed below:

1. An electron infusion to the semiconductor conduction band must be faster than dye de-excitation.
2. The dye rejuvenation time constant must be speedy enough to minimize any exhaustion effects.
3. Electron transport within TiO$_2$ must be faster than the electron recombination process.

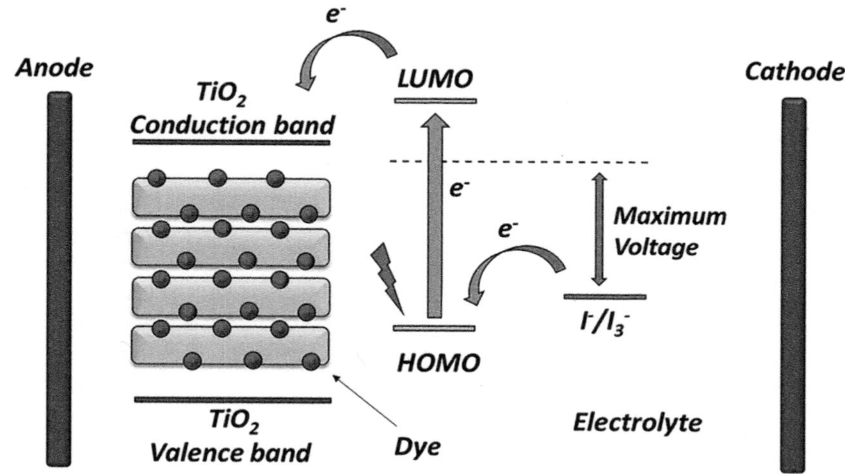

FIGURE 13.5   Schematic diagram of a DSSC.

Generally, in TiO_2 nanoparticles, grain boundaries, defects, surface states *etc.* create trapping sites and that obstruct the diffusion of electron flow [36,37]. On contrary, one-dimensional nanostructures, such as TNT can significantly enhance overall electron transport mechanism by reducing recombination due to limited inter-crystallite traps [26,38].

In DSSCs, TiO_2 serves solely to harvest electrons injected from dye. In spite of the fact that other oxides may be considered as an alternative, TiO_2 remains the most effective oxide until now. Eventually, the degree of crystallinity will have a major impact on TiO_2 DSSC performance. It was observed that upon rising the temperature leads to the dominance of the rutile phase which may collapse. Thus, titanium-based solar cells are most efficient when anatase is used, as it is the most photoactive phase.

13.5.1.2   Photo/Photoelectron Catalytic Application

Globalisation and industrial revolution have created a demand for sustainable energy sources and pollution free environment. Evidentially, future energy demands can be met with hydrogen because it can be produced sustainably, and its consumption does not pollute the environment. Solar energy has been harnessed for photocatalytic hydrogen generation via water splitting. Besides, photo/photo-electro catalytic can be potent to eradicate the pollutant through oxidation process. Mechanistically, a semiconductor is exposed to UV light, which promotes electrons to the conduction band from the valence band; charge carriers (e^-/h^+ pairs) are created and separated, reaching the semiconductor-environment interface, and reacting with appropriate redox species [39,40]. A number of reactions are caused due to the charge exchange process *e.g.* $OH(H_2O + h^+ \rightarrow OH\bullet)$ originated at the valance band and O^-_2 ($O_2 + e^- \rightarrow O_2$) in the conductional band from the surrounding water molecules. These reactive species can potentially oxidised all toxic organic chemicals and convert into CO_2 and H_2O.

Therefore, for the effective employment of photo/photo-electro catalysts the following condition must be satisfied:

i. Desirably, small band gap material to harvest maximum energy from solar spectrum and convert these photons to well disconnected charge carriers.
ii. Chemical steadiness in aqueous electrolytes is one of the vital points to address.
iii. Cost-effectivity of the earth abundant material is important in terms of industrial applications.

A semiconducting material for solar water splitting should meet the conditions shown in Figure 13.6. In this respect, TiO$_2$ has become apparent to be an ideal candidate to meet all these criteria. However, the low UV absorbance of TiO$_2$ and the fast recombination rate of its photo generated charge carriers make it an undesirable solar material. Photo generated charge carrier recombination time can be prolonged by modifying the morphology of TiO$_2$; 1-D TiO$_2$ nanotubes exhibit excellent photocatalytic activity by augmented charge carrier separation due to decoupling light absorption and carrier collection directions [41]. Furthermore, the band gap can be tailored *via* doping or alloying with different elements (metals and non-metals) to an extent where the visible light of the solar spectrum can be effectively harvested [12,42].

13.5.2 ENERGY STORAGE APPLICATIONS

13.5.2.1 Electrochromic Devices [43]

Fabrication of enduring energy storage devices, electrochemical capacitors are considered to be a prospective candidate that can store the energy in the form of electrical charges. An energy storage process based on redox reactions is known as the "faradaic process". The faradaic process occurs when electrolyte ions are adsorbing onto the electrode material surface and electrons are exchanging between the electrode material and the electrolyte after a fast reversible redox reaction takes place at

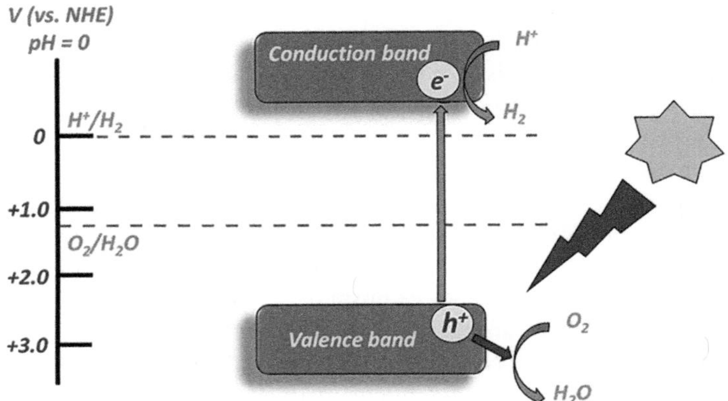

FIGURE 13.6 Semiconductor requirements for solar water splitting.

the electrode surface. Ions adsorption and charge trade are the vital factors in this process which is solely controlled by the surface of the materials used for. Thus, studying the material's morphology is important, since it impacts the capacitor's quality and its mechanism. In this aspect, TiO_2 can be utilized in redox reaction, but inadequate conductivity makes it a poor target for supercapacitor application. Till date, several modification techniques have already been successfully documented to improve the conductivity of TiO_2. Perceptively, TiO_2 nanotubes array coated with carbon is found to be a promising one as TNT arrays display a characteristic rectangular cyclic voltammogram (CV), suggesting pseudo-capacitive behaviour [9,44].

13.5.3 BIOLOGICAL APPLICATIONS

13.5.3.1 Drug Delivery

Exploiting physicochemical properties and biocompatibility, TiO_2 nanotubes have been employed to deal with the flaws of the conventional drug therapeutic solutions. Prognostically, TNT helps to find an alternative route to reach drug at its target site smoothly [45]. Drug release is actually administered by numerous factors, including the nanotubes' charge, dimensions, and surface chemistry, as well as the loaded drug's charge, size, and diffusion coefficient. Also, how the drug molecules interact with TiO_2 molecules plays an important role [46,47]. Notably, most drugs are released in a zero-order manner, which is characterized by constant release rates over the course of the release process. Incidentally, quite a few studies have attempted to alter the nano-tubular structure to set off the desired therapeutic outcome. Among these modifications are, adjusting their length, thickness, pore opening, or stimulating their release by polymeric coatings or external sources. It is worth noting that in order to have better controlled and sustain release profile TNT need to be triggered by external source, like electric field, magnetic field, ultrasonic wave etc.

13.5.3.2 Antibacterial Applications

Integrating photocatalytic activities, TNT can further make use of destroying poisonous microorganisms [48] and bacteria from water using solar radiation. Basically, the formation of hydroxyl radicals (OH) in photocatalysis mechanism is believed to be responsible for water disinfection. Essentially, holes generating in the valance band, combined with H_2O molecules to produce OH• ions which either get hydrolysed to form OH^- or react with bacterial membrane lipids. The radicals bring about severe damage on the extracellular medium of the bacteria, causes dangerous chemical/biomolecular transformations. Conversely, the electron fuses with the proton ions (H^+) in the same biological environment to complete the other half of the electrochemical reaction. Figure 13.7 outlined the mechanism of this process.

In one study, it has been discovered that hydrothermally formed TNT is the best choice in comparison to other nanostructures, to get highest deactivation rate of *E-coli* bacteria in dark and light conditions for 120 min [34] due to high abundance of OH^- functional groups on their surfaces. Additionally, mixed phase of rutile and anatase were demonstrated high surface area which leads to high inactivation rate.

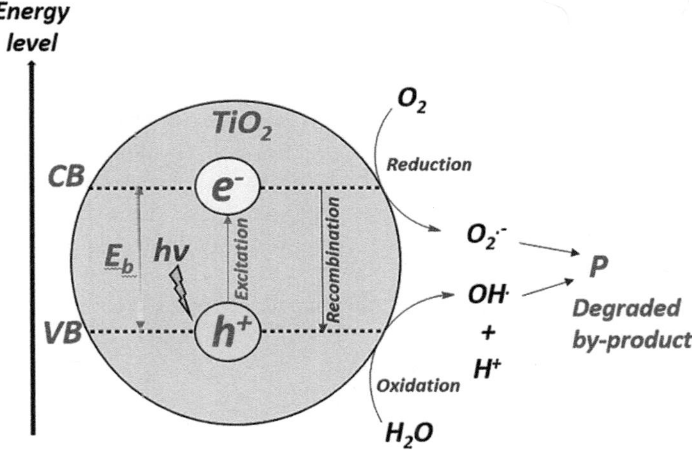

FIGURE 13.7 Mechanistic principle of water microbial disinfection using TNT.

13.5.4 SUPERHYDROPHOBIC COATING

The extreme repellent and self-cleaning surfaces, on which the contact angle (CA) value in respect of water droplet is more than 150° and roll off easily at very small perturbation are famously known as superhydrophobic surface [49]. Eventually, the blueprint for engineering this kind of superhydrophobic surface was based on imitating the natural phenomenon (*e.g.* lotus leaf wings of butterflies). Evidently, surface roughness is the distinctive feature in determining the wetting state of a solid surface. Henceforth, employments of different techniques have been espoused to grow the optimum roughness (micro/nano) on the surface to achieve high desire CA which will preferentially confer the Cassie-Baxter state [50]. In this model, large amount of air is trapped in the microgrooves and exert pressure in outward direction resulting the water droplets rest on the composite surface resist imbibition. Contextually, it has been found that TiO$_2$ nanotubes, whose length in the 1–2 µm range can effectively render the ideal roughness on the surface and make it apparently robust water-repelling surface with the help of conformal layer of low surface energy molecules [51–54] (Figure 13.8).

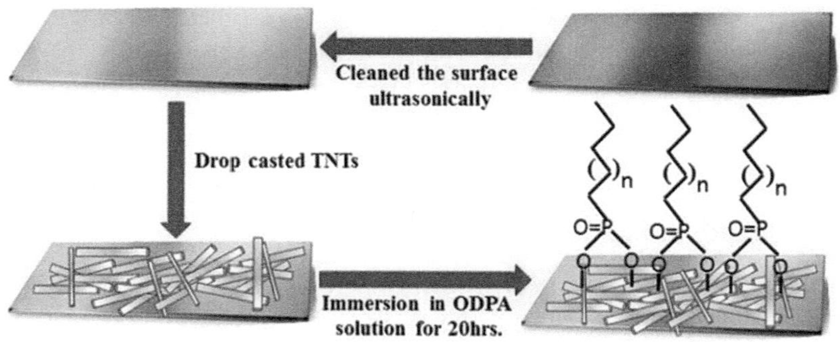

FIGURE 13.8 Stepwise schematic representation of passivation of the metal surface.

13.6 CONCLUSION

This chapter provides a comprehensive overview of the present state of knowledge regarding the growth mechanism that underlies the formation of TiO_2 nanotubes employing various techniques. The low-cost, non-hazardous, and simple two-electrode processing production approaches of TiO_2 nanotubes with tuneable properties like length, packing, and crystallinity have been thoroughly reviewed here. Above all, bottom-up methods have been the focus of the entire conversation rather than more expensive and time-consuming methods like lithography and vacuum deposition.

Owing to the distinct qualities over other morphologies (0-D and 2-D), tube-shaped nanomaterials have been popular for a variety of applications. Therefore, a number of key points emerge for different applications:

i. Apparently, lower recombination probability of the photogenerated excitons and faster electron transport in 1-D nanostructures such as TNT makes popular to efficiently harvesting solar energy and converting it to electrical energy.

ii. Chemical inertness, desirable band gap and cost-effective earth abundant materials are the key points to make a suitable choice photo/photo-electro catalyst application. In this respect, TiO_2 has emerged as a prime contender to fulfil each of these requirements. It is evident that photo generated charge carrier recombination time can be prolonged in 1-D TiO_2 nanotubes, results in increase the charge carrier separation and excellent photocatalytic activity. Nevertheless, this photocatalytic activity can be potentially employed in destroying poisonous microorganisms.

iii. Due to the physicochemical properties and biocompatibility, TNT has the potential to be utilised in drug delivery research and efficiently overcome the drawbacks of current drug therapy solutions. Actually, the course of drug release depends on several factors such as the nanotubes' charge, dimensions, and surface chemistry, and the loaded drug's charge, molecular size, and diffusion coefficient, as well as the type of interaction between the drug molecules and TiO_2 inner surface.

iv. The contact angle $\geq 150°$ and the sliding angle $< 10°$, are the fundamental requirement to achieve a superhydrophobic surface. It appears that, in order to achieve superhydrophobic coating, a combination of micro-nano dimension roughness on the surface is preferred. Contextually, TNT has emerged as an ideal candidate to accomplish this optimum roughness on a surface. Nonetheless, the inertness feature of TiO_2 add-on the value in this regard.

13.7 FUTURE PERSPECTIVE

TiO_2 nanotubes have already demonstrated the capability of sensing of various gases, *e.g.*, CO, H_2 and NO_x. In solution phase, gold nanoparticles supported TNT exhibit excellent O_2 sensor performance. Similarly, it has been discovered that Pd/

Pt nanoparticles grafted on TNT can oxidised methanol efficiently. This finding is likely to be trending in methanol fuel cells. Moreover, exploiting the unique ability of sensing of variable concentration of H$_2$O$_2$, TNT can be further used for biosensor after modifying the surface horseradish peroxidase and thionine chloride. Nonetheless, the possibility of reducing oxide tube layers to metallic nanostructures has essentially gone unexplored. It would be interesting enough to examine the growth of the nanotubes in ionic liquids, in metallurgy perspective. Synthesizing mixed oxide nanotube powder to enhance the optical, electrical, and electrochemical performance of TiO$_2$ nanotubes would be one of the important future insights. At last, but not the least, another future trend could be the development of various protocols to dope the TiO$_2$ nanotube powder with foreign elements for various applications.

REFERENCES

1. P. Roy, R. Kisslinger, S. Farsinezhad, N. Mahdi, A. Bhatnagar, A. Hosseini, L. Bu, W Hua, B. D. Wiltshire, A. Eisenhawer, P. Kar, K. Shankar, *Chem. Eng. J.* **2018**, *351*, 482.
2. W. Hua, P. Kar, P. Roy, L. Bu, L. C. T. Shoute, P. Kumar, K. Shankar, *Nanomaterials* **2018**, *8*, 783.
3. D. V. Bavykin, S. N. Gordeev, A. V. Moskalenko, A. A. Lapkin, F. C. Walsh, *J. Phys. Chem. B* **2005**, *109*, 8565.
4. G. Pfaff, P. Reynders, *Chem. Rev.* **1999**, *99*, 1963.
5. J. Desilvestro, M. Gratzel, L. Kavan, J. Moser, J. Augustynski, *J. Am. Chem. Soc.* 1985, *107*, 2988.
6. B. O'Regan, *Nature* **1991**, *353*, 737.
7. M. Gratzel, *Inorg. Chem.* **2005**, *44*, 6841.
8. B. Gao, et al., *Thin Solid Films* **2015**, *584*, 61.
9. H. Xiao, W. Guo, B. Sun, M. Pei, G. Zhou, *Electrochim. Acta* **2016**, *190*, 104.
10. I. Abdullah, et al., *Anal. Methods* **2018**, *10*, 2526.
11. C. M. The, A. R. Mohamed, *J. Alloys Compd.* **2011**, *509*, 1648.
12. L. K. Preethi, R. P. Antony, T. Mathews, L. Walczak, C. S. Gopinath, *Sci. Rep.* **2017**, *7*, 14314.
13. A. Fujishima, K. Honda, *Nature* **1972**, *238*, 37.
14. H. Gerischer, H. Tributsch, *Ber. Bunsen-Ges.* **1968**, *72*, 437.
15. M. P. Dare-Edwards, J. B. Goodenough, A. Hamnett, K. R. Seddon, R. D. Wrigh, *Faraday Discuss. Chem. Soc.* **1980**, *70*, 285.
16. M. Grtzel, *Nature* **2001**, *414*, 338.
17. J. Huusko, V. Lantto, H. Torvela, *Sens. Actuators B* **1993**, *16*, 245.
18. A. L. Linsebigler, G. Lu, J. T. Yates, *Chem. Rev.* **1995**, *95*, 735.
19. S. Y. Huang, L. Kavan, I. Exnar, M. Graetzel, *J. Electrochem. Soc.* **1995**, *142*, L142.
20. R. Wang, K. Hashimoto, A. Fujishima, M. Chikuni, E. Kojima, A. Kitamura, M. Shimohigoshi, T. Watanabe, *Nature* **1997**, *388*, 431.
21. K. Satake, A. Katayama, H. Ohkoshi, T. Nakahara, T. Takeuchi, *Sens. Actuators B* **1994**, *20*, 111.
22. H. Gerischer, *Electrochim. Acta* **1995**, *40*, 1277.
23. M. Wagemaker, D. Lutzenkirchen-Hecht, A. A. vanWell, R. Frahm, *J. Phys. Chem. B* **2004**, *108*, 12456.
24. A. R. Armstrong, G. Armstrong, J. Canales, R. Garcia, P. G. Bruce, *Adv. Mater.* **2005**, *17*, 862.
25. P. Roy, D. Kim, K. Lee, E. Spiecker, P. Schmuki, *Nanoscale* **2010**, *2*, 45.

26. A. Ghicov, P. Schmuki, *Chem. Commun.* **2009**, 2791. DOI: 10.1039/b822726h

27. J. M. Macak, H. Tsuchiya, A. Ghicov, K. Yasuda, R. Hahn, S. Bauer, P. Schmuki, *Curr. Opin. Solid State Mater. Sci.* **2007**, *11*, 3.

28. J. Zhang, P. Zhou, J. Liu, J. Yu, *Phys. Chem. Chem. Phys.* **2014**, *16*, 20382.

29. D. Regonini, C.R. Bowen, A. Jaroenworaluck, R. Stevens, *Mater. Sci. Eng. R Rep.* **2013**, *76*, 377.

30. D.V. Bavykin, M. Carravetta, A.N. Kulak, F.C. Walsh, *Chem. Mater.* **2010**, *22*, 2458.

31. B. M. Rao, S. C. Roy, *J. Phys. Chem. C* **2014**, *118*, 1198.

32. P. Roy, S. Berger, P. Schmuki, *Angew. Chem. Int. Ed.* **2011**, *50*, 2904.

33. H. Wang, Y. Song, W. Liu, S. Yao, W. Zhang, *Mater. Lett.* **2013**, *93*, 319.

34. M. Ge, Q. Li, C. Cao, J. Huang, S. Li, S. Zhang, Z. Chen, K. Zhang, S. S. Al-Deyab, Y. Lai, *Adv. Sci.* **2017**, *4*, 1600152.

35. J. Desilvestro, M. Gratzel, L. Kavan, J. Moser, J. Augustynski, *J. Am. Chem. Soc.* **1985**, *107*, 2988.

36. J. Bisquert, *Phys. Rev. Lett.* **2003**, *91*, 010602.

37. J. Nelson, *Phys. Rev. B: Condens. Matter Mater. Phys.* **1999**, *59*, 15374.

38. N. K. Awad, E. A. Ashour, K. Allam, *J. Renew. Sustain. Energy* **2014**, *6*, 022702.

39. C. A. Grimes, O. K. Varghese, S. Ranjan, *Light* 2008. DOI: 10.1007/978-0-387-68238-9.

40. H. A. Hamedani, et al., *Adv. Funct. Mater.* **2014**, *24*, 6783.

41. S. Cho, J. W. Jang, K. H. Lee, J. S. Lee, APL Mater. **2014**, *2*, 010703.

42. J. Liang, C. Hao, K. Yu, Y. Li, *Nanomater. Nanotechnol.* **2016**, *6*, 1847980416680808.

43. X. Zhou, N. Liu, P. Schmuki, *ACS Catal.* **2017**, *7*, 3210.

44. M. Zhou, A. M. Glushenkov, O. Kartachova, Y. Li, Y. Chen, *J. Electrochem. Soc.* **2015**, *162*, A5065–A5069.

45. H. Wu, D. Li, X. Zhu, C. Yang, D. Liu, X. Chen, Y. Song, L. Lu, *Electrochim. Acta* **2014**, *116*, 129.

46. E. Ajami, K. F. Aguey-Zinsou, *J. Colloid Interface Sci.* **2012**, *385*, 258.

47. M. Gary-Bobo, et al., *Int. J. Pharm.* **2012**, *423*, 509.

48. D. Losic, M. S. Aw, A. Santos, K. Gulati, M. Bariana, *Expert Opin. Drug Delivery* **2015**, *12*, 103.

49. W. S. Lee, Y.-S. Park, Y.-K. Cho, *Analyst* **2015**, *140*, 616.

50. W. A. Abbas, M. Ramadan, A. Faid, A. Abdellah, A. Ouf, N. Moustafa, N. K. Allam, *Environ. Nanotechnol. Monit. Manag.* **2018**, *10*, 87.

51. A. M. Rather, U. Manna, *ACS Appl. Mater. Interfaces* **2018**, *10*, 23451.

52. R.N. Wenzel, *Ind. Eng. Chem.* **1936**, *28*, 988.

53. Q. Huang, Y. Yang, R. Hu, C. Lin, L. Sun, E.A. Vogler, *Colloids Surf. B Biointerfaces* **2015**, *125*, 134.

54. S. Nishimoto, Y. Sawai, Y. Kameshima, M. Miyake, *Chem. Lett.* **2014**, *43*, 518.

14 Two-Way Fluid-Structure Interaction Hydroelastic Simulation of Vibrating Membranes with Applications in Marine Renewable Wave Power

O. Anwar Bég, S. Kuharat, Tasveer A. Bég,
J. Pattison, Ali Kadir, Henry J. Leonard, and W.S. Jouri

14.1 INTRODUCTION

Compliant, i.e., deformable structures are becoming increasingly popular in marine applications [1]. With the advent of nanomaterials and other advanced composites [2], purely metallic structures are being replaced with a new generation of ocean engineering structural materials. Exciting new possibilities for revolutionary civil, energy, robotic and propulsion systems are therefore emerging. National University of Singapore (NUS) recently created MantaDroid, an aquatic robot that emulates the swimming locomotion of manta rays. The robotic manta ray, which swims at a speed of twice its body length per second and can operate for up to 10h, could potentially be employed for underwater surveillance in future. MIT's Computer Science and Artificial Intelligence Laboratory (CSAIL) unveiled a soft robotic fish dubbed "SoFi," which can swim unsuspecting regular tropical fish and provides a unique tool for marine biologists. It has recently been deployed very successfully in Fiji's Rainbow Reef at depths over 50 feet. Other examples of compliant structures include a hybrid Solar Hydrogen Island where several OTEC power plants could also be integrated with both wind and photovoltaic solar cells, novel composite-body marine craft developed at Izmir shipyard in Turkey, mega yachts and floating museums exploiting tension structure design.

As noted earlier, the robust design of these marine systems requires fluid-structure interaction (FSI) simulations. FSI is the coupling between fluid dynamics and structural mechanics. It is a computational challenge to solve such problems due to complex geometries, intricate fluid physics and complicated fluid-structure interactions, as elaborated by Degroote et al. [3]. FSI effects are extremely important as they can be encountered in a number of engineering complications, such as the stability and response of aircraft wings, the pulsating blood flow through arteries, the response of bridges or tall buildings in high winds and also the response of marine structures in interaction with the sea. Moreover, when a particular body is in interaction with a certain fluid, if that body experiences some type of deformation then the subject area of hydroelasticity is introduced. In general, hydroelasticity is the study of a flexible body moving through a fluid and has traditionally been associated with ship design, although approaches have usually employed linearized models to obtain solutions for velocities and deflections. The concept of fully coupled hydroelasticity however has emerged mainly in the late 20th and 21st century due to the refinements in algorithms and advances in computational hardware. In general, when analysing and categorizing what type of FSI is to be conducted, there are a number of characteristics that have to be taken into consideration. The main three categories are rigid body FSI, one-way FSI and two-way FSI. These are of increasing complexity in terms of deformation in the structure, respectively. In addition, each mode also increases in terms of the computational detail which is needed in the analysis. In naval design, most engineers have employed the MIT software, WAMIT [4], for analysing wave interactions with structures (it is based on the boundary integral equation method (BIEM), also known as the panel method) and interfaced this with, for example, Simulink® software for the dynamic and structural analysis. However, commercial CFD/FEA software such as ANSYS FLUENT [5], ADINA [3], etc., offer one tool for simulating the entire FSI problem.

In order to predict the full-scale, load-dependent deformation response and potential failure mechanisms of self-adaptive compliant marine structures, appropriate hydroelastic modelling is required. This may be done either via *hydrodynamic similarity relationships* (e.g. for traditional rigid, metallic marine rotors) or via CFD/FSI modelling. Hydroelastic scaling of, for example, surface-piercing propellers, applies only to isotropic metallic blades. For flexible composite naval rotors, scaling of the material is highly nontrivial, especially for the prediction of material failure. The effects of specimen size, material properties, stacking sequence, number of plies, and fibre orientation, among other characteristics, have been shown to have a significant effect on the failure strength of composite materials, as well as the failure mode of the test specimen. Many types of hydroelasticity can arise in marine vessels. In addition to commercial CFD/FEM software approaches, other researchers have developed a number of in-house codes for FSI hydroelastic modelling. This usually involves coupling the equations of motion for the fluid with that for the structure, as elaborated by Lee et al. [6]. Loukogeorgaki et al. [7] presented a "wet" hydroelastic analysis of a free, flexible, mat-shaped floating breakwater (FB) comprising a grid of flexible floating modules connected flexibly in both horizontal directions, in the frequency domain under the action of oblique incident waves. They used a three-component simulation, i.e., a 3-D structural model for an initial "dry" eigenvalue analysis, a 3-D hydrodynamic model for the hydroelastic analysis and an iterative procedure (based on natural frequencies) for a "wet" eigenvalue analysis. Wei et al. [8] deployed a time-domain hydroelastic method for very large floating structures (VLFS') in inhomogeneous waves using Cummins' equation. They discretized the continuous VLFS into rigid modules connected by elastic beam elements, thereby enabling inhomogeneous wave effects to be simulated via adopting different wave spectra over different regions of the VLFS. Frequency-domain hydrodynamic coefficients, (representing the hydrodynamic interactions between each floating module) were transformed into the time-domain hydroelastic model using Cummins' equation. They showed that inhomogeneity of waves exerted a substantial effect on the bending moments, shear forces and torsional moments of the structure, especially for a wave direction of 90°, in which larger forces may be induced compared with the homogeneous waves. Zhang and Lu [9] employed the Lu discrete-module-beam-bending-based hydroelasticity method (which discretizes the system into several rigid submodules that are connected by Euler-Bernoulli beam elements) to simulate flexible marine structures with complex geometric features. They determined the stiffness matrix in terms of the geometrical and physical properties of the flexible structure for each beam element in an analytical form by combining simpler geometries e.g. rectangular cross sections and employed a finite element method for nonlinear computations. Wang et al. [10] employed a fluid–structure interaction (FSI) methodology for simulating elastic bodies embedded and/or encapsulating viscous incompressible fluid, by combining a finite volume method (FVM) with large eddy simulation (LES) for turbulent flow and a structural dynamic solver based on the combined finite element method–discrete element method (FEM-DEM). The two solvers were interfaced using an immersed boundary method (IBM) iterative algorithm to improve information transfer and applied to submerged vegetation stems and blades of small-scale horizontal axis kinetic turbines. They showed that compliant marine kinetic turbine rotors have the potential of significantly reducing power production due to undesired twisting and bending of the blades. Lee et al. [11] studied the hydroelastic vibrations of composite marine propellers. Barbarit et al. [12] presented a linear mathematical and numerical model for analysing the dynamic response of a flexible electroactive wave energy converter (WEC) comprising a floating elastic tube filled with slightly pressurized sea water and fabricated from electroactive polymers (EAPs). They computed the vibration eigenmodes and employed spectral decomposition. They obtained a reasonable correlation with scale model experiments in Ecole Centrale de Nantes wave tank. Barbarit et al. [13] also employed a similar methodology to simulate the hydroelastic energy absorption performance of a fixed-bottom pressure-differential wave energy converter. Kang et al. [14] computed the coupled hydroelastic responses (deformations and resonance) of a multi-unit floating offshore wind turbine platform (MUFOWT) considering multiple dynamic excitation sources acting on the frame structure e.g. random water waves, mooring tension, wind loads, and wind turbine dynamics with time-domain analysis. These studies however neglected extensive visualization of the velocity, pressure and stress contours on the hydroelastic body. This is the objective of the current study, i.e., to present a more general analysis with detailed 3-D visualization of membrane structure hydroelastic response with ANSYS 18.2 FLUENT, which uses a similar approach to ADINA-FSI [15]. This approach provides a complete simulation for two-way FSI and deeper insight into actual structural response for engineering designers.

14.2 ANSYS FSI HYDROELASTIC METHODOLOGY

As previously mentioned, fluid-structure interaction (FSI) is a mutual interaction between a fluid-solid boundary or a gas-solid boundary. This results in a coupling between two domains and therefore must be solved in a coupled manner in order to obtain acceptable results. Complications occur in this field when the mechanical coupling between the two bodies acts in a number of directions at the surface of contact

(Fluid-Solid Interface), which in turn causes deformation in the solid structure resulting in a force being applied by the fluid flow, and then the new force from the fluid being applied to the structure once again, causing a continuing cycle until the physics changes, as summarized in Figure 14.1.

A coupling is considered to be strong when there are high levels of exchange between the two domains, e.g. the fluid has a large impact on the structure or vice versa. The coupling is weak when the effect of one domain dominates the other. In order to classify FSI problems, three dimensionless numbers have been suggested.

 i. The *mass* number, M_A. This number defines the ratio between the density of the fluid, ρ_f and the density of the structure, ρ_s:

$$M_A = \frac{\rho_f}{\rho_s} \tag{14.1}$$

This determines the consequence of the inertial effects of the fluid and the structure. If the value is close to one, the inertial effects of the fluid are comparable to the structure, and so must be taken into account when obtaining a robust numerical solution.

 ii. The *Cauchy* number, C_y. This number is the ratio between the dynamic pressure and the elasticity of the structure, which is quantified by the Young's Modulus, E, fluid density (ρ_f) and a characteristic velocity (V).

$$C_y = \frac{\rho_f V^2}{E} \tag{14.2}$$

This number determines the significance of the deformations induced by the flow. If this number is small, e.g. the structure does not move and the fluid's velocity is small, the structural deformations are said to be negligible.

 iii. The reduced velocity, V_r. This number is the ratio between the *characteristic flow velocity and the velocity of the wave propagation within the structure.*

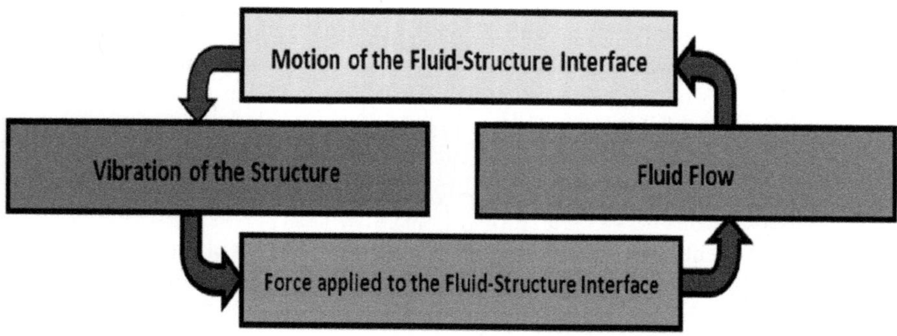

FIGURE 14.1   Fluid-structure coupling mechanism [15].

$$V_r = \frac{V}{\sqrt{\dfrac{E}{\rho_s}}} = \frac{V}{c_s} \qquad (14.3)$$

When this number is large, it can be said that the fluid dominates the structure in the perspective of time and the dynamics of the structure are negligible. However, if the dynamics of the structure increasingly dominate the number tends to zero. Moreover, if the number is close to 1, both dynamics of fluid and structure are taken into account within the problem.

In general, hydroelasticity deals with phenomena involving interaction between *inertial, hydrodynamic and elastic forces*. In ANSYS [16–19] the fluid domain is solved with ANSYS FLUENT CFD calculations, whilst the solid domain is simultaneously solved with ANSYS structural FEA calculations. The solutions obtained from the simulations complete the fluid-structure interaction analysis. Two-way interaction implies that we now need to send information between two codes. As a result, it is necessary to iterate in order to obtain adequate results. The type of coupling between the two codes must also now be established. This will be in the form of *weak coupling* and *strong coupling*. The weak coupling relates to a *steady-state solution* that can almost be regarded as a *one-way interaction* problem. Within these types of problems, the main interest is the steady-state solution. This means that in a typical algorithm, the CFD-solver will calculate the *pressure* of the fluid and export this information, whilst the FEA-solver imports that pressure and calculates the *deformations* in the structure due to that particular pressure. This cycle will continue until the steady-state condition is reached, usually until the velocity or pressure drops to zero. Typically, this coupling scheme is *explicit* as the previous step depends on the pressure calculated in that previous step. At this point, it can be said that weak coupling schemes do not take time scale of the problem into consideration. The strong coupling, however, takes *time scale* of the problem into consideration. Strong coupling relates to problems that are dynamic, where hydrodynamic and dynamic loads can change dramatically over time. In this type of coupling algorithm, the data is stored in the computer memory. In addition, the exchange between the two solvers will now occur at regular intervals, called the *coupling time step*. Depending on the particular problem it may be necessary to perform the exchange between the solvers several times during the time step in order to reach convergence and stability [20].

14.3 HYDROELASTIC VIBRATION OF RECTANGULAR PLATES

The ANSYS FSI simulation described in Section 14.5, involves a membrane immersed in a fluid- this specific membrane will be in the form of a rectangular plate. This important scenario has been analysed earlier by Kwak [21] in which the virtual mass effect on the natural frequencies and mode shapes of rectangular plates due to the presence of water on one side of the plate was addressed. The rectangular plate is immersed in water and is experiencing vibration due to an initial excitement. Due to this vibration, there will be an increase in kinetic energy which gives rise to motion in the water. As a result, the natural frequencies of the structure decrease significantly compared to the natural frequencies in a vacuum. The reason for this difference in natural frequencies mainly depends on the non-dimensionalized added virtual mass incremental (NAVMI) factor. This NAVMI factor Γ, reflects the *ratio of the kinetic energy of the water and the kinetic energy of the plate* [21]. This value is used in the approximate formula below to calculate the natural frequency of the plate in water; the determination of the NAVMI factor is the main area of interest as the natural frequency in water is calculated based on the natural frequency in a vacuum. In this study, the Rayleigh-Ritz method combined with the Green function method has been used. Two cases are considered for the outside boundary condition, one in which a plate is placed in an aperture of the rigid wall and the other independently resting on a free surface.

$$f_w = \frac{f_a}{\sqrt{1 + \beta\Gamma}} \tag{14.4}$$

Here f_w = natural frequency in water, f_a = natural frequency in a vacuum, $\beta = \rho_w a / \rho_p h$ = nondimensional parameter called thickness correction factor in which ρ_w and ρ_p are the mass densities of water and the plate, a = width for rectangular plate, Γ = NAVMI factor. The equation of motion is then obtained by combining the kinetic and potential energies of the plate and the kinetic energy of the water. The eigenvalue problem is solved based on the mass and stiffness matrices and the effect of water on mode shapes as well as the accuracy of the approximate formula is analysed, where the mass matrix consists of the mass matrix of the plate and the virtual mass matrix due to the presence of water. The results obtained show that the mode shapes differ slightly for lower modes but are different for higher modes when compared with experimental results. As a result, it is evident that the approximate formula guarantees very good accuracy for lower modes. This work provides a good reference for the ANSYS simulations to be conducted, although the natural frequencies of the plate (membrane) are not the primary focus of our work. In ANSYS, two-way FSI, careful computations are required since when the fluid and structure are tightly coupled, the deformation of the structure now influences the fluid flow within each time step, the *fluid mesh* now needs to be *automatically deformed and remeshed* to account for the deformation in the structural domain. In addition, both the FEA and CFD solvers are now needed to be coupled through co-simulation which requires close integration of the set-up and solution process. This type of analysis now needs powerful computational resources due to the requirement of performing both fluid and structural simulations simultaneously at each time step. ANSYS Workbench interface once again allows a relatively straightforward method of coupling the FEA and CFD solvers within a single GUI to carry out the two-way FSI. The integrated system allows the user to change the multiphysics model to propagate through the entire simulation automatically [21].

14.4   ANSYS STRUCTURAL AND CFD FORMULATION USED IN COUPLED TWO-WAY FSI ANALYSIS

Here we provide details on the formulations employed in ANSYS workbench for computing the flow variables (pressure and velocity) in ANSYS FLUENT [22] and the structural stresses, strains and deformations in ANSYS STRUCTURAL FEA (APDL) environment [23].

14.4.1 ANSYS FLUENT

This code has been deployed extensively by the authors in numerous multiphysical simulations including turbulent rocket channel cooling [24], thermal stress analysis of corroding gas turbine blade components [25], gold nanofluid solar collectors [26], rocket fuel tank sloshing [27], hydrogen/oxygen hybrid automotive fuel cells [28], aircraft wing moisture ingress FSI analysis [29], picodroplet microfluidic systems [30], aero-acoustic automotive broad band source simulation [31], gas turbine subsonic film cooling [32] and photovoltaic green systems [33]. To simulate incompressible, viscous fluid flow the following equations are used:

14.4.1.1 Mass Conservation Equation

The unsteady equation for mass conservation or continuity is written as follows:

$$\frac{\partial \rho}{\partial t} + \nabla \cdot \left(\rho \vec{v} \right) = 0 \tag{14.5}$$

The equation above is the general form of the unsteady mass conservation equation which is valid for both compressible and incompressible flows. Here $\left(\vec{v} \right)$ is the velocity vector in three dimensions, ρ is fluid density (kg/m³), t is time (s).

14.4.1.2 Momentum Conservation Equation

Conservation of momentum in an inertial (non-accelerating) reference frame takes the form.

$$\frac{\partial}{\partial t}\left(\rho \vec{v} \right) + \nabla \cdot \left(\rho \vec{v} \vec{v} \right) = -\nabla p + \nabla \cdot \left(\overline{\overline{\tau}} \right) + \rho \vec{g} + \vec{F} \tag{14.6}$$

Where p is the static pressure (Pa), $\overline{\overline{\tau}}$ is the stress tensor (N/m²), $\rho \vec{g}$ and $\vec{F}$ are the gravitational body force (N). $\vec{F}$ also has other model-dependent source terms that can be used for wave hydrodynamic forces in ANSYS FLUENT, although they are not considered in the present fully submerged membrane study since the structure is not surface-piercing. The $\overline{\overline{\tau}}$ term is given by:

$$\overline{\overline{\tau}} = \mu \left[\left(\nabla \vec{v} + \nabla \vec{v}^{T} \right) - \frac{2}{3} \nabla \cdot \vec{v} I \right] \tag{14.7}$$

Here μ is the dynamic viscosity (kg/(m s)), I is the unit tensor and the last term shown is the effect of volume dilation. No slip velocity conditions are applied on the hydroelastic membrane model (Section 14.5) at the rigid walls.

14.4.2 ANSYS STRUCTURAL

In ANSYS workbench, the stress analysis involves computations for equivalent stress, strain and total deformation. The principal stresses (σ_1, σ_2, σ_3) are calculated from the stress components by the cubic equation:

$$\begin{vmatrix} \sigma_x - \sigma_o & \sigma_{xy} & \sigma_{xz} \\ \sigma_{xy} & \sigma_y - \sigma_o & \sigma_{yz} \\ \sigma_{xz} & \sigma_{yz} & \sigma_z - \sigma_o \end{vmatrix} = 0 \qquad (14.8)$$

Here σ_o = principal stress (3 values). The three principal stresses are labelled σ_1, σ_2, and σ_3 (output quantities $S1$, $S2$, and $S3$). The principal stresses are ordered so that σ_1 is the most positive (tensile) and σ_3 is the most negative (compressive). The stress intensity σ_I (output as SINT) is the largest of the absolute values of $\sigma_1 - \sigma_2$, $\sigma_2 - \sigma_3$, or $\sigma_3 - \sigma_1$. That is:

$$\sigma_I = \mathrm{MAX}\left(|\sigma_1 - \sigma_{o2}||\sigma_2 - \sigma_3||\sigma_3 - \sigma_1|\right) \qquad (14.9)$$

The *von Mises* or equivalent stress σ_e (output as SEQV) is computed as follows:

$$\sigma_e \left(\frac{1}{2}\left[(\sigma_1 - \sigma_2)^2\right] + (\sigma_2 - \sigma_3)^2 + \left(\sigma_3 - \sigma_1\right)^2\right]\right) 1/2 \qquad (14.10)$$

or

$$\sigma_e \left(\frac{1}{2}\left[(\sigma_x - \sigma_y)^2 + (\sigma_y - \sigma_z)^2 + \left(\sigma_z - \sigma_x\right)^2 + 6\left(\sigma_{xy}^2 + \sigma_{yz}^2 + \sigma_{xz}^2\right)\right]\right)^{\frac{1}{2}} \qquad (14.11)$$

When $v' = v$ (*Poisson ratio* is input as PRXY or NUXY on MP command), the equivalent stress is related to the equivalent strain through the following relation:

$$\sigma_e = E\varepsilon_e \qquad (14.12)$$

Here E = Young's modulus (which is input as EX on MP command). It is important to note that the concept of von Mises stress is derived from distortion energy failure theory. It applies to test the isotropic and ductile materials such as metals and composites and to determine whether it will yield when directed towards complex loading conditions. Equivalent stress (*von Mises stress*) is often used in marine structural and naval design since it permits any arbitrary three-dimensional stress state to be represented as a single positive stress value. Equivalent stress is part of the maximum equivalent stress failure theory used to predict yielding in a ductile material. The *von Mises or equivalent strain* ε_e is computed as:

$$\varepsilon_e = \frac{1}{1+v'}\left(\frac{1}{2}\left[(\varepsilon_1 - \varepsilon_2)^2 + (\varepsilon_2 - \varepsilon_3)^2 + (\varepsilon_3 - \varepsilon_1)^2\right]\right)^{\frac{1}{2}} \qquad (14.13)$$

Here v' = effective Poisson's ratio which is defined as the material Poisson's ratio for elastic and thermal strains computed at the reference temperature of the body. The principal strains are calculated from the strain components by the 3-D matrix equation:

$$\begin{vmatrix} \varepsilon_x - \varepsilon_o & \dfrac{1}{2}\varepsilon_{xy} & \dfrac{1}{2}\varepsilon_{xz} \\[2mm] \dfrac{1}{2}\varepsilon_{xy} & \varepsilon_y - \varepsilon_o & \dfrac{1}{2}\varepsilon_{yz} \\[2mm] \dfrac{1}{2}\varepsilon_{xz} & \dfrac{1}{2}\varepsilon_{yz} & \varepsilon_z - \varepsilon_o \end{vmatrix} = 0 \tag{14.14}$$

Here ε_o = principal strain (3 values). The three principal strains are labelled ε_1, ε_2, and ε_3 (output as 1, 2, and 3 with strain items such as EPEL). The principal strains are ordered so that ε_1 is the most positive and ε_3 is the most negative. The strain intensity ε_1 (output as INT with strain items such as EPEL) is the largest of the absolute values of $\varepsilon_1 - \varepsilon_2$, $\varepsilon_2 - \varepsilon_3$, or $\varepsilon_3 - \varepsilon_1$. That is:

$$\varepsilon_1 = \text{MAX}\left(|\varepsilon_1 - \varepsilon_2|, |\varepsilon_2 - \varepsilon_3|, |\varepsilon_3 - \varepsilon_1|\right) \tag{14.15}$$

The *von Mises* or equivalent strain ε_e (output as EQV with strain items such as EPEL) is computed as:

$$\varepsilon_e = \frac{1}{1+v'}\left(\frac{1}{2}\left[(\varepsilon_1 - \varepsilon_2)^2 + (\varepsilon_2 - \varepsilon_3)^2 + (\varepsilon_3 - \varepsilon_1)^2\right]\right)^{\frac{1}{2}} \tag{14.16}$$

14.5 ANSYS FSI HYDROELASTIC MEMBRANE SIMULATION PROCEDURE

The specific simulation executed illustrates an example of an oscillating plate (membrane) two-way hydroelastic fluid-structure interaction. The structural physics is set up in the *transient structural*, the fluid physics set up in the fluid flow (ANSYS FLUENT) analysis system, and then the two are solved simultaneously under the solution of the fluid system. Throughout the simulation, coupling between the two analysis systems is needed in order to model the interaction between the structural and fluid systems correctly. The physical problem motivating this simulation is a flexible tidal energy generator system. To simplify the hydrodynamics, the membrane is considered to be fully submerged and simulated as a 2-dimensional immersed thin plate membrane which is anchored at the base. The simplified geometrical model is shown in Figure 14.2 with the dimensions of the plate membrane utilised. The actual problem we are dealing with here involves applying an initial pressure of 100 Pa to one side of the plate in order to distort it. Once this pressure is released the plate will begin to oscillate backward and forwards as it attempts to regain equilibrium. The fluid that is surrounding the plate will then damp the plate oscillations, which in turn will decrease the oscillations with time ("hydrodynamic damping"); this simulation constitutes the first stage in the FSI analysis of the more complex, multi-plate structure practical system which will be fully addressed in subsequent investigations.

The first step in carrying out the two-way FSI hydroelastic analysis for the vibrating membrane is to select the analysis systems. This was carried out in the ANSYS workbench by setting up a coupled analysis systems, i.e., a transient structural system and fluid flow (CFX) system. The important parameters for the material that will be used are Young's Modulus of 2.4×10^6 Pa and a Poisson's ratio of 0.34. These are then prescribed in the structural system by selecting the density and isotropic elasticity. Once the material has been selected for the plate, which will act as the membrane (Figure 14.3), a mesh needs to be generated around the structure. This is achieved by first selecting the 'Model' tab in the structural system. The first step from this point is to suppress the fluid body so that the membrane is the only body that appears in the model window. By selecting the 'solid' part, the material is then changed to 'VL_Plate' which was created earlier. Once this is complete attention can then be taken to the 'Mesh' tab. First of all, this tab was right clicked, and 'Sizing' was selected. This then allowed to select the number of divisions where the mesh would be created. For this particular mesh, a 10 by 4 quadrilateral mesh (grid) is designed as shown in Figure 14.4. At this stage, the next important parameter was the type of analysis that will be carried out for this simulation. This was carried out whilst still in 'Model' for the structural system and by selecting the 'Analysis Settings' tab within the 'Transient' tab. Once this was completed there were a number of parameters that had to be altered. First of all, the 'Step End Time' was set to 10 s, this acts as the total time for the simulation. Secondly, 'Auto Time Stepping' was then turned off. Finally, the analysis was then defined by 'substeps' and the 'Number of Substeps' was set to 1. The final step in terms of the physics regarding the mechanical/structural application was applying loads at the relevant locations. This involved fixed support, fluid-solid interface and finally pressure loads. This was carried out by right-clicking the 'Transient' tab and selecting the relevant loads and the face related to specific load. For the fixed support and fluid-solid interface, this stage is relatively simple. However, when selecting the pressure being applied to the plate, minor errors can be made. First of all, the magnitude of the pressure is set to 'Tabular Data'. The values that are inputted into this section can be seen below in Figure 14.5.

Once the mechanical/structural system has been completed, we turn our attention to the selected closed fluid medium. By selecting named selections for the fluid body (Figure 14.6) the geometry, faces can be easily grouped and analysed more efficiently later in the simulation. This is executed by selecting the face, right-clicking and selecting 'Named Selection'. The way in which they were named is self-explanatory and these can be seen below, the area in which is name 'wall_deforming' is the walls where the structural solid body is located. These named selections are

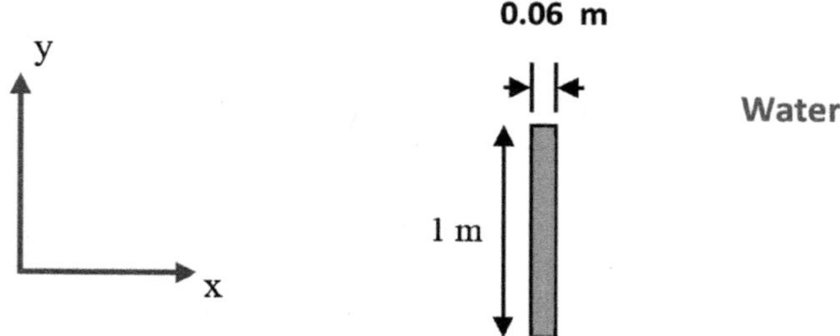

FIGURE 14.2   2D simplified geometric model of fully immersed membrane plate in water.

FIGURE 14.3 ANSYS thin membrane immersed.

deployed subsequently for aspects such as deforming mesh zones. Pending the naming of the selections, a mesh may be generated for the fluid system. This was carried out by selecting the 'Mesh' tab within the Fluid system. Once this is selected, the next tab that is selected is also 'Mesh'. Similar to the structural system the first step is to input *the sizing of the mesh* that will be used, therefore 'Sizing' constitutes the next selection. The sizing selections that were made can be seen below in Figure 14.7.

The selections for the method and mesh type are then selected, as shown in Figure 14.8. At this point, the mesh can be generated. After an extensive mesh independence study, the finalized mesh selected is shown in Figure 14.9 and comprises 2,500 surface quadrilateral elements engulfing the structural membrane.

The next stage of the simulation is to set up the simulation (solver, i.e., processor) in Fluent. Time is set to 'Transient'. Material properties for the fluid within the cavity are water (density 1,000 kg/m^3 and viscosity 1.0016 mPas). The next stage is then to analyse the areas which will be considered to be *dynamic mesh* zones. The appropriate selections are shown in Figure 14.10

Next the *dynamic mesh zones* are selected. Each face of the hydroelastic model is designated a different dynamic mesh zone name, as shown in Figure 14.11. The appropriate settings for the 'Solution Method' are then prescribed and summarized in Figure 14.12.

The final stage in Fluent is then to select the 'Solution Initialization' and select the 'Initialize' button and then close Fluent. The next stage of the hydroelastic simulation is to complete the set-up. This is opened by selecting the 'Setup' tab in the

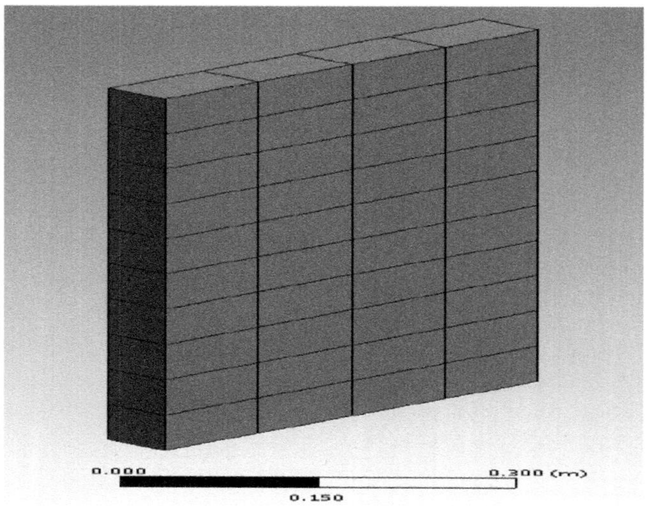

FIGURE 14.4 Generated mesh for membrane structure geometry [34].

Tabular Data

	Steps	Time [s]	✔ Pressure [Pa]
1	1	0.	100.
2	1	0.5	100.
3	1	0.51	0.
4	1	10.	0.

FIGURE 14.5 Tabulated pressure data.

'System Coupling' system. The first step is to select 'Analysis Settings' and change the Analysis type to Transient. The next is then to change the end time to 10 and step size to 0.1. Once this is complete, the next stage is to create data transfers. This is achieved by right-clicking the 'wall_deforming' region and selecting 'Create Data Transfer'. This is done for both the fluid and structural components. The selection for each data transfer is provided in Figures 14.13 and 14.14.

The next stage is to select the 'Intermediate Restart Data Output' and select 'At Step Interval' for the output frequency and then change the step interval to 5. Once this is complete, attention can be directed to the 'Solution' tab on the system coupling system right-clicked and updated. This completes the simulation process. Extensive contour visualizations of the hydrodynamic and structural results are provided in the next section.

FIGURE 14.6 Named selections for fluid body.

Sizing	
Size Function	Curvature
Relevance Center	Coarse
Initial Size Seed	Active Assembly
Smoothing	Medium
Transition	Slow
Span Angle Center	Fine
☐ Curvature Normal Angle	Default (18.0 °)
☐ Min Size	6.e-002 m
☐ Max Face Size	0.20 m
☐ Max Tet Size	Default (2.08760 m)
☐ Growth Rate	Default (1.20)
Automatic Mesh Based Defeaturing	On
☐ Defeature Size	Default (3.e-002 m)
Minimum Edge Length	6.e-002 m

FIGURE 14.7 Size selections for fluid mesh.

Scope	
Scoping Method	Geometry Selection
Geometry	1 Body
Definition	
Suppressed	No
Method	Sweep
Element Midside Nodes	Use Global Setting
Src/Trg Selection	Manual Source
Source	1 Face
Target	Program Controlled
Free Face Mesh Type	All Quad
Type	Number of Divisions
☐ Sweep Num Divs	1
Element Option	Solid
Advanced	
Sweep Bias Type	No Bias

FIGURE 14.8 Mesh method and mesh type for fluid.

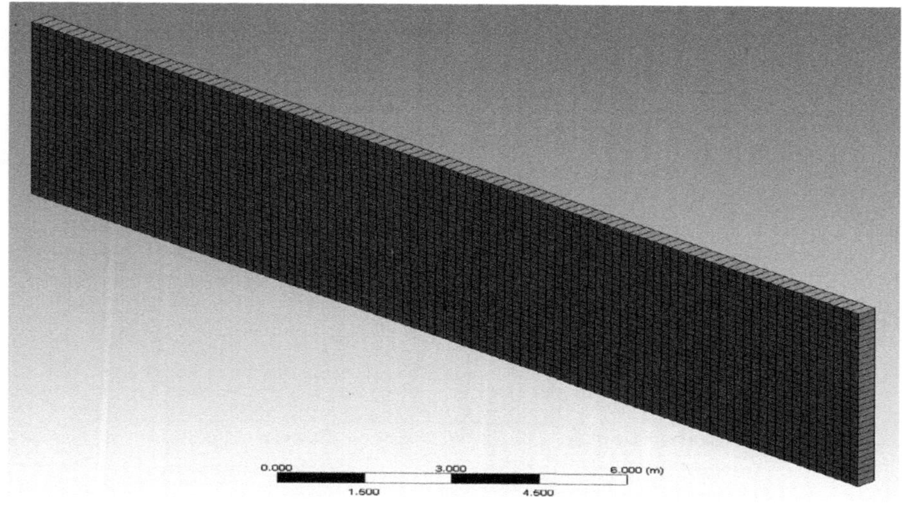

FIGURE 14.9 Mesh for fluid system.

FIGURE 14.10 Mesh method settings (ANSYS FLUENT).

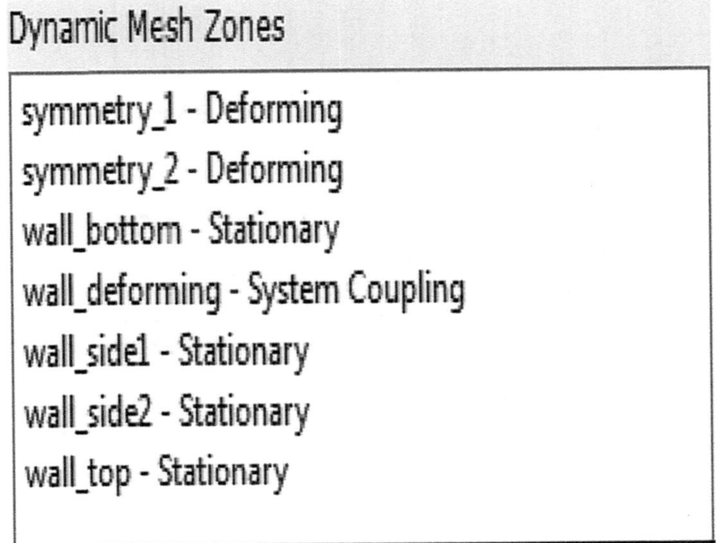

FIGURE 14.11 Dynamic mesh zones.

FIGURE 14.12 'Run calculation' selections.

	A	B
1	Property	Value
2	⊟ Source	
3	Participant	Structural
4	Region	Fluid Solid Interface
5	Variable	Incremental Displacement
6	⊟ Target	
7	Participant	Fluid
8	Region	wall_deforming
9	Variable	displacement
10	⊟ Data Transfer Control	
11	Transfer At	Start Of Iteration
12	Under Relaxation Factor	1
13	RMS Convergence Target	0.01
14	Ramping	None

FIGURE 14.13   FSI data transfer 1.

	A	B
1	Property	Value
2	⊟ Source	
3	Participant	Fluid
4	Region	wall_deforming
5	Variable	force
6	⊟ Target	
7	Participant	Structural
8	Region	Fluid Solid Interface
9	Variable	Force
10	⊟ Data Transfer Control	
11	Transfer At	Start Of Iteration
12	Under Relaxation Factor	1
13	RMS Convergence Target	0.01
14	Ramping	None

FIGURE 14.14   FSI data transfer 2.

14.6 ANSYS FSI HYDROELASTIC MEMBRANE SIMULATION RESULTS

ANSYS FSI provides an excellent 'CFD Post' facility for visualizing results for both the fluid and structural systems. The key results in the present simulations are Von Mises (equivalent) stress on the plate membrane (mathematically defined earlier in Section 14.4), the velocity and the pressure distributions.

14.6.1 VON MISES STRESS DISTRIBUTION

As previously discussed, an initial pressure was applied to the plate and then released so that the plate can oscillate. Therefore, it is essential to evaluate the stress distribution on the plate throughout the 10 s simulation- the appropriate contour plots (isochrones) are given below in 1 s intervals (i.e., $t = 1,2,3....10\,s$), with a supplementary visual at 0.2 s, in Figures 14.15–14.25.

When looking at the simulation the characteristics discussed earlier are evident as the membrane (plate) within the immersed fluid domain deforms significantly and influences the fluid surrounding it. Inspection of the von Mises stress plots (Figures 14.15–14.25) acting upon the plate membrane reveals that the material is experiencing elastic deformation and the main area of stress is localized near the base of the plate where it is fixed to the cavity. Out of the ten results displayed here, it can be seen that the highest amount of stress upon the plate was evident at 1 s at a value of 6.498 Pa. It is important to note here that this is also the time at which the plate tip experiences the largest displacement, as shown in Figures 14.26 and 14.27. In addition, the area at which the stress in concentrated on the thin plate is also of interest when analysing the properties of the plate and how the program uses the 'fixed support' function in the simulation setup. This is also confirmed by referring to the graph of tip deflection (Figure 14.28). The solution for this graph has been validated with COMSOL multiphysics finite element simulation which also provides a robust FSI methodology [35]. COMSOL offers two types of solvers for fluid-structure interaction problems (as well as other multiphysics problems). The first is the fully coupled solver, or *monolithic* solver as it is sometimes called in literature, and the second is the segregated solver (or *partitioned* solver). Having both solvers enables optimal solver selection for a wide range of FSI problems. The default solver settings work well for most problems; however, there are also a lot of solver functionalities for

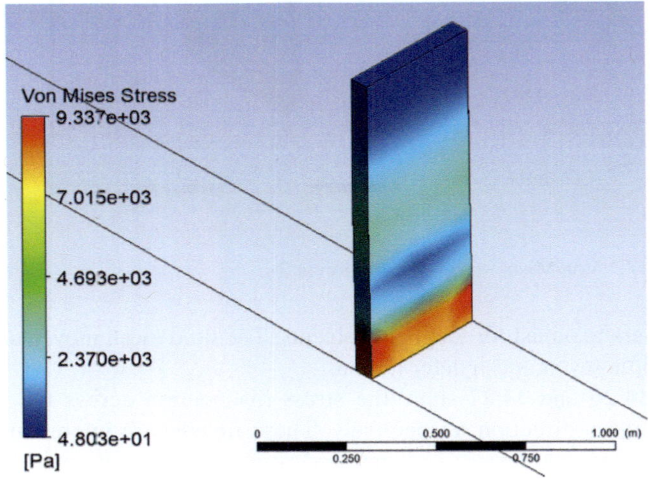

FIGURE 14.15 Von Mises stress distribution at 0.2 s.

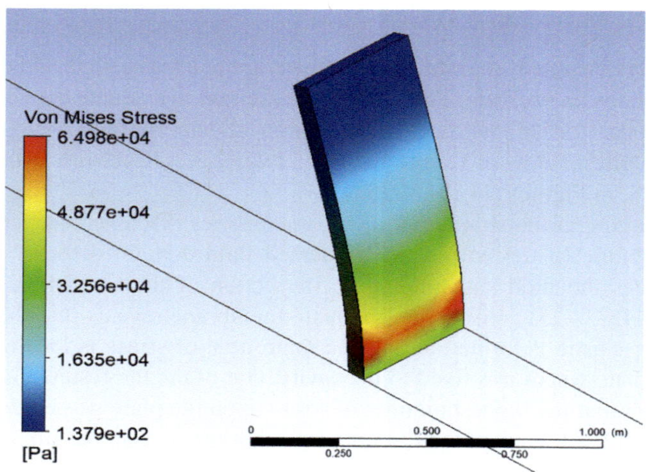

FIGURE 14.16   Von Mises stress distribution at 1 s.

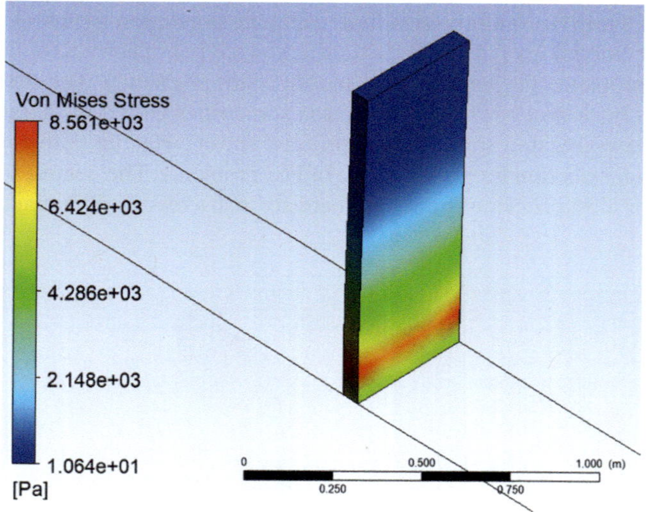

FIGURE 14.17   Von Mises stress distribution at 2 s.

advanced users to adjust for tougher problems. The fluid mesh movement algorithm can also handle severe mesh deformation.

Figures 14.26 and 14.27 show the stress distribution across the plate in the x-direction and y-direction, respectively. These images are interesting as the red areas show where the largest concentrated stress, therefore it could be suggested that this is the point at which the plate would fail and possibly experience plastic deformation. By carrying out some simple measurements this point is 0.112 m from

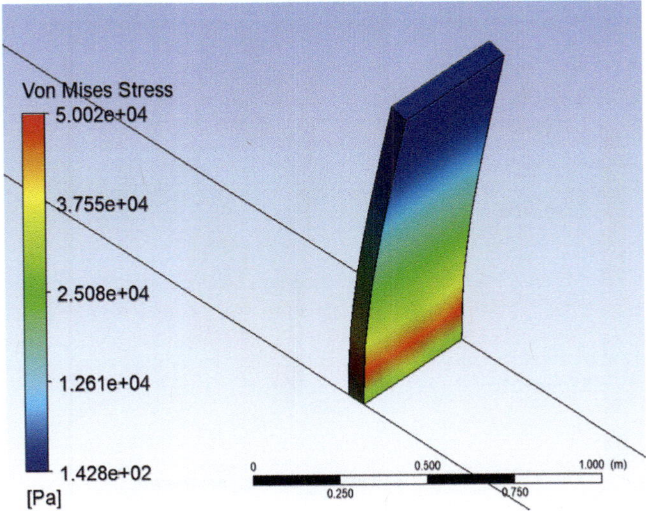

FIGURE 14.18   Von Mises stress distribution at 3 s.

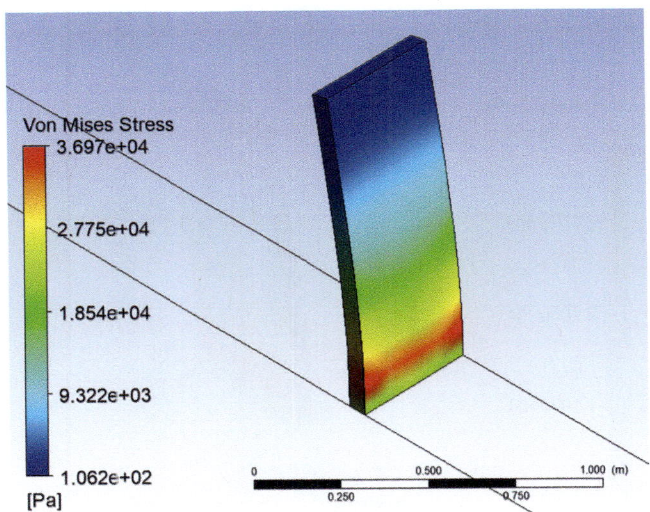

FIGURE 14.19   Von Mises stress distribution at 4 s.

the base of the membrane plate where it is fixed. Moreover, it is also interesting that from the top of the plate, the stress increases uniformly up until this point, and then, the stress reduces to roughly 50%.

In order to understand how much the plate membrane has oscillated and how much damping is present within the fluid, the tip deflection is computed in ANSYS and plotted against time. Figure 14.28 is generated by analysing how much the mesh has deformed within the simulation. Figure 14.28 shows via the tip deflection time

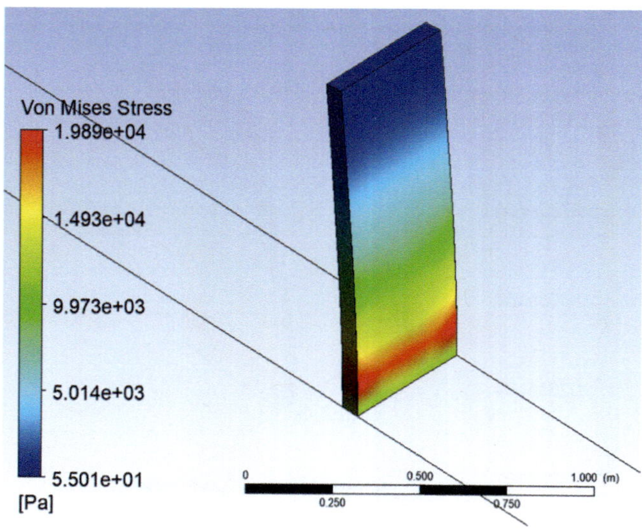

FIGURE 14.20   Von Mises stress distribution at 5 s.

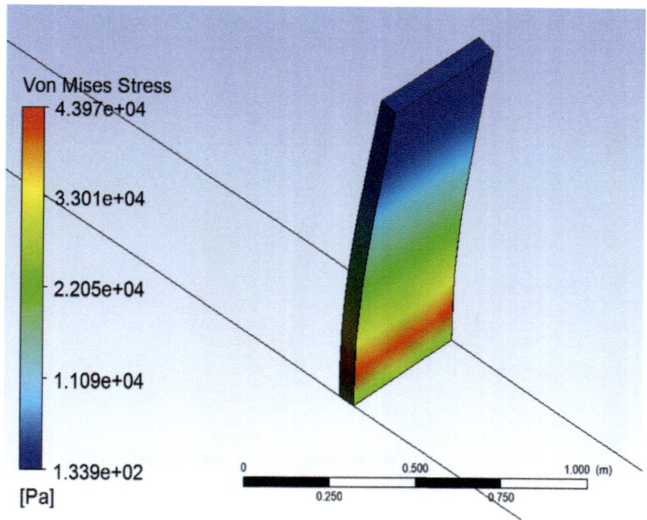

FIGURE 14.21   Von Mises stress distribution at 6 s.

history, that the plate oscillates in a fashion that would be expected over the 10 s period that simulation took place. Over the 10 s, the plate oscillated side to side with the tip deflection gradually decreasing over that time. It could be said that if the simulation was executed for a longer time period the plate would become stationary and return to its original position. By closely looking at this graph and the rate at which the displacement declines, it could be estimated from the damping that the

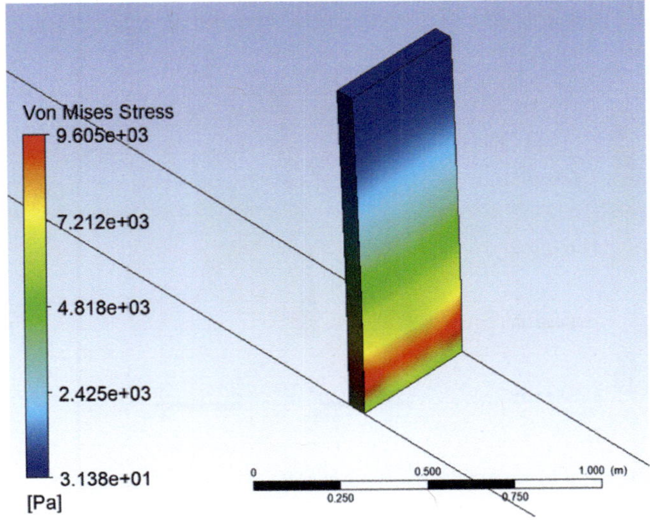

FIGURE 14.22 Von Mises stress distribution at 7 s.

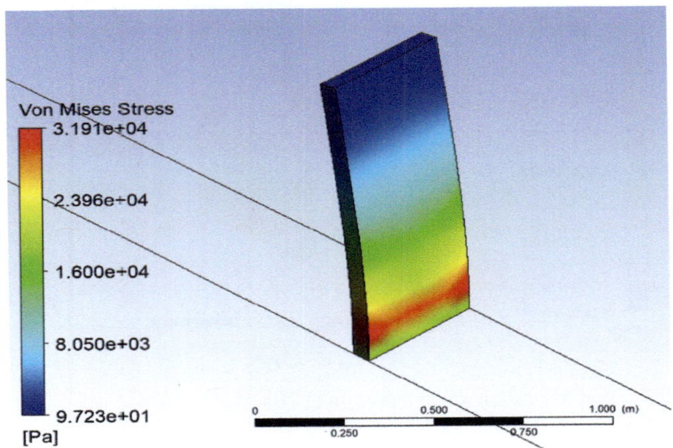

FIGURE 14.23 Von Mises stress distribution at 8 s.

plate would return to its original position after 20 s. The COMSOL finite element simulation concurs closely with the ANSYS FSI tip deflection results, adding further confidence to the mesh design and solution settings employed in the latter. 2,750 quadrilateral 6-node elements were employed in the COMSOL simulation. Figure 14.28 further shows that as the tip generally oscillates from one side of the axis to the other, hydroelasticity is correctly captured. As time progresses, the displacement decreases as the oscillations are damped due to the fluid and the elastic properties of the plate. Enhanced damping is evidently possible with larger stiffness of the plate; however, this is counter-productive for tidal membrane energy generation since compliance of the structure is essential for absorbing energy and harnessing it.

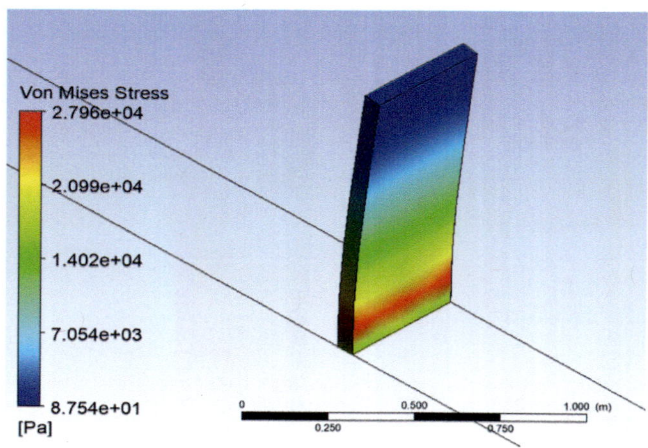

FIGURE 14.24 Von Mises stress distribution at 9 s.

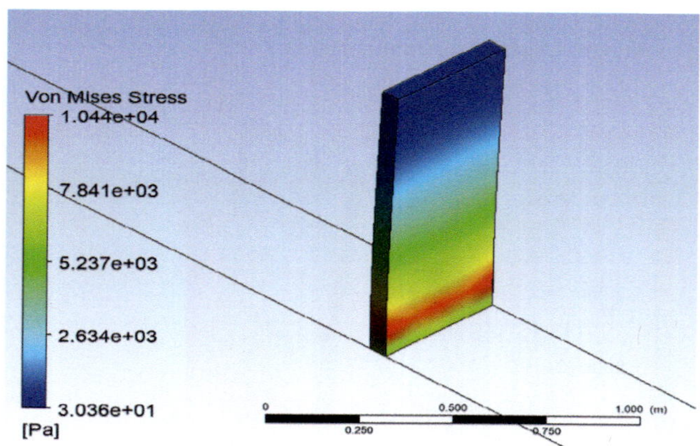

FIGURE 14.25 Von Mises stress distribution at 10 s.

14.6.2 Velocity Distribution

The next parameter that can be evaluated is the velocity distribution in terms of a vertex in the fluid domain. As discussed earlier, once the pressure is applied to the plate, the plate will deform causing a disturbance and results in movement of the fluid. The way in which this disturbance can be measured and analysed is by assessing both the velocity of the fluid as well as the pressure. The velocity vectors around the plate (membrane) can be seen below, once again in 1 s intervals, as illustrated in Figures 14.29–14.39.

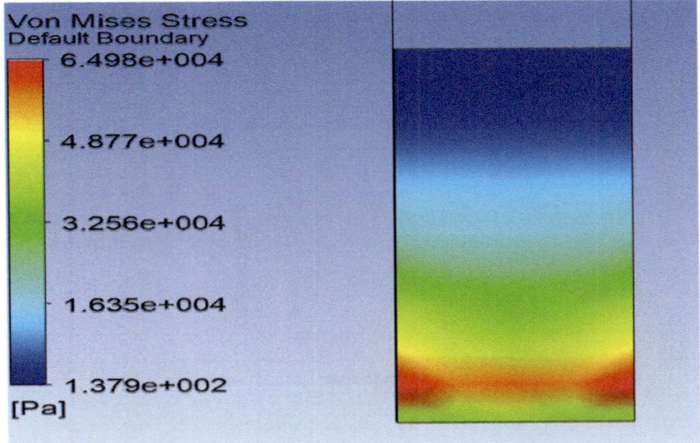

FIGURE 14.26　Von Mises stress at 1 s (front view, *x*-direction).

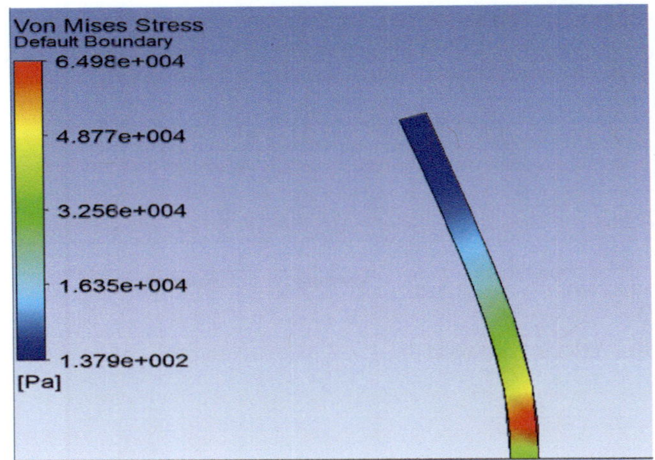

FIGURE 14.27　Von Mises stress at 1 s (side view, *y*-direction).

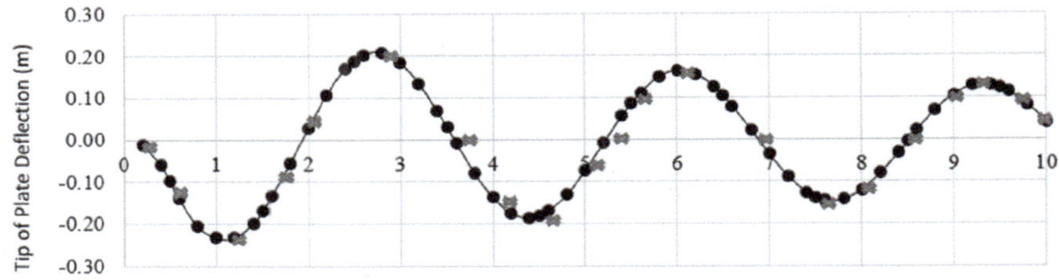

FIGURE 14.28　ANSYS FSI versus COMSOL FEM solution for tip deflection over 10 s.

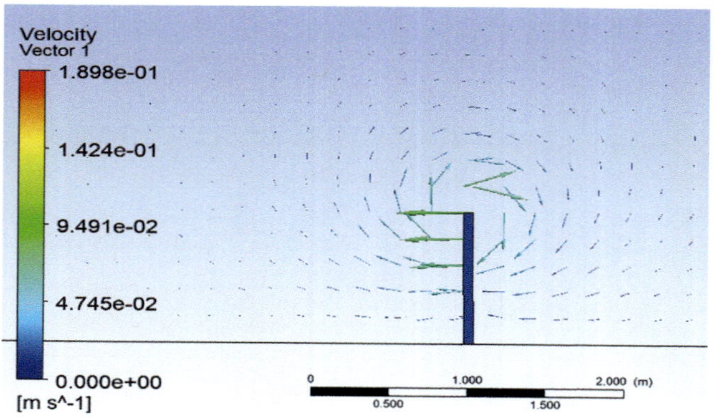

FIGURE 14.29   Velocity vectors at 0.2 s.

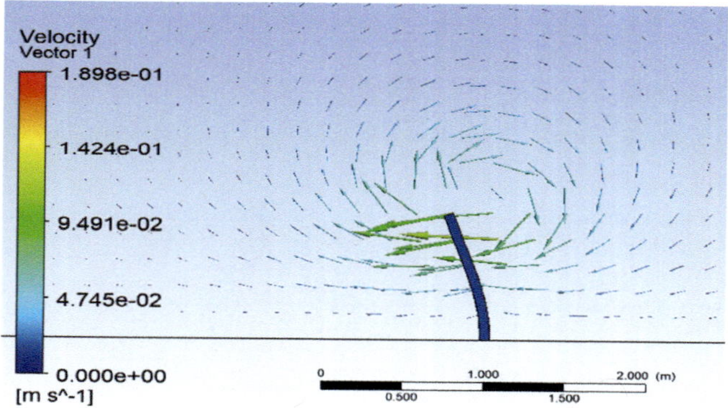

FIGURE 14.30   Velocity vectors at 1 s.

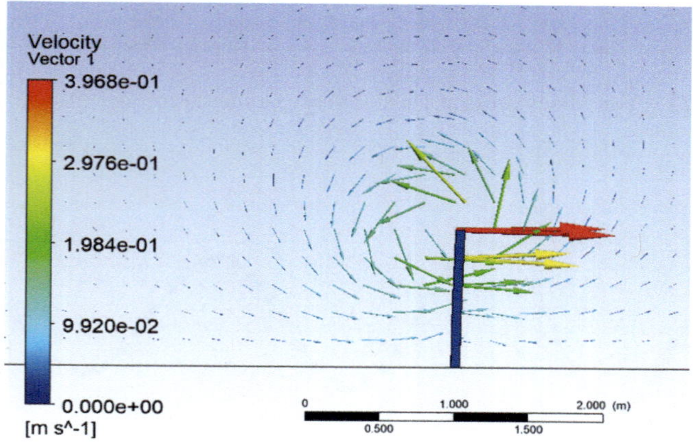

FIGURE 14.31   Velocity vectors at 2 s.

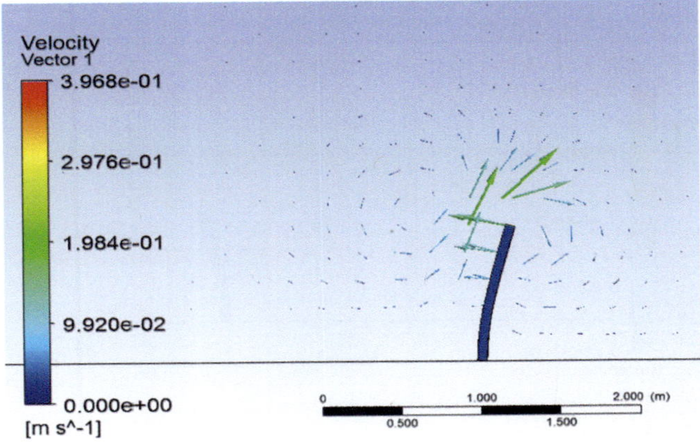

FIGURE 14.32 Velocity vectors at 3 s.

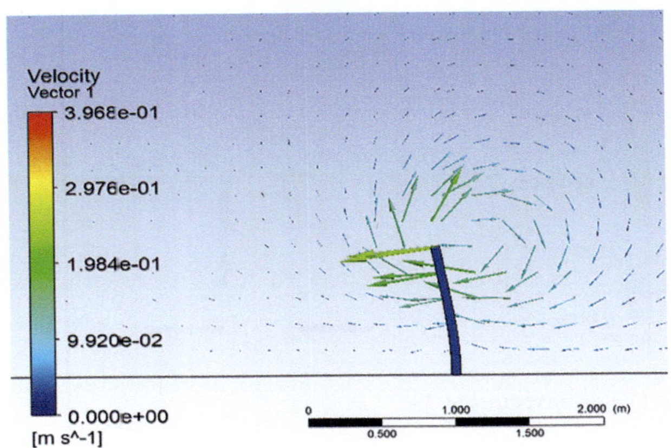

FIGURE 14.33 Velocity vectors at 4 s.

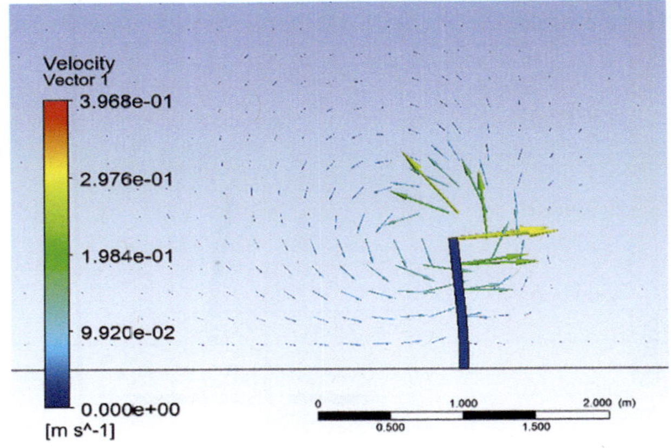

FIGURE 14.34 Velocity vectors at 5 s.

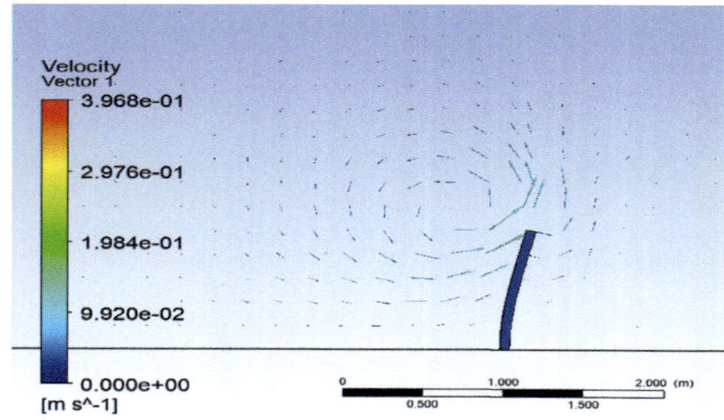

FIGURE 14.35   Velocity vectors at 6 s.

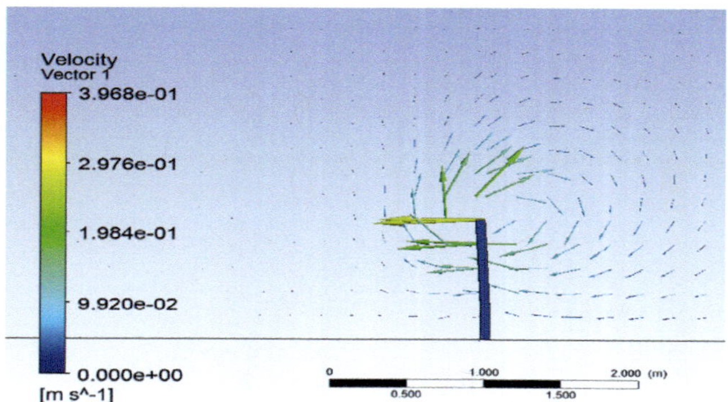

FIGURE 14.36   Velocity vectors at 7 s.

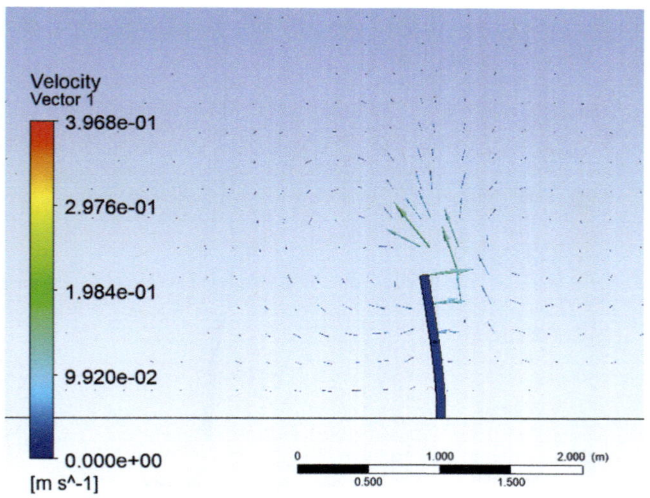

FIGURE 14.37   Velocity vectors at 8 s.

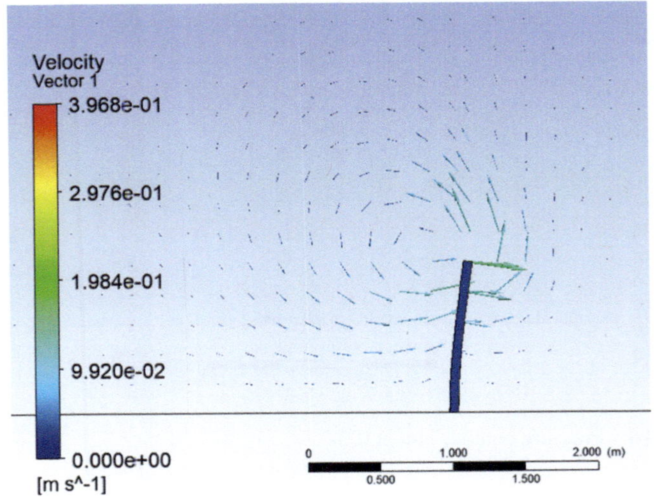

FIGURE 14.38   Velocity vectors at 9 s.

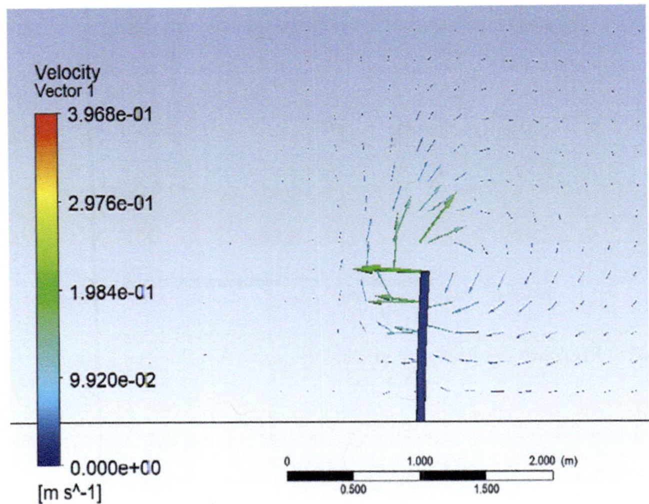

FIGURE 14.39   Velocity vectors at 10 s.

14.6.3   PRESSURE CONTOUR PLOTS

As discussed previously, due to the two-way FSI coupling, when the structural system deflects, a resulting deflection in the fluid will occur. As a result, a new pressure will be present as the plate oscillates throughout the simulation. The results below show the pressure contours around the membrane plate, once again at 1 s intervals (Figures 14.40–14.50).

Inspection of the velocity contours (Figures 14.29–14.39) provide a good insight to the distortion in the fluid interfacing with the structure. The point at which the

 Eco-Materials and Green Energy for a Sustainable Future

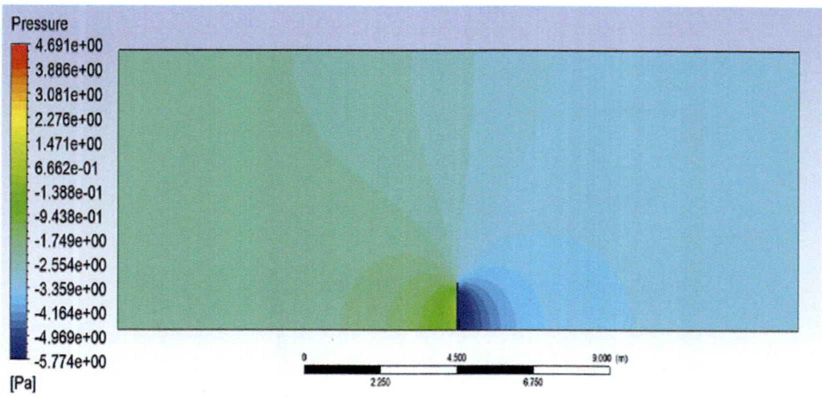

FIGURE 14.40 Pressure contour plot at 0.2 s.

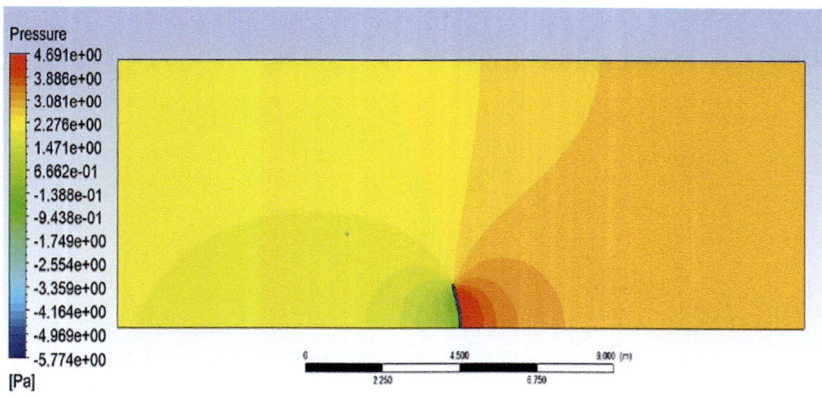

FIGURE 14.41 Pressure contour plot at 1 s.

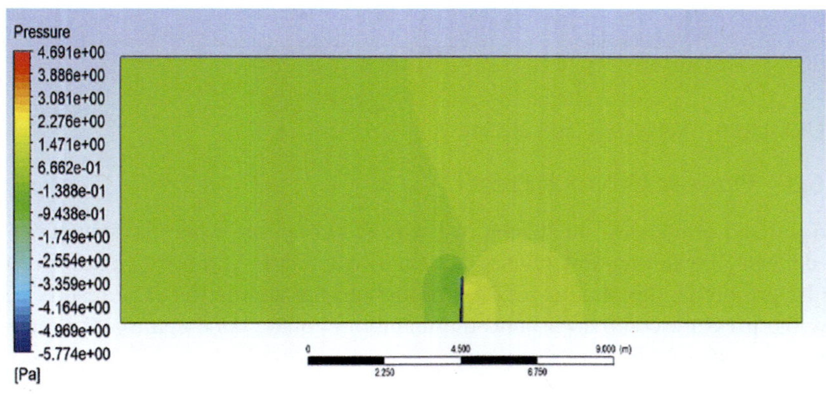

FIGURE 14.42 Pressure contour plot at 2 s.

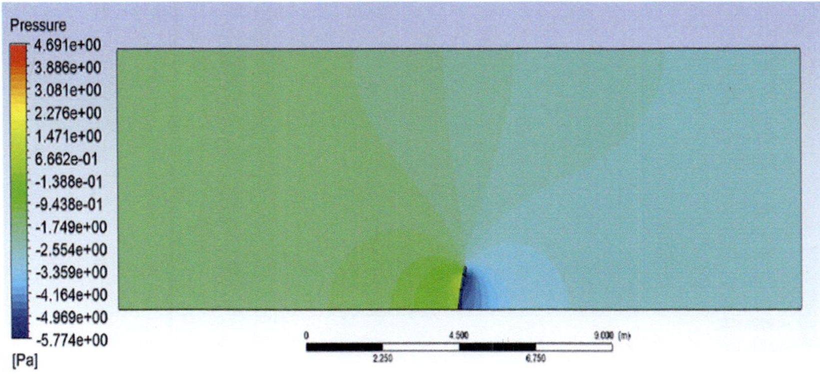

FIGURE 14.43 Pressure contour plot at 3 s.

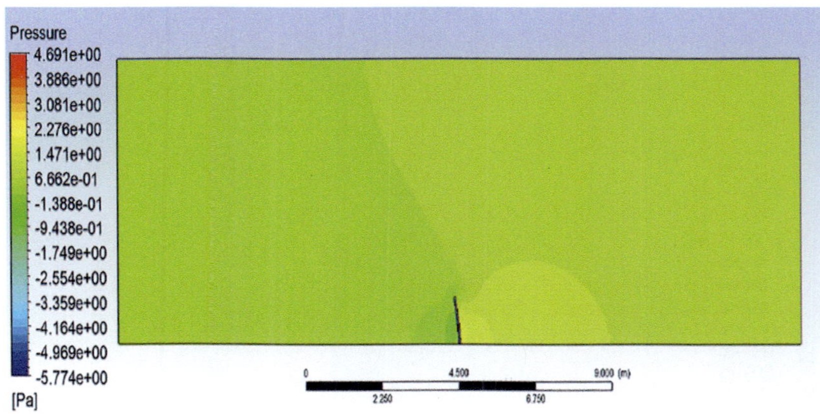

FIGURE 14.44 Pressure contour plot at 4 s.

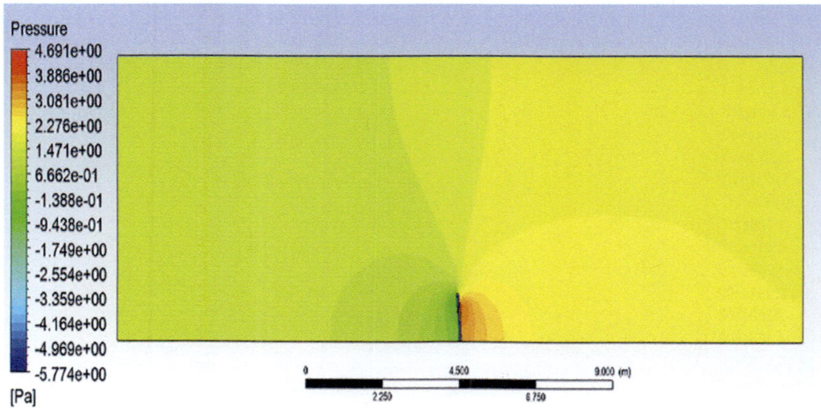

FIGURE 14.45 Pressure contour plot at 5 s.

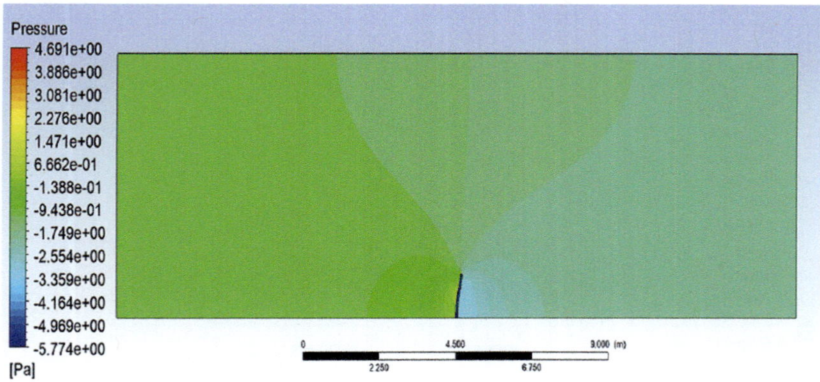

FIGURE 14.46 Pressure contour plot at 6 s.

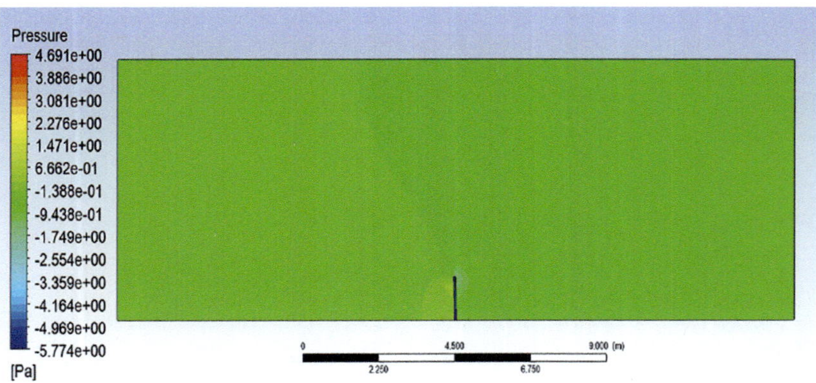

FIGURE 14.47 Pressure contour plot at 7 s.

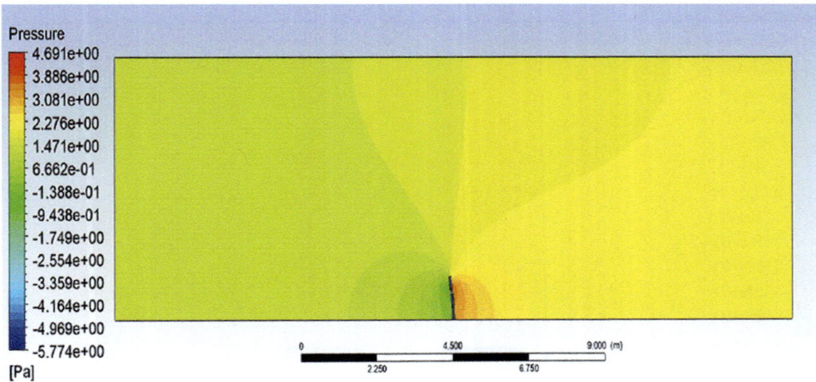

FIGURE 14.48 Pressure contour plot at 8 s.

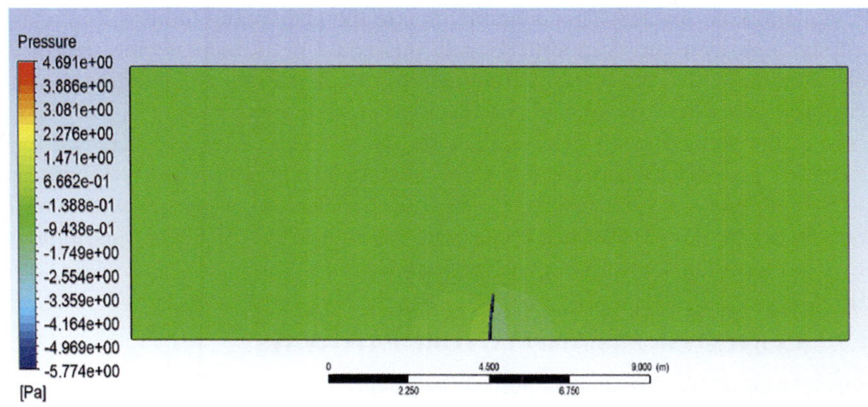

FIGURE 14.49 Pressure contour plot at 9 s.

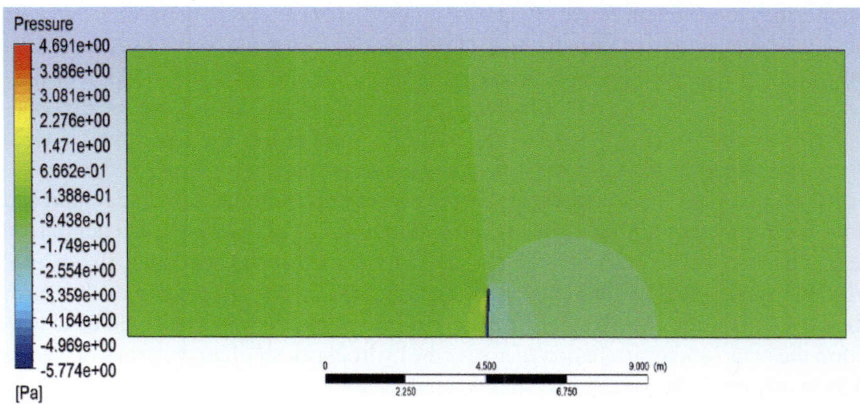

FIGURE 14.50 Pressure contour plot at 10 s.

largest velocity occurs in this section is at 2 s, at a velocity of 0.3968 m/s. This is noteworthy since it arises *after the first oscillation,* as the plate is acting as a spring action upon the fluid in the cavity. It is also important to note here the circular motion in the fluid where the tip of the plate is the source of this shape. This vertex shape remains the same throughout the simulation, however, it progressively weakens as time elapses due to the oscillatory motion being damped.

Examining the pressure contour plots within the fluid domain is also conducted throughout the entire simulation in increments of 1 s (Figures 14.40–14.50). As expected, significantly high pressure is computed at the very start of the simulation, shortly after the initial pressure is applied to one side of the plate membrane, the pressure being 5.774 Pa. The next value for the highest pressure is evaluated at 1.2 s and has a magnitude of 6.087 Pa. In addition, when further analysing the results, it is evident that major changes in these contours can only be identified in close proximity to the membrane surface. This is mainly due to the plate being finite within a large hydrodynamic domain. A modification to this could be to increase the size of the plate or decrease the size of the hydrodynamic domain in order to gain more in-depth analysis. Earlier the vibration of the rectangular plate membrane was analysed in terms of the natural frequencies and mode shapes. The difference between these parameters is attributable to the non-dimensionalized added virtual mass incremental (NAVMI) factor. As a result, it is possible for the natural frequencies and the mode shapes to be computed and then compared with ANSYS FSI analysis, although this is not considered here due to space restrictions.

14.7 CONCLUSIONS AND FUTURE PATHWAYS

In the present work, a detailed analysis of the two-way fluid structure interaction (FSI) behaviour of a vertical membrane (steel) structure is conducted in two-dimensions as a simulation of a tidal energy membrane system. The analysis is performed on a thin plate acting as a membrane, experiencing under damped oscillatory motion within a still marine environment (wave effects are ignored). Linear elastic material behaviour is considered and extensive visualization of pressure, velocity and Von Mises (equivalent) stress contours are provided. Mesh independence is conducted and validation of the tip deflection with COMSOL multiphysics software is included. Interesting localization of high pressure and Von Mises stress at certain time stages has been identified. Velocity vector visualization has also shown distinct patterns in the near-surface membrane hydrodynamic field. The hydroelastic model successfully simulates underdamped oscillatory motion within a finite hydrodynamic domain. The simulations also demonstrate the excellent capability of ANSYS FSI to execute efficiently and accurately two-way FSI analysis via the ANSYS Workbench and accelerated coupling capabilities. To further refine the simulations, there are a number of different parameters that could be changed within the setup of the simulation to assess the hydroelasticity characteristics of the plate (membrane) structure and hydrodynamic characteristics:

- **Geometry**: inspection of the ANSYS results indicates that the plate acting as a membrane submerged in the fluid within the closed cavity was of a finite nature. As a result, the plate membrane influence on the fluid within the hydrodynamic domain is restricted to the area closely surrounding the plate tip. Therefore, in the future the geometry of this plate could be changed to enhance the displacement in the fluid region. This could be achieved via increase the size of the membrane or completely changing the morphology of the plate to for example a wavy surface (corrugated) or staggered design (as shown in the actual energy capturing mechanism).
- **Load Conditions**: A relatively low initial pressure of 100 Pa was applied to the face of the plate membrane, to simulate steady tidal conditions. This value is selected to initiate the simulation as it is sufficient to induce hydroelastic vibration and for the membrane to experience the oscillatory behaviour. This also acted as "wave exciter". However, since the membrane is not surface-piercing, wind wave effects are not incorporated. There are many linear and nonlinear hydrodynamic wave models (e.g. Stokes second-order

theory, cnoidal wave theory, etc.) available in ANSYS physics which could generalize the hydroelastic modelling including the effects of slamming. This could be executed by partially submerging the plate within the fluid and applying a larger pressure onto one side of the plate, as noted by Newman [36].

- **Modification in Elastic Membrane Properties**: To simulate *composite* materials for the energy harvesting system, a multi-layered (laminated) membrane model is required and anisotropic elastic properties are necessary. This is currently under investigation for novel ocean "wave park" energy devices, pioneered by Oregon State University, USA [37] which also feature magnetohydrodynamic (MHD) coupling [38–41] to enhance efficiency. The *MHD thermo-hydroelastic system* requires coupling between ANSYS FLUENT, ANSYS Structural and ANSYS MAXWELL electromagnetics and efforts in this direction will be communicated imminently.

REFERENCES

1. M.H. Patel and J.A. Witz, *Compliant Offshore Structures*, Elsevier, London (1991).
2. https://amc-catamaran.com/ (2020).
3. J. Degroote, K.-J. Bathe, and J. Vierendeels, Performance of a new partitioned procedure versus a monolithic procedure in fluid-structure interaction, *Computers and Structures*, 87 (11–12), 793–801 (2009).
4. J. van Rij et al., Structural loads analysis for wave energy converters, *Proceedings of the 36th International Conference on Ocean, Offshore & Arctic Engineering OMAE2017*, Trondheim, Norway, June 25–30 (2017).
5. P. Krawczyk, A. Beyene, and D. MacPhee, Fluid structure interaction of a morphed wind turbine blade, *International Journal of Energy Research*, 37, 1784–1793 (2013).
6. K-H. Lee et al., Hydroelastic analysis of floating structures with liquid tanks and comparison with experimental tests, *Applied Ocean Research*, 52, 167–187 (2015).
7. E. Loukogeorgaki et al., Hydroelastic analysis of a flexible mat-shaped floating breakwater under oblique wave action, *Journal of Fluids and Structures*, 31, 103–124 (2012).
8. W. Wei et al., A time-domain method for hydroelasticity of very large floating structures in inhomogeneous sea conditions, *Marine Structures*, 57, 180–192 (2018).
9. X. Zhang and D. Lu, An extension of a discrete-module-beam-bending-based hydroelasticity method for a flexible structure with complex geometric features, *Ocean Engineering*, 163, 22–28 (2018).
10. M. Wang et al., Fluid-structure interaction of flexible submerged vegetation stems and kinetic turbine blades, *Computational Particle Mechanics*, 4, 1–10 (2019).
11. H. Lee et al., A numerical study on the hydro-elastic behavior of composite marine propeller, *Fourth International Symposium on Marine Propulsors, Austin*, Texas, USA, June (2015).
12. A. Barbarit et al., A linear numerical model for analysing the hydroelastic response of a flexible electroactive wave energy converter, *Journal of Fluids and Structures*, 74, 356–384 (2017).
13. A. Barbarit et al., Investigation on the energy absorption performance of a fixed-bottom pressure-differential wave energy converter, *Applied Ocean Research*, 65, 90–101 (2017).
14. H.Y. Kang et al., Hydroelastic analysis of multi-unit floating offshore wind turbine platform (MUFOWT), *27th ISOPE International Ocean and Polar Engineering Conference*, San Francisco, California, USA, 25–30 June (2017).

15. K.J. Bathe and G. Ledezma, Benchmark problems for incompressible fluid flows with structural interactions, *Computers and Structures*, 85 (11–14), 628–644 (2007).

16. D. Rubinstein et al., Riser and J-Tube VIV Analysis Using CFD and FSI for platform retrofit application, *ASME 2013 32nd International Conference on Ocean, Offshore and Arctic Engineering*, Nantes, France, June 9–14 (2013).

17. T. Liaghat, Two-way fluid-structure coupling in vibration and damping analysis of an oscillating hydrofoil. *MSc Thesis*, Mechanical Engineering, École Polytechnique de Montréal, Canada, April (2014).

18. W.Z. Lim and R.Y. Xiao, Fluid-structure interaction analysis of gravity-based structure (GBS) offshore platform with partitioned coupling method, *Ocean Engineering*, 114, 1–9 (2016).

19. C-H. Lee et al., FSI analysis of deformation along offshore pile structure for tidal current power, *Renewable Energy*, 54, 248–252 (2013).

20. https://www.ansys.com/products/connect (2020).

21. M.K. Kwak, Hydroelastic vibration of rectangular plates, *ASME Journal of Applied Mechanics*, 63(1), 110–120 (1996).

22. *ANSYS FLUENT; Theory Manual*, Version 18.1, Lebanon, New Hampshire, USA (2018).

23. *ANSYS Structural; Theory Manual*, Version 18.1, Lebanon, New Hampshire, USA (2018).

24. O.A. Bég, A. Zubair, S. Kuharat, and M. Babaie, CFD simulation of turbulent convective heat transfer in rectangular mini-channels for rocket cooling applications, *ICHTFM 2018: 20th International Conference on Heat Transfer and Fluid Mechanics, WASET, Istanbul*, Turkey, August 16–17 (2018).

25. A. Kadir, O.A. Bég, M. E. El Gendy, T.A. Bég, and M. Shamshuddin, Computational fluid dynamic and thermal stress analysis of coatings for high-temperature corrosion protection of aerospace gas turbine blades, *Heat Transfer*, (2019) (25 pages). doi: 10.1002/htj.21493.

26. S. Kuharat, O.A. Bég, A. Kadir, B. Vasu, T. A. Bég, and W.S. Jouri, Computation of gold-water nanofluid natural convection in a three-dimensional tilted prismatic solar enclosure with aspect ratio and volume fraction effects, *Nanoscience and Technology-An International Journal*, 11 (2), 141–167 (2020).

27. O.A. Bég, M. Valter, A. Kadir, W.S. Jouri, T.A. Bég, and H.J. Leonard, Finite volume and smoothed particle hydrodynamic numerical simulation of rocket fuel tank sloshing with baffles, *ICSMA 2021: 15TH International Conference on Structural Mechanics and Analysis*, Istanbul, May 6–7 (2021).

28. O.A. Bég, H. Javaid, T.A. Bég, V.R. Prasad, S. Kuharat, A. Kadir, H.J. Leonard, W.S. Jouri, and Z. Ozturk, Computational fluid dynamic simulation of thermal convection in green fuel cells with finite volume and lattice Boltzmann methods, In *"Energy Conversion and Green Energy Storage"*, CRC/Taylor and Francis, Boca Raton, FL (A. Soni, D. Tripathi, J. Sahariya & K. N. Sharma, Eds.), September, pp. 1–38 (2022).

29. O.A. Bég, B. Islam, MD. Shamshuddin, and T.A. Bég, Computational fluid dynamics analysis of moisture ingress in aircraft structural composite materials, *Arabian Journal of Science & Engineering*, (2019) (23 pages). doi: 10.1007/s13369-019-03917-4.

30. O.A. Bég, T.A. Bég, W.S. Jouri, A. Kadir, G. Umesh, G.J. Reddy, and M.N. Satyanarayan, ANSYS (VOF) interface tracing method simulation of dynamic wetting and trapping efficiency of picolitre droplets in biomicrofluidic channels, *ICMEAMT 2020: 14th International Conference on Mechanical Engineering, Applied Mechanics and Technology, Cairo*, Egypt, December 14–15 (2020).

31. O.A. Bég, M. Kitende, S. Kuharat, T.A. Bég, A. Kadir, M. El Gendy, B. Vasu, A. Sohail, H.J. Leonard and W.S. Jouri, Aero-acoustic turbulent CFD simulation of a generic automotive body with multiple broadband source models, *ICMEAMT 2020: 14th International Conf. Mechanical Engineering, Applied Mechanics and Technology*, Cairo, Egypt, December 14–15 (2020).

32. H.A. Daud, Q. Li, O.A. Bég, and S.A. Ghani, CFD modeling of blowing ratio effects on 3-D skewed gas turbine film cooling, *11th International Conference on Advanced Computational Methods and Experimental Measurements in Heat Transfer*, Tallinn, Estonia, 14–16 July (2010).

33. C.N.P. Sze, B.R. Hughes, and O.A. Bég, Computational study of improving the efficiency of photovoltaic panels in the UAE, *ICFDT 2011-International Conference on Fluid Dynamics and Thermodynamics*, Dubai, United Arab Emirates, January 25–27 (2011).

34. J. Pattison, Two-way fluid-structure interaction hydroelastic analysis of vibrating membranes, *MSc Dissertation*, Aerospace Engineering, University of Salford, Manchester, UK, September (2018).

35. *COMSOL Theory Manual- FSI Applications*. https://www.comsol.de/multiphysics/fluid-structure-interaction, Stockholm, Sweden (2020).

36. J.N. Newman, *Marine Hydrodynamics*, MIT Press, Cambridge, Massachusetts (1977).

37. O.A. Bég, T.A. Bég, I. Karim, M.S. Khan, M.M. Alam, M Ferdows, and M. Shamshuddin, Numerical study of magneto-convective heat and mass transfer from inclined surface with Soret diffusion and heat generation effects: A model for ocean magnetohydrodynamic energy generator fluid dynamics, *Chinese Journal of Physics*, 60, 167–179 (2019).

38. S.R. Sheri, O.A. Bég, P. Modugula, and A. Kadir, Computation of transient radiative reactive thermo-solutal magneto-hydrodynamic convection in inclined MHD Hall generator flow with dissipation and cross diffusion, *Computational Thermal Sciences*, 11(6) 541–563 (2019).

39. O.A. Bég, T.A. Bég, S.R. Munjam, and S. Jangili, Homotopy and Adomian semi-numerical solutions for oscillatory flow of partially ionized dielectric hydrogen gas in a rotating MHD energy generator duct, *International Journal of Hydrogen Energy*, 46, 17677–17696 (2021).

40. M.D. Shamshuddin, N. Akkurt, O.A. Bég, H.J. Leonard, and T.A. Bég, Analysis of unsteady rotating thermo-solutal MoS2-EO Brinkman electro-conductive nanofluid transport with heat source, radiative and chemical reaction effects: Modelling a hybrid rotating nanofluid Hall MHD generator, *Partial Differential Equations in Applied Mathematics*, 7 (2023) 100525 (13 pages). doi: 10.1016/j.padiff.2023.100525.

41. O.A. Bég, M. Ferdows, M.E. Karim, M.M. Hasan, T.A. Bég, M.D. Shamshuddin, and A. Kadir, Computation of non-isothermal thermo-convective micropolar fluid dynamics in a Hall MHD generator system with non-linear distending wall, *International Journal of Applied and Computational Mathematics*, 6, 1–44 (2020).

15 Schiff Bases and Their Transition Metal Complexes Composites as Energy Storage Materials

Nisha Yadav and Gyandshwar Kumar Rao

15.1 INTRODUCTION

The energy crisis is being observed worldwide since fossil fuel reserves are running out, and hence, researchers are looking for sustainable energy options [1]. The utilization of available energy resources by human beings led to an increasing reliance on new energy sources in order to meet future energy requirements and promote a more environmental friendly atmosphere [2,3]. It is a difficult task to create renewable energy systems that ensure the effective utilization of energy in the future. Renewable sources, like solar power, hold the potential to offer substantial energy and might serve as a feasible solution to address our energy requirement. Considerable research is directed toward the development of renewable resources and the advancements in technologies for energy storage and interconversions. One area of extensive research is storage devices, which include technologies such as batteries and super-capacitors [4]. The energy storage process in batteries is the transfer of electrons at a particular electrochemical potential from a cathode, which has a high reduction potential, to an anode, which has a lower reduction potential. This means that electrical energy at a constant potential can be produced whenever needed. On the other hand, super-capacitors use the movement of electrons between conductive materials to store electrical energy. Graphite-based materials are commonly used as a conductive counterpart for energy storage devices [5]. Further efforts have been made for the search of new materials having higher lithium storage capacity per unit mass than graphite [6] which in turn can produce lighter electronic devices. Such materials exhibit notable versatility in their chemical properties, allowing for the straightforward introduction of various functional groups through a simple condensation process [7]. In addition, condensation results in the formation of Schiff-base materials having a significantly high surface area, which allows for the integration of redox-active counterparts to other components for their utilization as super-capacitors [8,9]. Schiff bases can be synthesized through the condensation of a primary amine with an active carbonyl

DOI: 10.1201/9781003473749-18

molecule, resulting in the creation of an azomethine (—R—C=N—) group [10]. Such compounds have attracted research interest due to their facile synthesis and the ability to form complexes with metals. Some Schiff bases have also been tailored in the form of organic polymers [5] and nitrogen-doped carbon compounds [11,12]. Also, the lower abundance of lithium as compared to other alkali metals is another reason why scientists are looking for other battery designs, like sodium-ion batteries [13–15]. Thus, the synthesis and utility of Schiff bases and their metal complexes in diverse energy-related applications, including their effectiveness in organic light-emitting diodes (OLEDs), batteries, and carbon-based materials in energy storage have been summarized in the present chapter.

15.2 GENERAL SYNTHESIS OF SCHIFF BASES

The one-step condensation reaction between primary amines and aldehydes has arisen in a new and easy way for the synthesis of a new class of compounds, known as Schiff bases [16]. Initially, in 1864, the reaction of aniline and various aldehydes was carried out by Hugo Schiff [17,18]. The reaction proceeded by generating water and experiencing a temperature rise, indicating a condensation mechanism. The resulting imines, also known as Schiff bases, can also be easily produced when different primary amines react with different aldehydes or ketones at elevated temperature.

15.2.1 MECHANISM OF FORMATION OF SCHIFF BASES

The bond between amines and carbonyl groups is formed when the primary amine nucleophile attacks the carbon of the carbonyl group of aldehyde or ketone (Figure 15.1). This results in the formation of hemiaminal (having a secondary amine and hydroxyl group at the carbon center) due to proton transfer. The addition of a proton to the hydroxyl group promotes the removal of water, leading to the formation of the iminium ion. The imine was generated through the removal of a proton from the iminium ion, which generates a double bond between carbon and nitrogen [19]. Imines can undergo hydrolysis in aqueous or humid environments, especially with

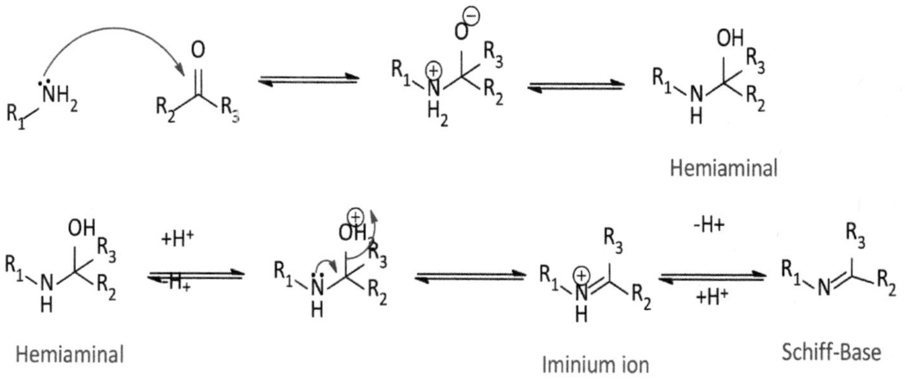

FIGURE 15.1 Mechanism of synthesis of Schiff bases.

aliphatic aldehydes or ketones. The continuous removal of water produced during reaction or present in solvent using Dean-Stark apparatus gives rise to the desired imine in higher yield. Aromatic aldehydes react more easily due to steric effects and lower electron density on aldehyde carbon compared to ketone carbon. Therefore, the impact of each individual reaction step on the total rate is highly influenced by the proton activity during condensation. The reversibility of all stages involved in imine synthesis presents a practical issue. Aqueous syntheses, although uncommon, may result in a lower yield of product. Recently, mechanochemistry has proven to be a simple technique to produce Schiff bases by simply grinding precursors. The melt synthesis involves neat heating of one precursor component above its melting point in the absence of solvent [20–22].

15.2.2 N-DOPED CARBON MATERIALS SYNTHESIZED USING SCHIFF BASES

While the imine bond is extremely strong and durable in a water-free condition, the practical application of Schiff-base compounds is sometimes restricted due to their hydrolysis in the presence of water. Since alkali-ion batteries have been found to work in water-free conditions and hence they present an attractive platform for the integration of Schiff bases. The carbonaceous compounds derived from Schiff bases showed better electrochemical energy storage properties as compared to pure Schiff bases. Carbon-based N-doped materials have attracted a lot of interest as super-capacitor electrodes [23–25] and anodes [26–28] for alkali-metal batteries. Starting materials do not require sublimation or evaporation for the preparation of such carbon-based materials. Since polymers tend to condense, carbonize, and then evaporate, they are frequently used for this purpose. However numerous, sustainable polymers frequently lack nitrogen, an essential heteroatom for improving the electrochemical characteristics of the material. Nitrogen doping in carbon has been made possible using organic chemistry. It can be achieved by adding heteroatoms as functional groups either on the surface or inside the carbon framework. Further, such materials could be used to produce N-doped carbon-based systems at lower temperatures under the N_2 atmosphere. The electrochemical characteristics of such materials have been found to be enhanced due to N-doping which alters the band positions and electron density distribution. Notably, by introducing nitrogen in carbon materials, the adsorption abilities and wetting capacity can be enhanced [7]. On the other hand, amine-carbonyl condensation using tiny molecules with amine and carbonyl groups produces polymers that result in naturally nitrogen-doped carbons because of a high density of imine connections. Covalent organic frameworks (COFs) consist of interconnected organic building units through covalent bonds, displaying crystallinity due to the reversibility of the reactions used for their synthesis. Yaghi et al. pioneered COFs through reversible condensation of diboric acid, expanding the possibilities with the introduction of Schiff-base chemistry [29]. Kuhn et al. reported nitrile trimerization for COF formation, leading to covalent triazine frameworks (CTFs) [30]. The challenge of Schiff base chemistry in COFs is the imine bond instability in atmospheric conditions. Banerjee et al. addressed this by combining reversible and irreversible reactions, yielding stable COFs (TpCOFs) without imine bonds through irreversible tautomerization [31]. "ImCOFs" (imine COFs), emphasize the

COFs properties derived from Schiff-base materials. As a substitute example, a class of materials produced via solvothermal condensation is known as melamine-based SNWs [32]. The process involves the reaction of melamine with di-/tri-aldehydes in dimethyl sulfoxide in the absence of a catalyst, which is supersizing as it requires a catalyst. A further modification of the same precursors has resulted in the development of melamine-based conjugated polymers (MCPs) (Figure 15.2). The MCPs are synthesized from the same source materials as SNWs, but at lower temperature (between 150°C and 250°C for melamine-terephthalaldehyde networks), in an inert atmosphere and without the use of a solvent [22]. Materials like carbon, COF, and microporous materials that are derived from the Schiff base have been used in energy-storing devices.

15.3 TRANSITION METAL COMPLEXES (TMCS) OF SCHIFF BASES LIGANDS

TMCs with exciting optoelectronic characteristics are being used at a larger scale in the advancement of innovative functional molecular materials [33–37]. The few benefits of metalation are as follows: (i) Such systems showed improved redox activity that increases the charge transportation due to the recombination of redox activity of both the metal atoms and organic ligand; (ii) It enables easy adjustment of the energy gap between the highest occupied molecular orbital (HOMO) and the lowest unoccupied molecular orbital (LUMO) by interacting the d-orbitals of the transition metal with the HOMO or LUMO of the ligand; (iii) It offers a wide range of molecular frameworks due to the coordination number, geometry, and valence shell of the chosen metal atom and (iv) It enhances the solubility and makes easier to change intermolecular π-stacking and/or metallophilicity in solid state. Such complexes have received significant attention because of their appealing physical and chemical characteristics, as well as their numerous applications in several sectors [38–40]. A few examples such as nickel(II), platinum(II), and zinc(II) of Schiff base ligands used as phosphorescent materials for energy generating and energy storage materials, dye-sensitized solar cells, and potential conductive thermoelectric material (Figure 15.3).

15.4 UTILIZATION OF SCHIFF BASES AS ENERGY STORAGE MATERIALS

The energy that is being produced can be stored in a class of materials known as energy storage materials. This stored energy could be further utilized for different kinds of purposes. Such materials have been classified as batteries (sodium/lithium-ion batteries), carbon-containing materials, super-capacitors, LEDs, etc. The energy storage materials based on either Schiff bases or their complexes have been discussed in the following subsections.

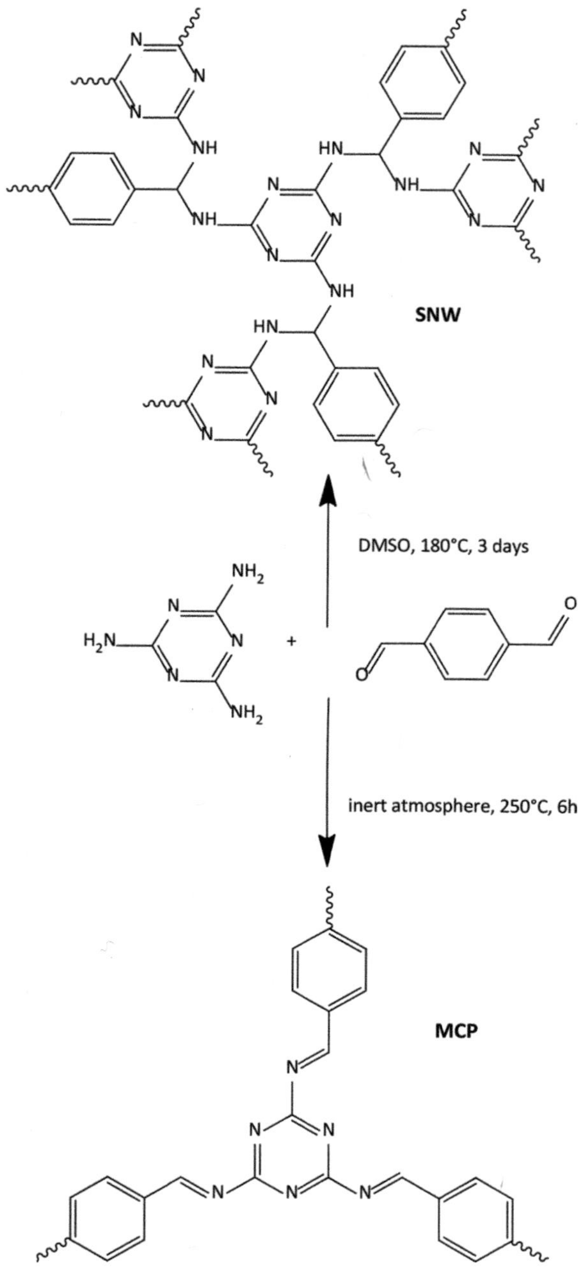

FIGURE 15.2 Synthesis of a melamine-based Schiff-base network (SNW) and a melamine-based conjugated polymer (MCP).

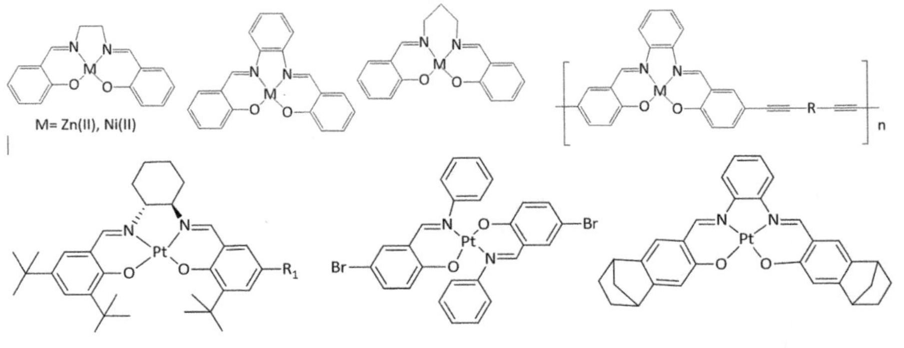

FIGURE 15.3 Examples of transition metal complexes of Schiff base ligands used as energy storage materials.

15.4.1 Batteries

The stored chemical energy can be easily converted into electrical energy by using an electrochemical process (oxidation-reduction) that involves the utilization of instruments known as batteries. Earlier, batteries were based on transition metal-metal oxide systems, but nowadays lithium-ion/sodium ion batteries are more common since these do not require aqueous electrolytes. These are more efficient and easily rechargeable; they take a longer time to discharge, and the energy densities of such batteries are quite high. Such batteries utilized materials derived from Schiff bases and their complexes.

15.4.1.1 Sodium Ion Storage

It was reported by Armand et al. that aromatic diamines and aliphatic dialdehydes can be combined to make linear polymeric Schiff bases. These Schiff bases exhibited promising performance as sodium-ion battery anodes, achieving capacities of up to 350 mAh g^{-1} at 26 mA g^{-1} [41]. The Schiff base was synthesized by refluxing reagents in toluene through azeotropic water elimination and characterized using infrared spectroscopy. The conjugated polymeric Schiff base has excellent efficacy when used with both aromatic dialdehydes and diamines. However, non-conjugated Schiff bases did not demonstrate the same level of effectiveness. It is also interesting to mention that Schiff-bases derived from aromatic dialdehydes and aliphatic diamines showed superior performance in comparison to those derived from aromatic diamines and aliphatic dialdehydes. The role of the —N=CH—Ar—CH=N— units at low voltages was to preserve sodium ions. The impact of functionalization was explored, revealing that certain modifications maintained performance, while steric hindrance from four methyl groups led to reduced efficiency of the resulting system [41]. This study highlights the capacity of Schiff-base materials to function as redox-active anodes, demonstrating their suitability to be used in sodium-ion batteries. Researchers conducted additional investigations using terephthalaldehyde, p-phenylenediamine, and polyether amine blocks to create polymeric Schiff bases. These Schiff bases demonstrate remarkable adhesive properties and function as binders for sodium-ion anodes with enhanced redox activity [42]. The study explores various materials based on Tp linkers as anodes for sodium-ion batteries, even though these materials don't have any imine bonds, and thus, they are directly related to Schiff-bases (keto-enol tautomerism). Lu et al. showed that a sodium anode TpCOF which contains anthraquinone can store up to 182 mAh g^{-1} at 50 mA g^{-1} [43]. Notably, the substance showed interesting redox activity at low voltages, even though it didn't have any imine bonds, which are usually thought to be necessary for storing energy at low voltages. Spectroscopic studies and DFT calculations were used together to figure out the storage kinetics of sodium. It has been proposed that the anthraquinone units went through a two-phase reduction process at about 1.5 V, which created a radical intermediate by moving the first electron around. The Tp-linker's keto-form has lower voltages (about 0.5 V) through a process with radical intermediates that happened in several steps. This process shows that TpCOFs are different from polymers with imine links in terms of energy storage. Furthermore, the exploitation improved the performance of the TpCOFs, increasing their ability to over 500 mAh g^{-1} at 50 mA g^{-1} [44] (Figure 15.4).

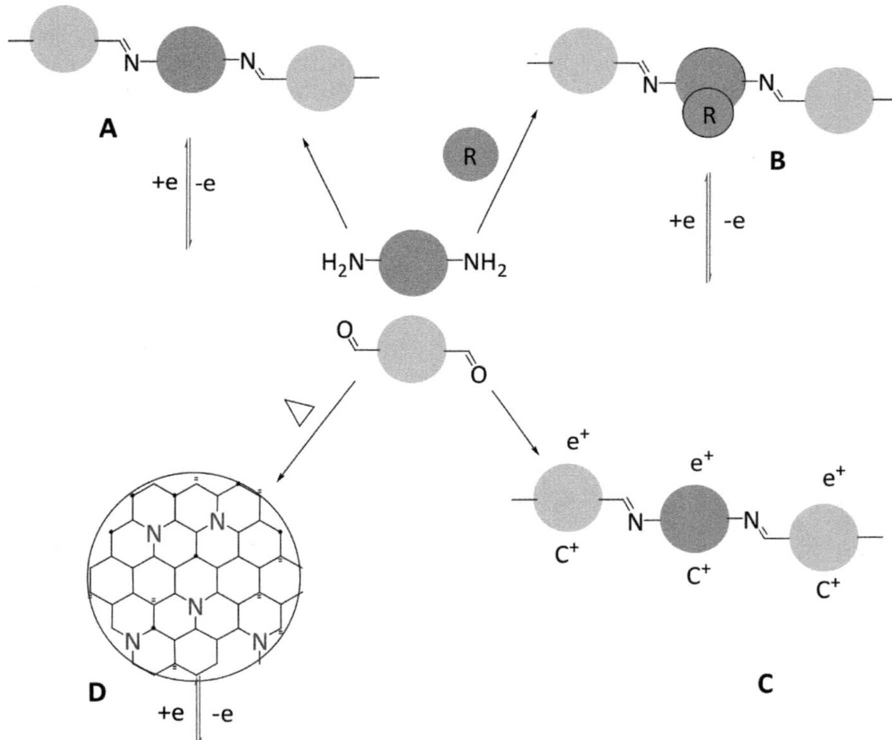

FIGURE 15.4 Electrochemical energy storage using different strategies utilizing Schiff base materials. (A) Schiff-base materials for organic anodes; (B) Schiff-base materials as redox-active species; (C) Microporous Schiff-based networks for supercapacitors and (D) Carbonized Schiff-base materials for super-capacitors and battery anodes [7].

15.4.1.2   Lithium-Ion Storage

Based on a study done on lithium-ion batteries, many new materials for sodium-ion batteries have been made [45,46]. The unique shape of Schiff-base anodes makes them stand out for their use in sodium-ion systems [41]. Nevertheless, Schiff-based materials have exceptional performance when used as anodes in lithium-ion batteries. Carbon-based organic frameworks (COFs) are becoming recognized as very advantageous materials. Zhao et al. synthesized an imine covalent organic framework (ImCOF) by reacting triformylbenzene (Tb) with 4,4′-diamino biphenyl through a condensation reaction [47]. The resulting material had a higher capacity of over 700 mAh g^{-1} at 1 Ag^{-1}, and its capacity is more than 80% higher even after 500 cycles. Additionally, Zhang et al. reported another example of an ImCOF that worked similarly to the anode material in a lithium-ion battery. The chemicals terephthalaldehyde and 2,4,6-triaminopyrimidine were used to make this ImCOF [48]. Interestingly, a direct synthesis method was employed; the reactants were pulverized using a mortar and pestle followed by adding p-toluene sulfonic acid as a catalyst, and the reaction was accelerated by adding a small quantity of water. Subsequently, the mixture was heated at a temperature of 120°C, leading to the formation of the product after the washing process. Thus, the narrative emphasizes that the ligands can only be synthesized in multiple steps has been simplified by using an efficient synthesis technique.

15.4.1.3   Schiff-Base Materials with an Added Functional Group

Recent research has shown that Schiff-based materials hold great promise in the development of anodes for sodium and lithium-ion batteries. The presence of the —N=CH—Ar—CH=N— moiety is responsible for the low redox voltage. However, there were instances where additional quinone units were incorporated, which underwent redox reactions at elevated redox potentials. While the Schiff-base groups did not actively take part in the electrochemical process, they served as a vital structural framework for integrating these innovative redox-active groups into polymeric networks and frameworks [45–48].

15.4.2   Redox Activity in Aqueous Media with Added Electrolytes

Triazine-based Covalent Organic Frameworks (TpCOFs) are stable in water, unlike materials that contain Schiff-base. These TpCOFs with redox functions have been used to store energy in wet conditions, especially in 1 M H_2SO_4. Dichtel et al. [54] synthesized a TpCOF having anthraquinone group and can store energy about 40 Fg^{-1}. In contrast, a standard TpCOF did not exhibit any redox activity and had a limited storage capacity of 15 Fg^{-1} at 0.1 Ag^{-1}. Banerjee et al. discussed a TpCOF with smaller quinone functions that were stabilized by intra-COF hydrogen bonding. This version had a higher capacity of 416 Fg^{-1} at 0.5 Ag^{-1} [49]. Furthermore, it exhibited excellent performance as a cathode material in liquid zinc-ion batteries, achieving a remarkable capacity of 276 mAh g^{-1} at 125 mA g^{-1} [50]. It's important to note that many redox-active Schiff base materials might not exactly fit the description of super-capacitors, even though they are often called that in the press. Super-capacitors utilize either an electric double layer or pseudo-capacitance to efficiently store electrons. There is a common misconception that pseudo-capacitance is limited to metal oxides and does not occur with redox-active organic polymers [51,52].

15.4.3 Redox Activity for Group I Ions in Batteries

Bu et al. showed that imine-based COFs can be used as cathode materials in lithium-ion batteries [53]. The COFs were constructed using a redox-active component, specifically the phthalimide moiety, in combination with diamine and a trialdehyde binder (Tp or Tb). This resulted in the formation of TpCOF and ImCOF. Interestingly, ImCOF proved to be more effective as compared to TpCOF [54]. This is due to its unique structure that facilitates the movement of electrons within the materials. This discovery holds significant implications as it demonstrates that the stability of TpCOFs is much higher in water-free systems. It indicates the potential of Schiff-base materials in alkaline-ion batteries even in dry conditions. Just like how peeling can enhance the anode performance, it can also enhance the performance of cathodes in COF-based lithium-ion batteries. Wang et al. exfoliate a TpCOF that contained anthraquinone using a ball mill. At low current levels (50 mA g^{-1}), the exfoliated COF exhibited similar performance to the pristine COF. However, when the currents were increased, noticeable distinctions became apparent [55].

15.4.4 Schiff Bases Used in Super-Capacitor Materials

Super-capacitors are devices that can store a huge amount of charge and are used as electrochemical devices. It has components such as a cathode, anode and a separator. These capacitors generally use an electrolyte solution which has ions (positively and negatively charged) separated by dielectric materials.

15.4.4.1 Microporous Schiff Base Material for Super-Capacitor

COFs and SNWs have large surface areas that make them appealing to be used as super-capacitors, however, they are not useful because they don't carry electricity well. The problem with conductivity can be fixed by growing ImCOF on amine-functionalized graphene. This creates a more conductive combination of materials. Graphene is an integrated current collector that lets electrons move quickly through the low-conductivity COFs to get to the highly conductive graphene sheets. This method effectively addresses the issue of low conductivity. Remarkably, the hybrid material demonstrated exceptional performance as a super-capacitor in a 1 M sodium sulfate water solution. It exhibited a capacitance of 533 Fg^{-1} at 0.2 Ag^{-1} [56].

Wang et al. made a COF by mixing a triamine with 2,5-dihydroxyterephthalaldehyde. This framework was held together by hydrogen bonds between molecules, which was explained in more detail earlier [57,58]. It was found that peeling off COFs helped the improvement of the efficiency of super-capacitors. This is shown by a mesoporous COF where peeling off shows unique rectangular cyclic voltammograms, which is a characteristic of super-capacitors.

15.4.5 Schiff Base Polymer as the Precursor for Carbon-Based Materials

Rechargeable batteries made up of carbon-containing materials for example carbon nanotubes are potential candidates for energy storage devices. A few carbon-based materials are discussed in this section which have higher efficiency as an electrode. Zhi et al. discussed the utilization of carbon materials derived from the Schiff base and supported by CNTs as cathode components in lithium-sulfur batteries [59]. A solution was prepared by mixing melamine and terephthalaldehyde with CNTs. The authors demonstrated the creation of nitrogen-rich areas on the carbon nanotube (CNT) structure through polymerization and carbonization. This approach ensures electrical conductivity and effectively serves as a cathode material. In 2021, Lin et al. conducted a noteworthy study [60]. They produced precursor compounds that contained imine links including triazine. Following the infusion of cobalt acetate, the precursor underwent pyrolysis, leading to the formation of cobalt-doped nitrogen compounds.

15.4.6 Schiff Base Complexes as Organic Light-Emitting Diode (OLEDs)

LED works on the principle of electroluminescence in which LED behaves as a semiconductor that emits light when current passes through the diode. Research on OLEDs has primarily concentrated on thin-film devices that incorporate organometallic molecules that can directly transform electrical energy into light [61–63]. OLEDs offer several intrinsic benefits for flat-panel display applications, including self-emission, increased luminous efficiency, full-color capability, broad viewing angles, superior contrast, lower power consumption, lightweight construction, the possibility for large-area color displays, and flexibility [64]. Furthermore, OLEDs are energy efficient and have lower costs in comparison to liquid crystal displays (LCDs). Transition metal complexes have been found to enhance the overall performance of OLEDs, however, it is essential to optimize their efficiency on the three basic properties; (i) Charge injection, (ii) Charge transport, and (iii) Electron-hole recombination. A few examples of the Schiff base transition metal complexes are discussed in this section which is being used in OLED. Schiff base Pt(II) complexes (Figure 15.3) constitute an attractive group of phosphorescent materials. These complexes can be easily synthesized, adaptable to structural changes, thermally stable, and possess potential importance in OLED applications. Che et al. have pioneered the investigation of the electroluminescent (EL) characteristics of phosphorescent d^8-Pt(II)-Schiff base complexes, proposing their suitability for high-performance OLEDs [65]. The electroluminescent (EL) properties of Zn(II) Schiff base complexes have been examined since 1993. Hamada et al. played a key role in the development of Zn(II) complexes that incorporate 2:1 bidentate azomethine and 1:1 tetradentate N_2O_2 Schiff base ligands, for their potential use in crafting blue OLEDs [66]. Lepnev et al. have extensively investigated the photoluminescent (PL) and electroluminescent (EL) properties of a series of Zn(II) Schiff base complexes (Figure 15.3) [67]. Their study involved a comprehensive analysis of the luminescence instability observed in thin films and OLEDs fabricated using Zn-complex [68]. The results indicate that prolonged exposure to intense UV light and extended heating can result in irreversible degradation. However, exposure to weak UV light does not exhibit noticeable degradation. This finding offers promise for the reliable operation of OLEDs in daylight conditions. To ensure the stability of OLED operation, precautions are necessary to prevent humidity during both the fabrication and encapsulation processes.

15.5 CONCLUSIONS

Energy storage materials are essential for effectively generating, converting, and utilizing various sources of energy. Materials containing Schiff-base groups exhibit significant promise for developing customized organic electrode materials for batteries and super-capacitors. The intrinsic propensity of these materials to combine and their significant molecular weight contribute to their exceptional ability to dissolve in electrolytes, thus maintaining overall stability. Due to the direct synthesis of Schiff bases, which can be done without the use of solvents or with ecologically friendly solvents such as water, these materials are well-suited for sustainable energy storage applications in the future. Metal Schiff base complexes are attractive for practical use because of their facile synthesis from Schiff base ligands, resilient chemical properties, and the convenience of large-scale synthesis. The composite derived from Schiff bases, or their metal complexes have been proven a great track record for their utilization in batteries and super-capacitors. Since a limited number of composites of Schiff bases and their complexes are explored for the applications mentioned above, thus there is a need to design and synthesize composites based on Schiff bases or their complexes of improved efficiency for their commercial application.

REFERENCES

1. Eisengberg, R.; Nocera, D. G. Preface: Overview of the forum on solar and renewable energy. *Inorg. Chem.*, **2005**, 44, 6799–6801.
2. Turner, J. A. A realizable renewable energy future. *Science*, **1999**, 285, 687–689.
3. Baranoff, E.; Yum, J.-H.; Grätzel, M.; Nazeeruddin, M. K. Cyclometallated iridium complexes for conversion of light into electricity and electricity into light. *J. Organomet. Chem.*, **2009**, 694, 2661–2670.
4. Larcher, D.; Tarascon, J. M. Towards greener and more sustainable batteries for electrical energy storage, *Nat. Chem.*, **2015**, 7, 19.
5. Whittell, G. R.; Hager, M. D.; Schubert, U. S.; Manners, I. Functional soft materials from metallopolymers and metallosupramolecular polymers. *Nat. Mater.*, **2011**, 10, 176–188.
6. Xiao, B.; Rojo, T.; Li, X. Hard carbon as sodium-ion battery anodes: progress and challenges. *ChemSusChem.*, **2019**, 12, 133.
7. Troschke, E.; Oschatz, M.; Ilic, I. K. Schiff-bases for sustainable battery and supercapacitor electrodes. *Exploration*, **2021**, 1, 20210128.
8. Yan, R.; Antonietti, M.; Oschatz, M. Toward the experimental understanding of the energy storage mechanism and ion dynamics in ionic liquid-based supercapacitors. *Adv. Energy Mater.*, **2018**, 8, 1800026.
9. Chen, L. F.; Zhang, X. D.; Liang, H. W.; Kong, M.; Guan, Q. F.; Chen, P.; Wu, Z. Y.; Yu, S. H. Synthesis of nitrogen-doped porous carbon nanofibers as an efficient electrode material for supercapacitors. *ACS Nano*, **2012**, 6, 7092.
10. Wang, Q.; Yan, J.; Fan, Z. Carbon materials for high volumetric performance supercapacitors: design, progress, challenges and opportunities. *Energy Environ. Sci.*, **2016**, 9, 729.

11. Mao, Y.; Duan, H.; Xu, B.; Zhang, L.; Hu, Y.; Zhao, C.; Wang, Z., Chen, L.; Yang, Y. Lithium storage in nitrogen-rich mesoporous carbon materials. *Energy Environ. Sci.*, **2012**, 5, 7950.

12. Wang, H.; Zhang, C.; Liu, Z.; Wang, L.; Han, P.; Xu, H.; Zhang, K.; Dong, S.; Yao, J.; Cui, G. Nitrogen-doped graphene nanosheets with excellent lithium storage properties. *J. Mater. Chem.*, **2011**, 21, 5430.

13. Slater, M.; Kim, D.; Lee, E.; Johnson, C. S. Sodium-ion batteries. *Adv. Funct. Mater.*, **2013**, 23, 947.

14. Kim, S. W.; Seo, D. H.; Ma, X.; Ceder, G.; Kang, K. Electrode materials for rechargeable sodium-ion batteries: potential alternatives to current lithium-ion batteries. *Adv. Energy Mater.*, **2012**, 2, 710.

15. Palomares, V.; Serras, P.; Villaluenga, I.; Hueso, K. B.; Carretero-González, J.; Rojo, T. Na-ion batteries, recent advances and present challenges to become low cost energy storage systems. *Energy Environ. Sci.*, **2012**, 5, 5884.

16. Tidwell, T. T. Hugo (Ugo) Schiff, Schiff bases, and a century of beta-lactam synthesis. *Angew. Chem. Int. Ed.*, **2008**, 47, 1016.

17. Schiff, H. *Justus Liebigs Ann. Chem.*, **1864**, 131, 118.

18. Schiff, U.; *G. di Sci. Nat. ed Econ.*, **1866**, 2, 201.

19. Clayden, J.; Greeves, N.; Warren, S.; *Organic Chemistry*, 2nd ed., Oxford University Press, Oxford, **2012**.

20. Saggiomo, V.; Lüning, U. On the formation of imines in water comparison. *Tetrahedron Lett.*, **2009**, 50, 4663.

21. Schmeyers, J.; Toda, F.; Boy, J.; Kaupp, G. Quantitative solid-solid synthesis of azomethines. *J. Chem. Soc.*, **1988**, 4, 989.

22. Zheng, H.; Dong, W.; Huang, X.; Wang, G.; Wu, Z. A sustainable method toward melamine-based conjugated polymer semiconductors for efficient photocatalytic hydrogen production under visible light, *Green Chem.*, **2017**, 20, 664.

23. Frackowiak, E.; Beguin, F. Carbon materials for the electrochemical storage of energy in capacitors. *Carbon*, **2001**, 39, 937.

24. Simon, P.; Gogotsi, Y. Materials for electrochemical capacitors. *Nat. Mater.*, **2010**, 7, 845.

25. Borchardt, L.; Oschatz, L.; Kaskel, S. Tailoring porosity in carbon materials for supercapacitor applications. *Mater. Horiz.*, **2014**, 1, 157.

26. Dahn, J. R.; Zheng, T.; Liu, Y.; Xue, J. S. Mechanisms for lithium insertion in carbonaceous materials. *Science*, **1995**, 270, 590.

27. Stevens, D. A.; Dahn, J. R. The Mechanisms of lithium and sodium insertion in carbon materials. *J. Electrochem. Soc.*, **2001**, 148, 803.

28. Saurel, D.; Orayech, B.; Xiao, B.; Carriazo, D.; Li, X.; Rojo, T. From Charge storage mechanism to performance: A roadmap toward high specific energy sodium-ion batteries through carbon anode optimization. *Adv. Energy Mater.*, **2018**, 8, 1703268.

29. Cote, A. P.; Benin, A. I.; Ockwig, N. W.; O'Keeffe, M.; Matzger, A. J.; Yaghi, O. M. Porous, crystalline, covalent organic frameworks. *Science*, **2005**, 310, 1166.

30. Kuhn, P.; Antonietti, M.; Thomas, A. Porous, covalent triazine-based frameworks prepared by ionothermal synthesis. *Angew. Chem. Int. Ed.*, **2008**, 47, 3450.

31. Kandambeth, S.; Mallick, A.; Lukose, B.; Mane, M. V.; Heine, T.; Banerjee, R. Construction of crystalline 2D covalent organic frameworks with remarkable chemical (acid/base) stability via a combined reversible and irreversible route. *J. Am. Chem. Soc.*, **2012**, 134, 19524.

32. Schwab, M. G.; Fassbender, B.; Spiess, H. W.; Thomas, A.; Feng, X.; Mullen, K. Catalyst-free preparation of melamine-based microporous polymer networks through Schiff base chemistry. *J. Am. Chem. Soc.*, **2009**, 131, 7216.

33. Whittell, G. R.; Hager, M. D.; Schubert, U. S.; Manners, I. Functional soft materials from metallopolymers and metallosupramolecular polymers. *Nat. Mater.*, **2011**, 10, 176–188.

34. Wild, A.; Winter, A.; Schlutter, F.; Schubert, U. S. Advances in the field of π-conjugated 2,2':6',2"-terpyridines. *Chem. Soc. Rev.*, **2011**, 40, 1459–1511.

35. Kingsborough, R. P.; Swager, T. M. Transition Metals in Polymeric π-Conjugated Organic Frameworks. *Inorg. Chem.*, **1999**, 48, 123–231.

36. Green, K. A.; Cifuentes, M. P.; Samoc, M.; Humphrey, M. G. Metal alkynyl complexes as switchable NLO systems. *Coord. Chem. Rev.*, **2011**, 255, 2530–2541.

37. Ho, C.-L.; Wong, W.-Y. Metal-containing polymers: Facile tuning of photophysical traits and emerging applications in organic electronics and photonics. *Coord. Chem. Rev.*, **2011**, 255, 2469–2502.

38. Katsuki, T. Catalytic asymmetric oxidations using optically active (salen) manganese(III) complexes as catalysts. *Coord. Chem. Rev.*, **1995**, 140, 189–214.

39. Li, K.; Tong, G. S. M.; Wan, Q.; Cheng, G.; Tong, W.-Y.; Ang, W.-H.; Kwong, W.-L.; Che, C.- M. Highly phosphorescent platinum(ii) emitters: photophysics, materials and biological applications. *Chem. Sci.*, **2016**, 7, 1653–1673.

40. Leung, A. C. W.; MacLachlan, M. J. Schiff base complex in macromolecules. *J. Inorg. Organomet. Polym. Mater.*, **2007**, 17, 57–89.

41. Castillo-Martínez, E.; Carretero-González, J.; Armand, M. Polymeric schiff bases as low-voltage redox centers for sodium-ion batteries. *Angew. Chem. Int. Ed.*, **2014**, 53, 5341.

42. Fernández, N.; Sánchez-Fontecoba, P.; Castillo-Martínez, E.; Carretero-González, J.; Rojo, T.; Armand, M. Polymeric redox-active electrodes for sodium-ion batteries. *ChemSusChem.*, **2018**, 11, 311.

43. Gu, S.; Wu, S.; Cao, L.; Li, M.; Qin, N.; Zhu, J.; Wang, Z.; Li, Y.; Li, Z.; Chen, J.; Lu, Z. Tunable redox chemistry and stability of radical intermediates in 2D covalent organic frameworks for high performance sodium ion batteries. *J. Am. Chem. Soc.*, **2019**, 141, 9623.

44. Haldar, S.; Kaleeswaran, D.; Rase, D.; Roy, K.; Ogale, S.; Vaidhyanathan, R. Tuning the electronic energy level of covalent organic frameworks for crafting high-rate Na-ion battery anode. *Nanoscale Horiz.*, **2020**, 5, 1264.

45. Nayak, P. K.; Yang, L.; Brehm, W.; Adelhelm, P.; From lithium-ion to sodium-ion batteries: advantages, challenges, and surprises. *Angew. Chem. Int. Ed.*, **2018**, 56, 2.

46. Adelhelm, P.; Hartmann, P.; Bender, C. L.; Busche, M.; Eufinger, C.; Janek, J. From lithium to sodium: cell chemistry of room temperature sodium-air and sodium-sulfur batteries. *J. Beilstein J. Nanotechnol.*, **2015**, 6, 1016.

47. Bai, L.; Gao, Q.; Zhao, Y. Two fully conjugated covalent organic frameworks as anode materials for lithium-ion batteries. *J. Mater. Chem. A*, **2016**, 4, 14106.

48. Chen, H.; Zhang, Y.; Xu, C.; Cao, M.; Dou, H.; Zhang, X. Two π-conjugated covalent organic frameworks with long-term cyclability at high current density for lithium-ion battery. *Chem. Eur. J.*, **2019**, 25, 15472.

49. Chandra, S.; Roy Chowdhury, D.; Addicoat, M.; Heine, T.; Paul, A.; Banerjee, A. Molecular level control of the capacitance of two-dimensional covalent organic frameworks: role of hydrogen bonding in energy storage materials. *Chem. Mater.*, **2017**, 29, 2074.

50. Khayum, A. M.; Ghosh, M.; Vijayakumar, V.; Halder, A.; Nurhuda, M.; Kumar, S.; Addicoat, M.; Kurungot, S.; Banerjee, R. Zinc ion interactions in a two-dimensional covalent organic framework based aqueous zinc ion battery. *Chem. Sci.*, 2019, 10, 8889.

51. Lukatskaya, M. R.; Dunn, B.; Gogotsi, Y. Multidimensional materials and device architectures for future hybrid energy storage. *Nat. Commun.*, **2016**, 7, 12647.

52. Brousse, T.; Bélanger, D.; Long, J. W. To be or not to be pseudocapacitive? *J. Electrochem. Soc.*, **2015**, 162, A5185.

53. Yang, D. H.; Yao, Z. Q.; Wu, D.; Zhang, Y. H.; Zhou, Z.; Bu, X. H. Structure-modulated crystalline covalent organic frameworks as high-rate cathodes for Li-ion batteries. *J. Mater. Chem. A*, **2016**, 4, 18621.

54. Vitaku, E.; Gannett, C. N.; Carpenter, K. L.; Shen, L.; Abruña, H. D.; Dichtel, W. R. Phenazine-Based covalent organic framework cathode materials with high energy and power densities. *J. Am. Chem. Soc.*, **2020**,142, 16.

55. Wang, S.; Wang, Q.; Shao, P.; Han, Y.; Gao, X.; Ma, L.; Yuan, S.; Ma, X.; Zhou, J.; Feng, X.; Wang, B. Exfoliation of covalent organic frameworks into few-layer redox-active nanosheets as cathode materials for lithium-ion batteries. *J. Am. Chem. Soc.*, **2017**, 139, 4258.

56. Wang, P.; Wu, Q.; Han, L.; Wang, S.; Fang, S.; Zhang, Z.; Sun, S. Synthesis of conjugated covalent organic frameworks/graphene composite for supercapacitor electrodes. *RSC Adv.*, **2015**, 5, 27290.

57. Kandambeth, S.; Venkatesh, V.; Shinde, D. B.; Kumari, S.; Halder, A.; Verma, S.; Banerjee, R. Self-templated chemically stable hollow spherical covalent organic framework. *Nat. Commun.*, **2015**, 6, 6786.

58. Sun, B.; Liu, B.; Cao, A.; Song, W.; Wang, D. Interfacial synthesis of ordered and stable covalent organic frameworks on amino-functionalized carbon nanotubes with enhanced electrochemical performance. *Chem. Commun.*, **2017**, 53, 6303.

59. Xiao, Z.; Kong, D.; Song, Q.; Zhou, S.; Zhang, Y.; Badshah, A.; Liang, J.; Zhi, L. A facile schiff base chemical approach: towards molecular-scale engineering of N-C interface for high performance lithium-sulfur batteries. *Nano Energy*, **2018**, 46, 365.

60. Tan, M.; Xiao, Y.; Xi, W.; Lin, X.; Gao, B.; Chen, Y.; Zheng, Y.; Lin, B. Cobalt-nanoparticle impregnated nitrogen-doped porous carbon derived from Schiff-base polymer as excellent bifunctional oxygen electrocatalysts for rechargeable zinc-air batteries. *J. Power Sources*, **2021**, 490, 229570.

61. Baldo, M. A.; Thompson, M. E.; Forrest, S. R. Phosphorescent materials for application to organic light emitting devices. *Pure Appl. Chem.*, **1999**, 71, 2095–2106.

62. Yang, X.; Zhou, G.; Wong, W.-Y. Functionalization of phosphorescent emitters and their host materials by main-group elements for phosphorescent organic light-emitting devices. *Chem. Soc. Rev.*, **2015**, 44, 8484–8575.

63. Chi, Y.; Chou, P. T. Transition-metal phosphors with cyclometalating ligands: fundamentals and applications. *Chem. Soc. Rev.*, **2010**, 39, 638–655.

64. Wong, W.-Y.; Ho, C.-L. Functional metallophosphors for effective charge carrier injection/transport: new robust OLED materials with emerging applications. *J. Mater. Chem.*, **2009**, 19, 4457–4482.

65. Che, C. M.; Chan, S. C.; Xiang, H. F.; Chan, M. C. W.; Liu, Y.; Wang, Y. Tetradentate Schiff base platinum(II) complexes as new class of phosphorescent materials for high-efficiency and white-light electroluminescent devices. *Chem. Commun.*, **2004**, 1484–1485.

66. Sano, T.; Nishio, Y.; Hamada, Y.; Takahashi, H.; Usuki, T.; Shibata, K. Design of conjugated molecular materials for optoelectronics. *J. Mater. Chem.*, **2000**, 10, 157–161.

67. Vashchenko, A. A.; Lepnev, L. S.; Vitukhnovskii, A. G.; Kotova, O. V.; Eliseeva, S. V.; Kuzmina, N. P. Photo- and Electroluminescent properties of Zinc(II) complexes with tetradentate Schiff bases, derivatives of salicylic aldehyde. *Opt. Spectrosc.*, **2010**, 108, 463–465.

68. Lepnev, L. S.; Vaschenko, A. A.; Vitukhnovsky, A. G.; Eliseeva, S. V.; Kotova, O. V.; Kuzmina, N. P. Luminescence instability of films and new organic light-emitting diodes based on zinc complexes with tetradentate schiff bases: influence of heating and laser irradiation. *J. Russ. Laser Res.*, **2008**, 29, 497–503.

Index